Epicurean Meteorology

Philosophia Antiqua

A SERIES OF STUDIES ON ANCIENT PHILOSOPHY

Editorial Board

C.J. Rowe (*Durham*)
K.A. Algra (*Utrecht*)
F.A.J. de Haas (*Leiden*)
J. Mansfeld (*Utrecht*)
D.T. Runia (*Melbourne*)
Ch. Wildberg (*Princeton*)

Previous Editors

J.H. Waszink †
W.J. Verdenius †
J.C.M. Van Winden †

VOLUME 142

The titles published in this series are listed at *brill.com/pha*

Epicurean Meteorology

Sources, Method, Scope and Organization

By

Frederik A. Bakker

BRILL

LEIDEN | BOSTON

Library of Congress Cataloging-in-Publication Data

Names: Bakker, Frederik A., author.
Title: Epicurean meteorology : sources, method, scope and organization / by
   Frederik A. Bakker.
Other titles: Philosophia antiqua ; v. 142.
Description: Leiden ; Boston : BRILL, 2016. | Series: Philosophia antiqua ;
   volume 142 | Includes bibliographical references and index.
Identifiers: LCCN 2016014669 (print) | LCCN 2016016482 (ebook) | ISBN
   9789004321564 (hardback) : alk. paper) | ISBN 9789004321588 (e-book)
Subjects: LCSH: Epicureans (Greek philosophy) | Meteorology–History. |
   Philosophy, Ancient.
Classification: LCC B512 .B35 2016 (print) | LCC B512 (ebook) | DDC
   551.501–dc23
LC record available at https://lccn.loc.gov/2016014669

Want or need Open Access? Brill Open offers you the choice to make your research freely accessible online in exchange for a publication charge. Review your various options on brill.com/brill-open.

Typeface for the Latin, Greek, and Cyrillic scripts: "Brill". See and download: brill.com/brill-typeface.

ISSN 0079-1687
ISBN 978-90-04-32156-4 (hardback)
ISBN 978-90-04-32158-8 (e-book)

Copyright 2016 by Koninklijke Brill NV, Leiden, The Netherlands.
Koninklijke Brill NV incorporates the imprints Brill, Brill Hes & De Graaf, Brill Nijhoff, Brill Rodopi and Hotei Publishing.
All rights reserved. No part of this publication may be reproduced, translated, stored in a retrieval system, or transmitted in any form or by any means, electronic, mechanical, photocopying, recording or otherwise, without prior written permission from the publisher.
Authorization to photocopy items for internal or personal use is granted by Koninklijke Brill NV provided that the appropriate fees are paid directly to The Copyright Clearance Center, 222 Rosewood Drive, Suite 910, Danvers, MA 01923, USA. Fees are subject to change.

This book is printed on acid-free paper and produced in a sustainable manner.

Printed by Printforce, the Netherlands

Contents

Preface IX
List of Tables X
List of Illustrations XI
List of Abbreviations XII

1 **General Introduction** 1

2 **Multiple Explanations** 8
   2.1 Introduction 8
   2.2 Preliminary Observations 9
      2.2.1 *Causes and Explanations* 9
      2.2.2 *Variations in the Use of Multiple Explanations* 10
   2.3 Truth, Probability and Personal Preferences 12
      2.3.1 *Introduction* 12
      2.3.2 *Are All Alternative Explanations True?* 13
      2.3.3 *Contestation or Disagreement with Appearances* 32
      2.3.4 *Incompatibility with Explanations of Other Phenomena* 32
      2.3.5 *Non-Contestation and Analogy* 34
      2.3.6 *Degrees of Probability and Personal Preferences* 37
      2.3.7 *Lucretius' Supposed Preference for the Theories of Mathematical Astronomy* 42
   2.4 Multiple Explanations and Doxography 58
   2.5 The Sources of the Method of Multiple Explanations 62
      2.5.1 *Introduction* 62
      2.5.2 *Democritus* 63
      2.5.3 *Aristotle* 65
      2.5.4 *Theophrastus* 67
      2.5.5 *The* Syriac Meteorology 70
      2.5.6 *Conclusions about the Origins of the Method* 73
   2.6 Conclusions 74

3 **Range and Order of Subjects in Ancient Meteorology** 76
   3.1 Introduction 76
   3.2 Range, Delimitation and Subdivisions of Meteorology 78
      3.2.1 *Introduction* 78
      3.2.2 *The Texts* 79
      3.2.3 *The Table* 98

	3.2.4	*Some Observations* 99
	3.2.5	*Some Conclusions* 106
3.3	Terrestrial Phenomena Other Than Earthquakes 109	
	3.3.1	*Lucretius* 109
	3.3.2	*Parallels in Meteorology and Paradoxography* 113
	3.3.3	*Conclusion* 125
3.4	Order of Subjects 127	
	3.4.1	*Introduction* 127
	3.4.2	*The Table* 130
	3.4.3	*Some Observations* 130
	3.4.4	*Proposed Original Order of Subjects* 137
	3.4.5	*Deviations from the Proposed Original Order* 137
	3.4.6	*The Internal Structure of Chapters and Sections* 140
3.5	Relations between the Four Texts 143	
	3.5.1	*Epicurus'* Letter to Pythocles *and His "Other Meteorology"* 144
	3.5.2	*Lucretius* DRN VI *and Epicurus' "Other Meteorology"* 144
	3.5.3	*Authorship and Identity of the* Syriac Meteorology 145
	3.5.4	*Lucretius, Epicurus and the* Syriac Meteorology 153
	3.5.5	*Aëtius'* Placita *and Theophrastus'* Physical Opinions 155
	3.5.6	*Summary* 156
3.6	Conclusions 158	
3.7	Epilogue: Epicurean Cosmology and Astronomy 160	

4 The Shape of the Earth 162

4.1	Introduction 162	
4.2	Historical and Conceptual Context 165	
	4.2.1	*The Shape of the Earth in Antiquity: A Historical Overview* 165
	4.2.2	*Ancient Proofs of the Earth's Sphericity* 169
	4.2.3	*Epicurus' Ancient Critics* 175
	4.2.4	*The Direction of Natural Motion and the Shape of the Earth* 177
4.3	Discussion of Relevant Passages 180	
	4.3.1	*The Rejection of Centrifocal Natural Motion* (DRN *I 1052ff.*) 181
	4.3.2	*Downward Motion* (DRN *II 62–250*) 210
	4.3.3	*The Apparent Proximity of the Sun* (DRN *IV 404–413*) 220
	4.3.4	Climatic *Zones?* (DRN *V 204–205*) 221
	4.3.5	*Lucretius' Cosmogony* (DRN *V 449–508*) 223

	4.3.6	*Stability of the Earth* (DRN V 534–563) 235
	4.3.7	*The Size of the Sun* (DRN V 564–591) 236
	4.3.8	*Centrifocal Terminology* (DRN V 621–636) 239
	4.3.9	*Sunrise and Sunset* (DRN V 650–679) 241
	4.3.10	*The Earth's Conical Shadow* (DRN V 762–770) 242
	4.3.11	*The 'Limp' of the Cosmic Axis* (DRN VI 1107) 245
	4.3.12	*Philodemus and the Gnomon* (Phil. De sign. 47.3–8) 254
4.4	Conclusions 255	

5 **General Conclusions**   264

Appendix 1: Multiple Explanations in Epicurus' *Letter to Pythocles*   269
Appendix 2: Multiple Explanations in Lucretius' *DRN* V **and** VI   272
Appendix 3: General Structure of the *Syriac Meteorology*   274
Bibliography   276
Index Locorum   286
General Index   298

Preface

This book is a revised and updated version of my doctoral dissertation entitled *Three Studies in Epicurean Cosmology*, which was prepared at Utrecht University and defended there on 17 November 2010. The revision was conducted at Radboud University Nijmegen.

While I worked on this book and its predecessor many people have helped me. At the risk of doing injustice to many others who have also helped and supported me over the years, I wish to single out a few people for special thanks. First of all I want to thank Keimpe Algra, who first recognized the potential scholar in me and offered me the chance to write a dissertation, which he subsequently supervised with immense patience, trust and friendliness. Thanks are also due to Jaap Mansfeld and David Runia, who saw some merit in my work and encouraged me to press on, first with the dissertation and later with the present work, providing me with numerous useful suggestions on the way. In addition I would like to thank Han Baltussen for the many detailed and pertinent comments on my dissertation. I am also immensely grateful to my present colleagues at the Center for the History of Philosophy and Science at Radboud University, and especially Christoph Lüthy, for all the support and encouragement they have given me the last couple of years. I am also much obliged to Brill's anonymous referee for saving me from some potentially embarrassing omissions and oversights, but most of all for the useful suggestions as to the structure and unity of my book. Finally and most importantly, I thank my wife, Ruthie, for bearing with me all these years.

List of Tables

2.1 Place of the mathematical astronomers' explanations in the astronomical section of *DRN* V   46
2.2 Place of the astronomers' explanations in the astronomical passages of the *Letter to Pythocles*   55
3.1 Range of subjects and subdivisions in various ancient accounts of meteorology   100
3.2 Lucretius' account of terrestrial phenomena   111
3.3 *DRN* VI 608 ff. with parallels in meteorology and paradoxography   116
3.4 Order of subjects in Aëtius, the *Syriac meteorology*, Lucretius and Epicurus   128
3.5 Terrestrial subjects other than earthquakes in Aëtius and Lucretius   136
3.6 Proposed original order of subjects   138
3.7 *Syriac meteorology* and Lucretius VI on thunderbolts   141
3.8 Alternative explanations of thunder in the *Syriac meteorology* and *DRN* VI   143
4.1 Time-line of ancient theories on the shape of the earth   170
4.2 Ancient proofs of the earth's sphericity   174
4.3 Passages to be discussed   180

List of Illustrations

3.1   Possible relations between Epicurus, Lucretius and the *Syriac meteorology*   154
3.2   Possible relations between Aëtius, Epicurus, Lucretius and the *Syriac meteorology*   157
4.1   *Linear* or *parallel* versus *centrifocal* cosmology   179
4.2   Three climatic regions versus five climatic zones   223
4.3   Furley's calculation of the sun's distance on the assumption of a flat earth   238
4.4   The moon in the conical shadow of the earth   244
4.5   Lucr. VI 1106–1113: five places, four winds   252

List of Abbreviations

In this work the following abbreviations will be used:

Arr.	G. Arrighetti, *Epicuro, Opere*, Torino 1960, revised 1973.
B–L	J. Bollack & A. Laks, *Epicure à Pythoclès: Sur la cosmologie et les phénomènes météorologiques*, Cahiers de Philologie de Lille, vol. 3, Villeneuve d'Ascq 1978.
D–K	H. Diels / W. Kranz, *Fragmente der Vorsokratiker*, 3 vols., Berlin 1960[10].
D.L.	Diogenes Laërtius, *Lives of the philosophers*.
DRN	Lucretius, *De rerum natura*.
E–K	L. Edelstein & I.G. Kidd, *Posidonius, I. The Fragments*, Cambridge 1972, 1989².
FHS&G	W.W. Fortenbaugh, P.M. Huby, R.W. Sharples & D. Gutas (eds.) (1992), *Theophrastus of Eresus: Sources for his life, writings, thought and influence*, Pt. I (*Philosophia antiqua* vol. 54, 1), Leiden.
Hdt.	Epicurus, *Letter to Herodotus*.
M&R 2	J. Mansfeld & D.T. Runia (2009a), *Aëtiana, the Method and Intellectual Context of a Doxographer, vol. 2 (in two parts): The Compendium (Philosophia Antiqua* vol. 114), Leiden.
OLD	*Oxford Latin Dictionary*, Oxford 1968.
Pyth.	Epicurus, *Letter to Pythocles*.
SVF	J. von Arnim, *Stoicorum Veterum Fragmenta*, 3 vols., Leipzig 1903–1905; vol. 4 with indices by M. Adler, Leipzig 1924.
TLL	*Thesaurus Linguae Latinae*, Munich 1894-present.
Us.	H. Usener, *Epicurea*, Leipzig 1887, repr. Rome 1963.

CHAPTER 1

General Introduction

One of the few complete works that have come down under Epicurus' name is the *Letter to Pythocles*.[1] In this work Epicurus undertakes to provide a 'concise and well-outlined discussion concerning the μετέωρα (or *lofty phenomena*)'.[2] What precisely these 'lofty phenomena' are is nowhere clearly defined, but the letter subsequently deals with a number of cosmological, astronomical, atmospherical and even seismological problems and phenomena, which must therefore be assumed to be included in the term μετέωρα, even though—strictly speaking—not all of them are 'lofty'. What they do have in common, or at least most of them, is that they are accounted for by a number of alternative explanations.[3] If this is what defines μετέωρα for Epicurus, we can safely extend the term to include also those phenomena which, although absent from the *Letter to Pythocles*, are explained in a similar way and in a similar context in book VI of Lucretius' Epicurean poem *De rerum natura*. It is in this sense then—in order to stay close to the Epicureans' own terminology and so as not to destroy the unity of the subject matter, as they understood it—that I will speak of Epicurean meteorology, i.e. the study of the μετέωρα.

In the present work I will investigate Epicurean meteorology from three main angles, which seemed to be in need of further clarification or re-examination in the light of recent insights or—conversely—on account of the uncritical persistence of outdated views. The emphasis will be on historical interpretation: to understand Epicurean meteorology in its own historical context. After the present introduction, Chapter Two deals with the method of multiple explanations—the explanatory method generally applied in Epicurean meteorology. Chapter Three investigates the range and order of subjects in Epicurean meteorology as compared to other ancient meteorological texts. Chapter Four

---

1 I will disregard the claims against the *Letter*'s authenticity by e.g. Usener (1887) xxxvii–xli; Reitzenstein (1924) 36–43; Bailey (1926) 275 and Schmidt (1990) 34–37, which I think have been sufficiently refuted by Arrighetti (1973) 691–705 and Bollack & Laks (1978) 45–55. See also Mansfeld (1994) 29 n. 2, Sedley (1998a) 119 n. 65, Podolak (2010) 39–41, 66–72, and De Sanctis (2012) 95–96.
2 Epic. *Pyth.* 1 [84] "περὶ τῶν μετεώρων σύντομον καὶ εὐπερίγραφον διαλογισμὸν". Henceforward, unless indicated otherwise, all translations are mine.
3 See Bénatouïl (2003), 16–20. On the subject matter of ancient meteorology see also Chapter Three below.

is concerned with the Epicurean view of the shape of the earth, and how this view fits in with the Epicureans' approach to meteorological problems in general. Finally, in Chapter Five I will summarize the main findings and draw some general conclusions.

Before introducing the subject of the present study in some more detail, I will first sketch the outlines of Epicureanism, dealing briefly, first with Epicurus himself, then with a number of later Epicureans, such as Lucretius, and their value as sources for a reconstruction of Epicurean theory, and finally with the most relevant parts of Epicurus' philosophy.

Circa 305 BC, Epicurus (341–270 BC) established a new school of philosophy around his newly acquired estate in Athens. Like the competing school of the Stoics, founded around the same time, Epicurus and his followers were principally concerned with ethics, to which other branches of philosophy, such as physics, were subordinate. According to Epicurus, the life of man was oppressed by unfounded fears—of the gods, of illness, of death—which stood in the way of real happiness. It was in order to eradicate these fears and replace them with a true understanding of nature that physics came into play, which was largely based on the atomism of Democritus (ca. 460–370). Yet, despite its ultimately subservient role, physics was no small matter to Epicurus, who devoted the 37 books (!) of his *On nature* (Περὶ φύσεως) to it.

Today the bulk of Epicurus' writings is lost. Only three works have come down more or less complete, being quoted by Diogenes Laërtius as part of his account of Epicureanism in book x of his *Lives of the philosophers*. These three are the *Letter to Herodotus*, a summary of Epicurean physics; the *Letter to Pythocles*, a summary of Epicurean meteorology (in the sense indicated above); and the *Letter to Menoeceus*, an overview of Epicurean ethics. Of his other works only scorched fragments—dug up in Herculaneum, where they were buried during the famous eruption of Vesuvius in 79 AD—and a large number of quotations and paraphrases by later authors remain.

The fullest account of Epicurean physics we possess today is the Latin didactic poem *De rerum natura*, 'On the nature of things', by the Roman Epicurean Lucretius (ca. 99–55 BC). In six books, and over 7,000 verses, Lucretius deals with every aspect of the physical world, from the invisible smallness of atoms to the infinite extent of the universe, and from sex and conception to inevitable death, diversifying his account with ethical exhortations and eulogies of Epicurus.

Other Epicureans whose works we can still partly read are Philodemus of Gadara and Diogenes of Oenoanda. Philodemus, a Greek philosopher roughly contemporary with Lucretius, was based in Herculaneum in southern Italy, where many of his works, scorched and buried by the eruption of Vesuvius in 79 AD, were found during excavations in the 1750s, eventually resulting in a number of fairly readable, though fragmentary, editions. Diogenes of Oenoanda was a wealthy Greek from Oenoanda in Lycia (southern Turkey), living in the second century AD, who in his old age had a wall in the town centre inscribed with a summary of Epicurean philosophy, many fragments of which can still be read and are available in modern editions.[4]

Between Epicurus and these three followers considerable time had elapsed and it is only natural to ask whether their teachings were still the same as Epicurus'. Especially Lucretius has become the object of an ongoing debate, in which one side detects traces of later intellectual and philosophical developments throughout his poem, while the other camp with equal vigour interprets any such sign as a reference to philosophical positions that would have been known to Epicurus himself. In this work I will try to avoid both extremes. Lucretius' explicit claim to be following Epicurus' writings[5] should deter us from looking for later developments when we can do without them, but on the other hand we should not reject an obvious interpretation because it might conflict with Lucretius' supposed intellectual isolation.

The basic tenets of Epicurean physics are not difficult to summarize, and here at least Lucretius is in complete agreement with Epicurus. According to the Epicureans there are two basic entities—bodies and space, while everything else is either a property of these or non-existent. Bodies are either compounds or atoms. The latter are imperceptibly small, indivisible and incompressible particles out of which compounds are made and into which they are eventually dissolved again. Among compounds Epicurus also numbers souls and the gods. There are many types of atoms, differing from each other in shape, size and weight. Beside these, atoms possess none of the qualities that belong to perceptible bodies, such as colour, temperature or smell. The number of atoms and the extent of space are infinite. Atoms are forever moving at a constant but inconceivable speed. Left to their own devices the atoms move either downwards by the force of their own weight or they swerve ever so little from their course, but if they collide with other atoms they may rebound in any direction, while in compounds they are reduced to vibrating. Every compound body is constantly shedding imperceptibly thin membranes which preserve its outward shape and texture. When these membranes, which are called 'images' (εἴδωλα), enter our eye we *see* the compound body. The Epicureans also

---

4   See now Smith (1993) and (2003).
5   Lucr. *DRN* III 9–12, quoted on p. 144, n. 161 below. See also *DRN* V 55–56.

distinguished a kind of 'mental perception' which is transmitted by still finer images that impinge directly on the mind, and there produce a kind of 'mental picture' which we call a memory, a dream or a thought. Other forms of sense-perception, such as hearing and smell, are brought on in similar ways by other kinds of effluences. Although the images and these other effluences are capable, because of their extreme subtlety, to travel almost unimpeded through the air, they can become confused. Yet, by itself every perception is true, because it accurately reports the way it is affected by the external object. It is only when we start interpreting our perceptions and form opinions about the external object that falsehood may arise. For instance, when an oar standing in the water appears broken to us, this perception is *true* in the sense that the images really convey this impression to us, but when we infer from this that the oar is really broken, this inference may well be *false*. The only way to find out for sure is by making another observation (which in itself is no more true than the other one) under circumstances where we know impressions to generally correspond to their objects. We might, for instance, handle the oar, or pull it out of the water and look again.

Starting from these basic tenets Epicurus constructed a complex theory that could account for every aspect of reality, and throughout Epicurus' *Letters to Herodotus* and *Pythocles* and throughout Lucretius' *De rerum natura* we see these principles being put to work to explain specific phenomena. An important class of such phenomena are the so-called μετέωρα or 'lofty things', to which Epicurus devoted his *Letter to Pythocles*, and which comprise both astronomical and meteorological phenomena. Lucretius too deals with such phenomena in books V and VI of his *De rerum natura*. A characteristic feature of these μετέωρα is that they can only be observed from afar and do not admit of more reliable observations. In these circumstances it is impossible to falsify every hypothesis about them, and we are forced to accept every theory that does not somehow conflict with other observations or with the basic tenets of Epicurean physics. Far from deploring this conclusion, Epicurus and Lucretius actually embrace it, accounting for almost every astronomical and meteorological phenomenon with a list of alternative explanations.

In Chapter Two I will explore some general aspects of this method of multiple explanations. In the first place there is the epistemological point of view. Nobody will object to the view that a theory that cannot be falsified must be held *possible*. There are indications, however, that Epicurus went further than that and claimed that any such theory is *true*. This would imply that in these cases several, sometimes conflicting, *truths* exist side by side. The evidence for ascribing this position to Epicurus is not unambiguous, though, and modern scholarship is divided on whether Epicurus really made this claim and how it should be interpreted. I will critically review the relevant textual evidence as well as a number of modern interpretations and propose a compromise. Another epistemological problem concerns the claim, made by Diogenes of Oenoanda, that among alternative explanations some are more probable than others. I will argue that this claim finds no support in the writings of Epicurus and Lucretius and actually constitutes a departure from Epicurus' views. A related question concerns Bailey's observation that Lucretius, in his astronomical passages, usually placed the view of the astronomers first, 'as though he really preferred these.' In a section devoted to this question I will propose a different interpretation, which will be more in line with Lucretius' and Epicurus' statements concerning the use of multiple explanations, and with Epicurus' general attitude to mathematical astronomy.[6] The remainder of the chapter is devoted to the question of Epicurus' sources for the alternative explanations, firstly for the individual explanations, which are thought to have been largely drawn from *doxography*,[7] and secondly for the systematic use of multiple alternative explanations, partial anticipations of which are suspected in Democritus, Aristotle and Theophrastus. In this context I will also deal with the '*Syriac meteorology*', a meteorological treatise preserved in Syriac and Arabic, which consistently offers multiple alternative explanations. Although the manuscripts are unanimous in naming Theophrastus as its author, I think this ascription has been too readily accepted and the alternative hypothesis of an Epicurean origin, suggested by the earlier commentators, too readily rejected. Yet, the question of its authorship is important because the answer to this question largely determines our view as to the extent of Epicurus' dependence on Theophrastus. A thorough comparison of the way multiple explanations are employed in the *Syriac meteorology*, in Epicurus' *Letter to Pythocles* and Lucretius' *De rerum natura*, and in the undisputed writings of Theophrastus, may bring us closer to an answer.

Although at the beginning of this chapter I styled Epicurus' *Letter to Pythocles* a meteorological treatise, the range of its subject matter is rather different from that of other ancient meteorologies, such as for instance Aristotle's *Meteorology* or the *Syriac meteorology* ascribed to Theophrastus. While the most conspicuous difference is, of course, Epicurus' inclusion of cosmological and astronomical matters, which other authors preferred to assign to a separate branch of physical inquiry, a close comparison will reveal many other interesting dif-

---

6 See esp. Sedley (1976).
7 Doxography = genre of writings devoted to the systematic collection of *doxai* or philosophical opinions: for a quick reference see Runia (1997b) or Mansfeld (2013).

ferences as well as similarities, both in the range and the order of subjects. Chapter Three of this work will, therefore, be devoted to such a comparison, with a view to establishing the relations of these texts to one another. I will start with a thorough comparison of nine meteorological texts with regard to the range and subdivision of their subject matter, in order to distinguish the various traditions and the place of Epicurus and Lucretius therein. A characteristic feature of Lucretius' meteorology is the attention he pays to exceptional local phenomena, which are either absent or far less conspicuous in other meteorologies and belong more properly to the genre known as *paradoxography*.[8] I will therefore continue the investigation with a comparison of the latter part of Lucretius' book VI with a number of meteorological as well as paradoxographical works, with regard to the inclusion and treatment of exceptional local phenomena. Next I will deal with the order of meteorological subjects in Lucretius' book VI, Aëtius' book III, the *Syriac meteorology*, and Epicurus' *Letter to Pythocles*. The large degree of correspondence between the first three works has often been observed, but never thoroughly examined; and the not-so-obvious resemblance of Epicurus' *Letter* to the other three has generally been overlooked. I will therefore carry out a more thorough comparison of all four works with due attention to the similarities and differences in the order of their subjects. In this context I will also resume my investigation into the authorship and identity of the *Syriac meteorology* by comparing its theories with, on the one hand, Epicurus and Lucretius, and, on the other hand, Theophrastus and Aristotle.

While Chapters Two and Three each deal with Epicurean meteorology in general, Chapter Four is devoted to a more specific cosmological problem: the shape of the earth. Yet, this chapter has a wider application as well. As the problem of the shape of the earth is intimately connected with several other cosmological and astronomical problems, some of which are accounted for with single, others with multiple explanations, it is an excellent case to illustrate Epicurus' scientific method and its relation to orthodox astronomy, topics that will also be addressed in Chapter Two.

As for the actual subject of Chapter Four: although many scholars confidently claim that Epicurus and Lucretius believed the earth to be flat, even scorning them for having clung to such an antiquated idea, in reality no explicit statements about the shape of the earth can be found in their works. In this chapter I will consider this problem from various angles. I will re-examine the evidence that has been adduced so far for attributing a flat-earth cosmology

---

8 Paradoxography = genre of writings devoted to the collection of *paradoxa* or marvellous stories: see e.g. Ziegler (1949), Schepens & Delcroix (1996), Wenskus & Daston (2000).

to the Epicureans. In addition, I will discuss every passage in the works of Epicurus and Lucretius, as well as other Epicureans, that may be relevant to the question of the shape of the earth. I will also inventory the ancient proofs of the earth's sphericity and search for clues that the Epicureans may have known, and somehow responded to, these proofs. In the final section of the chapter I will conclude that the evidence is hopelessly ambiguous, that the Epicureans probably had no firm conviction about the shape of the earth at all, and that to try and wrest such a conviction from them goes against the scientific method they adopted when dealing with cosmological and astronomical problems.

CHAPTER 2

Multiple Explanations

2.1 Introduction

According to Epicurus and Lucretius, sunset and sunrise can be explained not only on the assumption that the sun passes unaltered below the earth and then emerges again, but equally well by its extinction and subsequent rekindling. Both explanations are retained because neither can be eliminated on the evidence of the senses, which is the Epicureans' principal criterion of truth.

This approach is typical of the way Epicurus, in the *Letter to Pythocles*, and Lucretius, in books v and vi of the *De rerum natura*, deal with astronomical, meteorological and terrestrial phenomena. Just like sunrise and sunset almost every one of these problems is accounted for with a number of alternative explanations, sometimes two, sometimes more. In defence of this method of multiple explanations Epicurus points out that in these fields of inquiry single explanations are neither possible nor necessary. They are not possible because the objects in question cannot be clearly observed on account of their distance, and they are not necessary because the main aim of studying celestial and atmospherical phenomena is to rule out divine interference, for which it is enough to show that every phenomenon can be explained physically, although absolute certainty as to the actual cause is not needed.[1]

The method of multiple explanations depends on earlier philosophy in various ways. It has long been known that Epicurus and Lucretius borrowed most of their alternative theories from earlier philosophers, and there is good reason to believe that they learned of these views not from the original works but from doxographies, thematically organized collections of opinions of earlier thinkers. Nor were Epicurus and Lucretius the first to apply multiple explanations to certain problems. One notable instance is found among the fragments of Democritus and occasional instances are found in the works of Aristotle and Theophrastus. The most complete parallel for the use of multiple explanations is a meteorological treatise ascribed to Theophrastus and preserved in Syriac and Arabic, which employs multiple explanations on a similar scale

---

[1] For the ethical motivation of multiple explanations and Epicurean science in general, see e.g. Wasserstein (1978) 493–494; Jürss (1994), esp. 240–251, and Bénatouïl (2003) 24–28.

as Epicurus and Lucretius, and which, if the ascription is correct, would make Theophrastus the real 'inventor' of the method of multiple explanations.

In this chapter I propose to investigate the method of multiple explanations from several angles. After some preliminary observations I will first deal with a number of epistemological problems concerning the method of multiple explanations. Then I will discuss the connection of multiple explanations with doxography, and finally I will look into the possible antecedents of Epicurus' method of multiple explanations.

2.2 Preliminary Observations

2.2.1 *Causes and Explanations*

In scholarly literature Epicurus' method is variously referred to as relating to multiple *causes* or multiple *explanations*.[2] Epicurus himself uses various expressions. In the *Letter to Herodotus*, the preferred term seems to be αἰτία,[3] a word whose general sense varies between 'cause' and 'explanation'.[4] The same word, αἰτία, also occurs sometimes in the *Letter to Pythocles*.[5] There, however, it may refer not just to individual causes/explanations, but also to the whole range of causes/explanations used to account for a particular phenomenon.[6] However, Epicurus' favourite term in the *Letter to Pythocles* is τρόπος. This word, too, is ambiguous. Its general meaning is 'manner' or 'way', which in context may refer either to the way a phenomenon is produced, i.e. its *cause*, or to the way a phenomenon is accounted for, i.e. its *explanation*. In addition, just like αἰτία, it may denote both individual causes/explanations and their combination.[7] In short, Epicurus' choice of words does little to resolve the

---

2 With 'cause' I mean the mechanism that is actually responsible for producing a certain phenomenon, and with 'explanation' the verbal statement of a possible cause. For the distinction between cause and explanation see also Hankinson (1999a) 4 & (1999b) 480, and Masi (2015) 39.

3 Αἰτία is used in *Hdt.* 50.1; 63.11; 64.3; 76.7; 78.1; 79.5; 79.8; 81.5. In *Hdt.* 79.8 *multiple causes/explanations* are referred to as πλείους αἰτίας.

4 See again Hankinson (1999a) 4 & (1999b) 480, and Masi (2015) 39. Verde (2013) 129 rightly observes that Epicurus himself does not distinguish between the two senses.

5 Αἰτία is used in *Pyth.* 2 [86.7]; 11 [95.4]; 31 [111.10]; 33 [113.8 & 11]; 37 [116.9].

6 The clearest instance of this use is *Pyth.* 2 [86.7]: πλεοναχὴν αἰτίαν (note the singular).

7 Τρόπος denotes individual causes/explanations in the expression τρόπους πλείους and similar expressions (*Pyth.* 17A [99.6] *et passim*: see also APPENDIX 1 on p. 269 below). The expression πλεοναχὸν τρόπον (*Pyth.* 2 [87.3] & 11 [95.3]), on the other hand, may refer to either the *method* or the *combination* of multiple causes/explanations.

ambiguity. Lucretius' language shows a similar ambiguity. His preferred terms in the relevant sections (*DRN* V 509–770 and VI) are *causa* ('cause'),[8] which seems clear enough, and *ratio*,[9] which—like the Greek τρόπος—may refer both to the *manner* in which a phenomenon is produced and to the *account* that is given of its production. However, in order not to run ahead of my argument, I will speak of multiple *explanations*, leaving open for now the question whether or not each explanation corresponds to an actual cause.[10]

2.2.2 Variations in the Use of Multiple Explanations

Although Epicurus' own methodological remarks on the use of multiple explanations (*Hdt.* 78–80 and *Pyth.* 2 [85–88]) make it sound as if we are dealing with a single and homogeneous method, in fact several variants can be discerned, depending on the number, the exhaustivity, the mutual exclusivity or, contrariwise, the subsidiarity, and the type-differentiation of the individual alternative explanations. Below I will briefly review these five kinds of variation and their possible significance for the present investigation.

1. Number. The total number of alternative explanations for a single problem varies from *two* to *nine*.[11] For sunset and sunrise, for instance, both Epicurus and Lucretius offer just *two* possible explanations.[12] Thunder, on the other hand, is accounted for with no less than *nine* different explanations by Lucretius, and *five* by Epicurus.[13]

A few of the astronomical, meteorological and terrestrial problems discussed by Epicurus and Lucretius, e.g. the size of the heavenly bodies,[14] the stability of the earth,[15] and the temperature fluctuation in wells,[16] are accounted for with only *one* explanation. In these cases we should distinguish between

---

8   V 509, 529, 531, 620, 753, VI 172, 204, 363, 577, 703, 706, 761, 1001.
9   V 590, 614, 643, VI 132, 279, 425, 475, 506, 535, 575, 639, 861, 866, 881, 902, 1000, 1090, 1131.
10 For a somewhat parallel account, see Masi (2015), who arrives at the conclusion that αἰτία for Epicurus denotes a 'causal possibility'—an actually existing potency that may or may not be operative in a particular case.
11 See APPENDICES 1 & 2 on p. 269 and p. 272 below.
12 Epic. *Pyth.* 7 [92]; Lucr. *DRN* V 650–679.
13 Lucr. *DRN* VI 96–159 (for the structure of this section see table 3.8 on p. 143 below) and Epic. *Pyth.* 18 [100].
14 Epic. *Pyth.* 6 [91] and Lucr. *DRN* V 564–591.
15 Lucr. *DRN* V 534–563.
16 Lucr. *DRN* VI 840–847.

problems that may be open to other explanations as well and those which exclude them (see the next entry).

2. *Exhaustivity.* In most cases the lists of alternative explanations are not exhaustive, but appear to be open to other options as well, as Epicurus often tells us explicitly.[17] Only rarely do the alternative explanations given seem to exhaust the entire range of possibilities,[18] as when the moon is said to shine either with its own or with reflected light,[19] or when the sun at sunset is said either to pass unaltered below the earth, or to be extinguished.[20]

This distinction of exhaustivity and inexhaustivity may perhaps be applied to single explanations as well (see above). Sometimes, although only one explanation is given, there is no need to suppose that this is the only possible one. This seems to be the case with a number of terrestrial phenomena discussed by Lucretius in the latter part of his book VI. Perhaps, because he was one of the first to discuss these problems, there was no anterior tradition from which he could borrow his explanations (see p. 125 ff. below), and he therefore contented himself with providing just one, and sometimes two. However, other instances of single explanations seem to exclude alternative views. Epicurus' and Lucretius' emphatic claim that the sun, the moon and the stars are the size they appear to be seems to rule out any other option,[21] and so does Epicurus' account of the formation of the heavenly bodies out of windy and fiery matter.[22] These two varieties of single explanations may perhaps be distinguished as *provisionally* single and *exclusively* single, only the latter being truly opposed to multiple explanations.

3. *Mutual exclusivity.* In some cases the alternatives offered seem to exclude each other: the sun is either extinguished at sunset or it is not, and the moon's light is either borrowed or its own.[23] In most cases, however, nothing prevents

---

17 See APPENDIX 1 on p. 269 below: a '+'-sign after the number indicates that the list is explicitly *in*exhaustive.
18 Masi (2015), 59, argues that the entire range of possible explanations, however extensive and varied it may be, must nevertheless be limited.
19 Epic. *Pyth.* 11 [94–95].
20 See n. 12 above.
21 See n. 14 above. For a different interpretation of this theory see n. 184 on p. 236 below.
22 Epic. *Pyth.* 5 [90–91].
23 See n. 12 and n. 19 above. This mutual exclusivity only holds with respect to one particular instance of a phenomenon: in the universe at large every possible explanation is true and may coexist with any other explanation (see § 2.3.2.1.6 on p. 21 below).

the alternative explanations from obtaining at the same time. Once, in *Pyth.* 13 [96–97] on solar and lunar eclipses, Epicurus even tells us so explicitly: eclipses may be caused by extinction or interposition of another body, or both at the same time.[24]

4. Subsidiarity. There are even some explanations that not merely allow, but actually require the simultaneous occurrence of another cause: a number of phenomena discussed in the latter part of *DRN* VI are accounted for with one principal explanation and one or two that are merely subsidiary to the first, and not capable of producing the desired effect on their own.[25] Eruptions of the Etna, for instance, are said to be caused by wind blowing in caverns beneath the mountain, catching fire and then violently escaping, an effect that is further strengthened by incursions of the sea into these caverns.[26]

5. Type-differentiation. While in most cases any instance of a certain phenomenon can be accounted for by any one of the alternative explanations, sometimes—especially in meteorology—different explanations seem to apply to different types of the same phenomenon.[27] Thunder, for instance, is explained according to the nature of its sound,[28] lightning according to the presence or absence of thunder,[29] and earthquakes according to their effects.[30]

Although Epicurus himself does not comment on the different ways multiple explanations can be used, but instead seems to present multiple explanations as a single method, it may be useful to keep these distinctions in mind.

2.3 Truth, Probability and Personal Preferences

2.3.1 *Introduction*

Epicurus' and Lucretius' consistent use of multiple explanations in astronomy and meteorology is epistemologically very interesting if not problematic. There

---

[24] Concerning the question whether the alternative explanations are mutually exclusive or not, see also Masi (2015) 48–56.
[25] See APPENDIX 2 on p. 272 below. Cf. Bailey (1947) pp. 1655, 1684, 1704.
[26] Lucr. *DRN* VI 639–702.
[27] Bailey (1947) 1567.
[28] Lucr. *DRN* VI 96–159; cf. Bailey (1947) 1575.
[29] Lucr. *DRN* VI 160–218; cf. Bailey (1947) 1578, 1586.
[30] Lucr. *DRN* VI 535–607. Seneca's report, in *NQ* VI 20, 7, that Epicurus considered wind the strongest cause of earthquakes may be another case in point: see p. 38 below.

is, for instance, a continuing debate on whether Epicurus and Lucretius considered all alternative explanations merely possible or actually true, and, if true, how these simultaneous truths should be conceived, and on what grounds Epicurus felt he could pronounce some explanations true and others false. Conversely, if alternative explanations are only possible, are they all equally possible or do they admit different degrees of probability, and is it permitted to prefer one explanation over another, and, if so, were certain explanations actually preferred over others? These are some of the problems I propose to deal with below.

2.3.2 Are All Alternative Explanations True?

It is often claimed that Epicurus considered all alternative explanations true. Although we do not have any explicit statement by Epicurus to this effect, it seems to follow logically from his use of *non-contestation* to confirm individual alternative theories and his claim that *non-contestation* establishes *truth*.[31] On the other hand, the simultaneous truth of several, often mutually exclusive, explanations seems to violate the principle of *non-contradiction*, to which Epicurus was also committed.[32]

2.3.2.1 Ancient Texts and Testimonies

The various modern views concerning the truth-value of Epicurus' multiple explanations are based on a limited number of ancient texts and testimonies, the most important of which will be presented below.

2.3.2.1.1 *The Principles of Epicurean Epistemology*

According to our sources,[33] Epicurus acknowledged three criteria of truth: perceptions (αἰσθήσεις), preconceptions (προλήψεις) and affections (πάθη). Perceptions are the raw data presented to us by the senses; in addition Epicurus distinguished a kind of 'mental perception', called φανταστικὴ ἐπιβολὴ τῆς διανοίας ('impressional projection of the mind'), which directly affects the mind,

---

31 On *non-contestation* and multiple explanations in Epicureanism see e.g. Striker (1974/1996); Sedley (1982) 263–272; Long & Sedley (1987), vol. I, 90–97; Asmis (1984) 178–180, 193–196, 211, 321–336; id. (1999) 285–294; and Allen (2001) 194–205 & 239–241; Fowler (2002) 191–192.

32 See Asmis (1984) 194. See also n. 35 below.

33 The most important sources for Epicurus' epistemology are Diog. Laërt. x 31–34 and Sext. Emp. *Math.* VII 203–211, which can be supplemented at crucial points with quotations from Epicurus himself.

without the mediation of the senses, such as we experience in dreams; later Epicureans made this into a separate criterion of truth. Preconceptions are general notions naturally formed in our minds in response to repeated perceptions, and affections are the primary emotions—pleasure and pain—by which choice and avoidance are determined. These three, or four, types of data, which are themselves incontrovertibly true, serve as the criteria by which the truth and falsity of opinions are established.

For the investigation of physical reality not all criteria are equally relevant. Affections are mainly of use in ethics. 'Mental perceptions' are useful insofar as they help us to form a preconception of the gods, including the realisation that the gods are in no way responsible for the occurrences in our world. Preconceptions provide us with the necessary notions to be able to formulate opinions at all: we can only investigate the causes of thunder if we have a clear notion of what thunder is. However, supposing that the necessary preconceptions are there, and that hypotheses concerning physical reality can be formulated, the criteria by which these are tested are perceptions and 'things perceived', i.e. the *appearances* (τὰ φαινόμενα).[34]

Opinions are either true or false.[35] An opinion is true when it is *attested* (ἐπιμαρτυρεῖται) or *not contested* (οὐκ ἀντιμαρτυρεῖται) by the appearances, and false when it is *not attested* (οὐκ ἐπιμαρτυρεῖται) or when it is *contested* (ἀντιμαρτυρεῖται) by the appearances.[36] Opinions to be tested fall into one of two categories: the προσμένοντα, those 'awaiting' confirmation by a closer and clearer observation, and the ἄδηλα, the 'unclear' or 'non-apparent', which do not allow closer observation.[37] The latter can be further subdivided into those whose objects can only be observed from afar, such as astronomical and meteorological phenomena (τὰ μετέωρα),[38] and those whose objects cannot be observed

---

34 I translate Greek φαινόμενα as 'appearances', which better conveys the essential characteristic of being observed than modern English 'phenomena'.

35 Diog. Laërt. x 34; Sext. Emp. *Math.* VII 211. See, however, Cic. *De fato* 19, 21, 37; *Acad. pr.* II 97; *De nat. deor.* I 70 (collected as Epic. fr.376 Us.), where it is stated that Epicurus refused to assign any truth-value to opinions *about the future*.

36 Epic. *Hdt.* 51: ἐὰν μὲν μὴ ἐπιμαρτυρηθῇ ἢ ἀντιμαρτυρηθῇ, τὸ ψεῦδος γίνεται· ἐὰν δὲ ἐπιμαρτυρηθῇ ἢ μὴ ἀντιμαρτυρηθῇ, τὸ ἀληθές.—"if it is not attested, or is contested, falsehood arises; but if it is attested or not contested, truth." Cf. Diog. Laërt. x 34; Sext. Emp. *Math.* VII 211 & 216. The translations of the technical terms are those of Sedley (1982), Long & Sedley (1987) and Allen (2001). Asmis (1984) & (1999) prefers *witnessing, no-counterwitnessing, no-witnessing, counterwitnessing*.

37 Epic. *Hdt.* 38; id. *RS* 24. Cf. Diog. Laërt. x 34.

38 Epic. *Hdt.* 80 (... αἰτιολογητέον ὑπέρ τε τῶν μετεώρων καὶ παντὸς τοῦ ἀδήλου, ...).

at all, such as the existence of atoms and void, and other theories fundamental to Epicurean physics.³⁹ Of the two categories mentioned above, the προσμένοντα are typically tested by *attestation* and *non-attestation*,⁴⁰ the ἄδηλα by *non-contestation* and *contestation*.⁴¹

2.3.2.1.2   Sextus Empiricus on Epicurus' Conditions of Truth and Falsehood

The only complete account of Epicurus' four methods of verification and falsification that has come down to us is provided by Sextus Empiricus in *Math.* VII 211–216.⁴² According to Sextus, *attestation* occurs when a hypothesis about something is confirmed by a closer observation of the object in question, e.g. when we see someone approaching from afar and hypothesize that it is Plato, this hypothesis is confirmed when he has come closer and is seen to be really Plato.⁴³ When, on the other hand, on approaching he turns out not to be Plato, the hypothesis is rejected by *non-attestation*.⁴⁴ Note that Sextus' use of the term *non-attestation* is more restrictive than the words themselves suggest: *non-attestation* seems to denote not merely the *negation or absence of attestation* but rather the *attestation of the negated hypothesis*.

Of *contestation* Sextus gives the following account (214):

Ἡ μέντοι ἀντιμαρτύρησις {…} ἦν γὰρ συνανασκευὴ τοῦ φαινομένου τῷ ὑποσταθέντι ἀδήλῳ, οἷον ὁ Στωικὸς λέγει μὴ εἶναι κενόν, ἄδηλόν τι ἀξιῶν, τούτῳ δὲ οὕτως ὑποσταθέντι ὀφείλει τὸ φαινόμενον συνανασκευάζεσθαι, φημὶ δ' ἡ κίνησις· μὴ ὄντος γὰρ κενοῦ κατ' ἀνάγκην οὐδὲ κίνησις γίγνεται …	*Contestation* {…} was the elimination of the appearance with the hypothesized non-evident fact, as, for instance, when the Stoic says there is no void, claiming something non-evident, the appearance (I mean motion) must be co-eliminated with what is thus hypothesised, for if there is no void, by necessity motion doesn't occur either …

---

39   Epic. *Hdt.* 38, introducing the fundamental theories of *Hdt.* 38–44.
40   Epic. *RS* 24. Cf. Diog. Laërt. X 34.
41   Epic. *Pyth.* 2 [88], 3 [88], 7 [92] (*non-contestation* applied to astronomical and meteorological phenomena); Sext. Emp. *Math.* VII 213–214 (*contestation* and *non-contestation* applied to what is by nature unobservable).
42   = Long & Sedley 18A = Usener 247 (part).
43   Sext. Emp. *Math.* VII 212.
44   Sext. Emp. *Math.* VII 215.

A hypothesis about something non-evident (ἄδηλον) is proved wrong by *contestation* when its acceptance would lead to the elimination or cancellation of an evident fact. The argument can be set out as follows:

¬v → ¬m   if there were no void, there would be no motion
   m   but there is motion
   ───
   v   therefore there is void

If the necessity of the first premise, and the evidence of the second are accepted, the conclusion must be true.

Non-contestation is explained by Sextus as follows (213–214):

Οὐκ ἀντιμαρτύρησις δέ ἐστιν ἀκολουθία τοῦ ὑποσταθέντος καὶ δοξασθέντος ἀδήλου τῷ φαινομένῳ, οἷον ὁ Ἐπίκουρος λέγων εἶναι κενόν, ὅπερ ἐστὶν ἄδηλον, πιστοῦται δι᾽ ἐναργοῦς πράγματος τοῦτο, τῆς κινήσεως· μὴ ὄντος γὰρ κενοῦ οὐδὲ κίνησις ὤφειλεν εἶναι, τόπον μὴ ἔχοντος τοῦ κινουμένου σώματος εἰς ὃν περιστήσεται διὰ τὸ πάντα εἶναι πλήρη καὶ ναστά, ὥστε τῷ δοξασθέντι ἀδήλῳ μὴ ἀντιμαρτυρεῖν τὸ φαινόμενον κινήσεως οὔσης.

Non-contestation is the attendance of the hypothesized and supposed non-evident fact upon the appearance, as for instance, when Epicurus says there is void, which is non-evident, this is proved by an evident thing, motion: for if there were no void, there shouldn't be motion either, as the moving body wouldn't have a place into which to come round, because everything would be full and packed; therefore the appearance does *not contest* the supposed non-evident thing, since there *is* motion.

A hypothesis about something non-evident (ἄδηλον) is proved right by *non-contestation* when it can be shown to 'attend upon', or be implied by, an evident appearance, as, for instance, the non-evident existence of void follows from the evident existence of motion. Sextus' general account of *non-contestation* suggests the following schema (*modus ponens*):

m → v   if there is motion, there is void
  m   there is motion
  ───────────────────────────────────────
  v   therefore there is void

This schema, however, fails to make clear the function of the negation in *non-contestation*. If we follow Sextus' example rather than his theoretical account, we find a different, though logically equivalent, schema (*modus tollens*):

¬v → ¬m	if there were no void, there would be no motion
m	but there is motion

v	therefore there is void

Now the schema becomes identical to that of *contestation*! The only difference is one of focus: *contestation* is about *falsifying* a hypothesis (viz. the nonexistence of void), whereas *non-contestation* is about *verifying* a hypothesis (viz. the existence of void), *by falsifying its negation* (viz. the *non*-existence of void). In other words, according to Sextus' account, *non-contestation* does not just denote—as the name would suggest—the mere absence of *contestation*, but *contestation* of the *negated* hypothesis, just as *non-attestation* was *attestation* of the *negated* hypothesis.

2.3.2.1.3 Epicurus' Account of the Fundamental Theories of Physics

In *Hdt.* 38–44, Epicurus discusses, under the general heading of ἄδηλα, i.e. *nonevident* things, the principal and fundamental tenets of his physical theory (e.g. that nothing comes from nothing, that the sum total of things is unchanging, that everything consists of bodies and void, etc.). One of the subjects discussed is the existence of void, which Epicurus sets out as follows (*Hdt.* 40):[45]

... εἰ ⟨δὲ⟩ μὴ ἦν ὃ κενὸν καὶ χώραν καὶ ἀναφῆ φύσιν ὀνομάζομεν, οὐκ ἂν εἶχε τὰ σώματα ὅπου ἦν οὐδὲ δι' οὗ ἐκινεῖτο, καθάπερ φαίνεται κινούμενα.	... and if there were not what we call void and space and intangible nature, bodies would have no place to be or through which to move, as they are observed to move.

Except for the suppressed conclusion, viz. that void exists, this argument is identical, both in subject and structure, to Sextus' example of *non-contestation* (§ 2.3.2.1.2 on p. 15 above), and most of Epicurus' other arguments in this section

---

45 Cf. Lucr. *DRN* I 334–345. On this argument see Allen (2001) 195–196; Furley (1971), and Furley's response to Schrijvers in Gigon (1978), 117–118.

have the same structure. Epicurus does not, however, in this context speak of *non-contestation*, nor does he provide any other name for the procedure.

A little further on Epicurus discusses the sizes of atoms (*Hdt.* 55–56):

Ἀλλὰ μὴν οὐδὲ δεῖ νομίζειν πᾶν μέγεθος ἐν ταῖς ἀτόμοις ὑπάρχειν, *ἵνα μὴ τὰ φαινόμενα ἀντιμαρτυρῇ*· {…} πᾶν δὲ μέγεθος ὑπάρχον οὔτε χρήσιμόν ἐστι πρὸς τὰς τῶν ποιοτήτων διαφοράς, ἀφῖχθαί τε ἅμ᾽ ἔδει καὶ πρὸς ἡμᾶς ὁρατὰς ἀτόμους· ὃ οὐ θεωρεῖται γινόμενον …	Nor, moreover, must we suppose that every size exists among the atoms, *lest the appearances contest this*, {…} but the existence of every size of atoms is not required for the differences of their qualities, and at the same time visible atoms would have to come within our ken, which is not observed to happen …

Here Epicurus has interwoven two different arguments. In the first place the existence of atoms of every size is unnecessary for his physical theory, and in the second place it would entail the existence of observable atoms,[46] which is in conflict with the evidence of the senses. The second part of the argument can be set out schematically as follows:

p → q if atoms could have every size, some atoms would be observable
¬q   but there are no observable atoms
―――――――――――――――――――――――――――――――――――――
¬p   therefore atoms cannot have every size

The argument closely matches Sextus' account of *contestation*, and—what is more—this time Epicurus himself refers to the argument by this very term: μὴ τὰ φαινόμενα ἀντιμαρτυρῇ—'lest the appearances *contest* this'. May we then conclude that Sextus has correctly reported Epicurus' views on *contestation* and *non-contestation*?

---

46 Epicurus speaks of 'visible atoms', but on his own theory (cf. *Hdt.* 46–50 and Lucr. IV 54–216) even enormous atoms would not be directly *visible*, because, being atomic, they would not be able to shed the necessary images. They would, however, be *indirectly visible*, by blocking other things from view, and also *tangible* since touch does not require the shedding of particles. Such atoms could therefore be said to be *observable*.

2.3.2.1.4 Epicurus' Use of Non-Contestation in Astronomy and Meteorology

Until now we have only seen *contestation* and *non-contestation* applied to fundamental physical theories. It remains to be seen how they are used in astronomy and meteorology.

In *Pyth.* 7 [92], Epicurus offers two alternative explanations to account for the risings and settings of the sun, the moon and the stars:

Ἀνατολὰς καὶ δύσεις ἡλίου καὶ σελήνης καὶ τῶν λοιπῶν ἄστρων

(1) καὶ κατὰ ἄναψιν γίνεσθαι δύνασθαι καὶ κατὰ σβέσιν, τοιαύτης οὔσης περιστάσεως καὶ καθ' ἑκατέρους τοὺς τόπους, ὥστε τὰ προειρημένα ἀποτελεῖσθαι· οὐδὲν γὰρ τῶν φαινομένων ἀντιμαρτυρεῖ,

(2) ⟨καὶ⟩ κατ' ἐκφάνειάν τε ὑπὲρ γῆς καὶ πάλιν ἐπιπροσθέτησιν τὸ προειρημένον δύναιτ' ἂν συντελεῖσθαι· οὐδὲ γάρ τι τῶν φαινομένων ἀντιμαρτυρεῖ.

Risings and settings of the sun and the moon and the other heavenly bodies

(1) may come about by kindling and extinction, the circumstances at both places [i.e. the places of rising and setting] being such as to produce the afore-mentioned results: *for nothing in the appearances contests this*,

(2) and by their appearance above the earth and again the (earth's) interposition the afore-mentioned result might be produced: *for nothing in the appearances contests this.*

According to Epicurus, the heavenly bodies are either repeatedly extinguished and then rekindled, or they pass unaltered below the earth and then emerge again.[47] Both options must be accepted, because '*nothing in the appearances contests*' either one of them. The same or similar terms, expressing either the absence of disagreement[48] or the presence of (positive) agreement with the appearances,[49] occur throughout the *Letter to Pythocles*, often to account for each one of a number of alternative explanations.

---

[47] Similarly Lucr. *DRN* V 650–679 on the causes of nightfall and dawn.
[48] *Pyth.* 2 [88] "οὐκ ἀντιμαρτυρεῖται", *Pyth.* 3 [88] "τῶν γὰρ φαινομένων οὐδὲν ἀντιμαρτυρεῖ", *Pyth.* 9 [93] "οὐθενὶ τῶν ἐναργημάτων διαφωνεῖ", *Pyth.* 11 [95] "οὐθὲν ἐμποδοστατεῖ τῶν ἐν τοῖς μετεώροις φαινομένων", *Pyth.* 16 [98] "οὐ μάχεται τοῖς φαινομένοις".
[49] *Pyth.* 2 [86] "τοῖς φαινομένοις συμφωνίαν" & "ταῖς αἰσθήσεσι σύμφωνον"; *Pyth.* 2 [87] "συμφώνως τοῖς φαινομένοις" and "σύμφωνον ὂν τῷ φαινομένῳ"; *Pyth.* 9 [93], 12 [95], 32 [112] "τὸ σύμφωνον τοῖς φαινομένοις".

Now, if, as Epicurus claims,[50] *non-contestation* establishes truth, then each one of the alternative explanations must be true. It should be noted, however, that Epicurus' argument in these passages is not at all like Sextus' account of *non-contestation* (§ 2.3.2.1.2 on p. 15 above); nowhere in the *Letter to Pythocles* do we encounter anything that resembles the syllogistic structure of *non-contestation* as set out by Sextus and as applied (though without this 'label' of *non-contestation*) by Epicurus in the *Letter to Herodotus* to prove his fundamental physical theories. In the present context *non-contestation* seems to be nothing more than the absence of *contestation* by the appearances, or—in other words—the (positive) agreement with the appearances.

2.3.2.1.5 Epicurus on the Distinction between Single and Multiple Explanations

It would seem then, that Epicurus dealt differently with the fundamental physical theories and the more specialised theories concerning astronomical and meteorological phenomena. Whereas the first are proved to be uniquely true by showing that their negation is contested by appearances, the second are accounted for with a number of alternative theories which must all be accepted because none of them is contested by appearances. Epicurus explicitly contrasts the two types of problems in *Pyth.* 2 [86]:

Μήτε τὸ ἀδύνατον [καὶ] παραβιάζεσθαι, μήτε ὁμοίαν κατὰ πάντα τὴν θεωρίαν ἔχειν ἢ τοῖς περὶ βίων λόγοις ἢ τοῖς κατὰ τὴν τῶν ἄλλων φυσικῶν προβλημάτων κάθαρσιν, οἷον ὅτι τὸ πᾶν σῶμα καὶ ἀναφὴς φύσις ἐστίν, ἢ ὅτι ἄτομα στοιχεῖα καὶ πάντα τὰ τοιαῦτα δὴ ὅσα μοναχὴν ἔχει τοῖς φαινομένοις συμφωνίαν· ὅπερ ἐπὶ τῶν μετεώρων οὐχ ὑπάρχει, ἀλλὰ ταῦτά γε πλεοναχὴν ἔχει καὶ τῆς γενέσεως αἰτίαν καὶ τῆς οὐσίας ταῖς αἰσθήσεσι σύμφωνον κατηγορίαν.	We must not try to force an impossible explanation, nor employ a method of inquiry similar in every respect to our reasoning either about the modes of life or with respect to the sorting-out of other physical problems, such as our statement that 'the universe consists of bodies and the intangible', or that 'the elements are indivisible', and all such statements which exhibit a singular agreement with appearances. For this is not so with the things above us: they admit of a plural account of their coming-into-being and a plural expression of their nature which agrees with our sensations.

---

50 See n. 36 on p. 14 above.

Although the fundamental physical theories (just like the theories concerning the modes of live) no less than astronomical and meteorological theories need to fulfil the requirement of agreement with the appearances, the method of inquiry by which they are approached is different. In other words: Epicurus explicitly recognizes the existence of two different methods of inquiry, one that is applied to such problems as admit only one answer (such as the fundamental physical theories), and one that is applied to those problems that admit several answers (such as we find in astronomy and meteorology).

2.3.2.1.6   Lucretius on the Truth of All Alternative Explanations (1)

Lucretius too offers multiple explanations to account for astronomical (*DRN* V 509–770) and meteorological (*DRN* VI) phenomena. Close to the beginning of his astronomical section, having just offered a number of alternative explanations for the movements of the stars, he states his view on the epistemological status of these explanations (V 526–533):[51]

nam quid in hoc mundo sit eorum ponere certum	for to state with certainty which of these causes holds in our world is
difficilest; sed quid possit fiatque per omne in variis mundis varia ratione creatis,	difficult; but what can and does happen throughout the universe in the
id doceo plurisque sequor disponere causas,	various worlds created in various ways,
530 motibus astrorum quae possint esse per omne;	this I teach, and I proceed to set forth several causes for the motions of the
e quibus una tamen siet hic quoque causa necessest,	stars, which may apply throughout the universe; one cause out of this number,
quae vegeat motum signis; sed quae sit earum	however, is necessarily the case here too, which gives force to the motion of
praecipere haudquaquamst pedetemptim progredientis.	the stars, but which of them it is, is not for them to lay down who proceed step by step.

Although in our world each explanation can at best be called possible, in the universe at large, given the infinity of space and matter and hence of worlds, any given possibility cannot fail to be realised (the 'principle of plenitude'),[52]

---

51   I will assume, with Striker (1974) 78, Sedley (1982) 270, Asmis (1984) 325 & (1999) 289, Allen (2001) 197, Bénatouïl (2003) 45, and others, that Lucretius' account may be used to supplement Epicurus', providing details not preserved in Epicurus' extant works. For a more cautious view see Verde (2013) 139–141, and Masi (2015) 48–50.

52   Cf. Lucr. *DRN* V 422–431 and Epic. fr.266 Us. (both referring to the infinity of *time* rather

and so every possible explanation is also 'true', if not here, then somewhere else. This may also explain why Epicurus in his *Letter to Pythocles* sometimes seems to speak of the alternative explanations as actual and coexisting rather than merely possible and mutually exclusive.[53]

2.3.2.1.7 Lucretius on the Truth of All Alternative Explanations (2)

There is a second passage where Lucretius deals with the method of multiple explanations. In VI 703–711, preceding his account of the summer flooding of the Nile, Lucretius writes:

Sunt aliquot quoque res quarum unam dicere causam non satis est, verum pluris, unde una tamen sit; corpus ut exanimum siquod procul ipse iacere conspicias hominis, fit ut omnis dicere causas conveniat leti, dicatur ut illius una; nam ⟨ne⟩que eum ferro nec frigore vincere possis interiisse neque a morbo neque forte veneno, 710 verum aliquid genere esse ex hoc quod contigit ei scimus. Item in multis hoc rebus dicere habemus.	There are also a number of cases for which naming one cause is not enough, but several, one of which is nevertheless the case; just as, if you should yourself see a person's dead body lying at a distance, it happens to be fitting to name all the causes of death, to make sure that the one cause of this death be named; for you could not prove that he died from the sword or from cold or from disease or perchance from poison, but we know that it was something of this sort that befell him. Similarly we must say this in many cases.

---

than matter and space). On the Epicurean use of the principle of plenitude see e.g. Sedley (1998a) 175, n. 29. A precursor of this principle (based on the infinity of space and matter) is described by Aristotle (*Phys.* III, 4, 203b25–30), who may be rendering a Democritean view: see Asmis (1984) 264–265.

53 In the *Letter to Pythocles* I have counted 34 cases where multiple explanations are offered. 20 of these are accompanied by some verb or expression denoting possibility (ἐνδέχεται, δύναται, οὐκ ἀδύνατον, etc.). Of the 14 remaining cases, 5 exhibit a purely *disjunctive* list of alternative explanations (ἤτοι A ἢ B ἢ Γ). The 9 remaining cases are either *conjunctive* (καὶ A καὶ B καὶ Γ) or *mixed* (καὶ A καὶ B ἢ Γ): in these 9 cases the language seems to suggest that several explanations are true at the same time. For more details see APPENDIX 1 on p. 269 below.

Although this passages is not incompatible with the earlier one, there is an interesting shift of focus. This time we hear nothing of the infinite number of worlds, nothing of the principle of plenitude, and nothing of the truth of all explanations, and, although the event is still viewed as a particular instant of a certain class of events, the emphasis seems to be on the particular instance, to which only one explanation applies (although we do not know which one), rather than the whole class of events, to which many explanations apply. Moreover, the chosen example, a dead body, seems strangely inappropriate. While multiple explanations are typically applied to things that can *only* be seen from a distance, there does not seem to be any cogent reason why the dead body could not be approached and examined more closely, so as to eliminate certain explanations and perhaps even arrive at the one true cause of death (concerning this example see also p. 30 below).

2.3.2.2   Three Modern Approaches

Having reviewed the available evidence we seem to end up with two different and seemingly incompatible conceptions of *non-contestation*. On the one hand there is *non-contestation* as presented by Sextus Empiricus (§ 2.3.2.1.2 on p. 15 ff. above). According to this account *non-contestation* is *contestation of the negated hypothesis*, which by means of a syllogism establishes the exclusive truth of its hypothesis. On the other hand there is *non-contestation* as applied by Epicurus himself in the *Letter to Pythocles* (§ 2.3.2.1.4 on p. 19 ff. above). Here *non-contestation* seems to mean nothing more than the *absence of contestation*, which only establishes the *possibility* of each of a number of alternative explanations, which can be called 'true' only by virtue of the principle of plenitude (§ 2.3.2.1.6 on p. 21 above). Various approaches to these two conceptions of *non-contestation* have been put forward, which can be divided into three main groups:

1. Both conceptions are correct: Epicurus sometimes used *non-contestation* in the stricter sense of *contestation of the negated hypothesis*, as explained by Sextus Empiricus, and sometimes in the wider and vaguer sense of *absence of contestation*, as in the *Letter to Pythocles*.
2. Sextus' account of *non-contestation* is correct, and Epicurus' use of *non-contestation* in the *Letter to Pythocles* must be interpreted accordingly: *non-contestation* does not apply to individual alternative explanations, but only to the entire disjunction of alternative explanations.
3. Sextus' account of *non-contestation* is incorrect or at least incomplete. Epicurus' own use of *non-contestation* in the *Letter to Pythocles* is our only certain guide to the working of *non-contestation* as conceived by Epicurus.

Below I will examine each of these three approaches more thoroughly.

Ad 1. The first approach, proposed by Gisela Striker,[54] is to simply acknowledge the existence, in Epicurean epistemology, of two different kinds of *non-contestation*. The first kind, which we may call *non-contestation in the strict sense*, is *contestation of the negated hypothesis*, the method that is described by Sextus Empiricus (§ 2.3.2.1.2 on p. 15 above) and repeatedly applied (though without the label of 'non-contestation') by Epicurus himself in the *Letter to Herodotus* (§ 2.3.2.1.3 on p. 17 above). It is only with reference to this type of *non-contestation* that Epicurus calls *non-contestation* a condition of truth (§ 2.3.2.1.1 on p. 13 above). The second kind, which may be called *non-contestation in the weak sense*, is simply the negation or absence of *contestation*, which is invoked several times in Epicurus' *Letter to Pythocles* to establish the possibility of each of a number of alternative explanations of astronomical and meteorological phenomena (§ 2.3.2.1.4 on p. 19 above). Only with recourse to the 'principle of plenitude' (§ 2.3.2.1.6 on p. 21 above) can each alternative explanation also be called true, if not in this world, then in another.[55]

This approach has the great advantage of preserving all the available evidence: it allows us to accept Sextus' account of *non-contestation* and at the same time do justice to Epicurus' own use of *non-contestation* in the *Letter to Pythocles*. The great drawback is that it leaves us with two different kinds of *non-contestation*, not distinguished by name, but having different truth-values, despite Epicurus' unqualified claim that *non-contestation* always establishes truth.

Ad 2. Another interesting solution to the problem is offered by Jim Hankinson.[56] According to Hankinson, an Epicurean explanation of an atmospherical or celestial phenomenon takes the form of a disjunction of possible explanations: "x occurs because either E_1 or E_2 or ... E_n. At most one of the E_i's can be the true explanation (cf. Lucretius 6. 703–704); but if the disjunction is sufficiently all-embracing, one of them will be: and that is all that is required." Tad Brennan,[57] elaborating on Hankinson's remark, adds: "the point could be strengthened by reflecting that the actual reference in DL 10.86 [*Pyth.* 2] does not mention multiple "*aitiai*", plural, but a "*pleonachên aitian*",[58] i.e. a single

---

54 Striker (1974) 76; id. (1996) 45. See also Fowler (2002) 191–192.
55 Striker (1974) 78–79; id. (1996) 47–48.
56 Hankinson (1999a) 221–223, and (1999b) 505–507.
57 Brennan (2000) commenting on Hankinson (1999b) 505–507.
58 See the text quoted in item E on p. 20 above.

explanation with a complex, manifold structure. This is why the assertion of the whole disjunction is safe but the isolated assertion of any one disjunct is not (DL 10.87) [*Pyth.* 2]." In other words, if complete, the whole disjunction, i.e. the entire range of possible explanations, can be called *true*, and could in principle be demonstrated to be so by *non-contestation* (as interpreted by Sextus Empiricus). In this context Brennan might also have quoted *Pyth.* 2 [88], where *non-contestation* seems to be applied to the fact of there being multiple explanations, not to any single explanation in particular:

Τὸ μέντοι φάντασμα ἑκάστου τηρητέον καὶ ἐπὶ τὰ συναπτόμενα τούτῳ διαιρετέον, ἃ οὐκ ἀντιμαρτυρεῖται τοῖς παρ' ἡμῖν γινομένοις πλεοναχῶς συντελεῖσθαι.	Yet the appearance of each phenomenon must be preserved, and, as regards what is associated with it, those things must be distinguished whose production in multiple ways is *not contested* by the phenomena here with us.

Although Hankinson's approach may seem to solve the observed incompatibility, I think it must be rejected for the following reasons:

a. Epicurus and Lucretius do *not* generally present their lists of alternative explanations as logical disjunctions; instead several of the lists in the *Letter to Pythocles* exhibit a conjunctive or a mixed structure,[59] and similar observations can be made with respect to Lucretius' lists of alternative explanations.
b. Several of Epicurus' lists of alternative explanations are explicitly open to the addition of further explanations,[60] a fact which—though not strictly fatal to Hankinson's interpretation—would turn the 'truth' of the disjunction into an empty truism.
c. The most important objection, however, is the fact that in the *Letter to Pythocles non-contestation* is sometimes explicitly invoked to prove each *one* of a number of alternative explanations rather than the disjunction as a whole.[61]

Ad 3. A third approach is to simply dismiss Sextus' account of *non-contestation* as incorrect or inaccurate. This has been, with minor variations, the approach

---

59   See p. 22 with n. 53 above. See also Bakker (2010) 20 with n. 52 and Masi (2015) 50 n. 37.
60   See p. 11 above. See also Bakker (2010) 10 and Masi (2015) 50 n. 37.
61   E.g. *Pyth.* 7 [92]: text quoted on p. 19 above. See also Bakker (2010) 24. A similar criticism of Hankinson's and Brennan's position is also voiced by Verde (2013) 131 n. 1.

of David Sedley, Elizabeth Asmis and James Allen.[62] According to this approach, there is only one kind of *non-contestation*, viz. the method which Epicurus himself employs several times in his *Letter to Pythocles*, where *non-contestation* amounts to nothing more specific than agreement or compatibility with the appearances, possibly resulting in the acceptance of several theories at the same time. It is to this procedure that Epicurus' claim that *non-contestation* establishes truth must be applied.

It cannot be denied that the method which Sextus Empiricus designates as '*non-contestation*' corresponds very well to the way Epicurus, in the *Letter to Herodotus*, proves many of his fundamental physical theories by showing that *their contradictories* are *contested* by the appearances (§ 2.3.2.1.3 on p. 17 above). It is also true, however, that in the *Letter to Pythocles* Epicurus makes the fundamental physical tenets, no less than astronomical and meteorological theories, subject to agreement with the appearances (§ 2.3.2.1.5 on p. 20 above). This apparent contradiction can be explained, basically, in two ways: either (a) Epicurus did not consider *contestation of the contradictory* a proof in its own right, despite his repeated use of this method in the *Letter to Herodotus*, believing that the real proof consisted only in *non-contestation* (as applied in the *Letter to Pythocles*), i.e. agreement with the appearances, which—curiously—he often fails to invoke in the *Letter to Herodotus*, or (b) he though that *contestation of the contradictory* by itself somehow implied, and therefore could be subsumed under, *non-contestation*.

The first option is chosen by Sedley, who minimizes the importance of *contestation of the contradictory* for Epicurus, pointing out that the logical implication on which this procedure rests, e.g. 'if there is no void, there is no motion', must itself be proved by *non-contestation*, i.e. agreement with the appearances,[63] and therefore cannot count as a condition of truth. Against this position I would like to stress the following two points: firstly that Epicurus in the *Letter to Herodotus* (see § 2.3.2.1.3 on p. 17 above) repeatedly uses *contestation of the contradictory* as a proof in its own right, presenting the underlying implications as self-evident, rather than requiring further proof; and secondly that Epicurus in the *Letter to Pythocles* (2 [86]: see § 2.3.2.1.5 on p. 20 above) explicitly distinguishes *two* methods of inquiry, one applied to the fundamental physical problems and resulting in single explanations, the other applied to meteorological and astronomical problems and resulting in multiple explanations.

---

62 Sedley (1982) 263–272; Long & Sedley (1987), vol. I, 90–97; Asmis (1984) 178–180, 193–196, 211, 321–336; id. (1999) 285–294; and Allen (2001) 194–205 & 239–241.

63 Sedley (1982) 269 with n. 70; cf. Allen (2001) 203.

The second option, or something like it, is advocated by Asmis and Allen. Asmis acknowledges the importance of *contestation of the contradictory*, but generally refers to it as a kind of (positive) *contestation* (rather than *non-contestation* as Sextus does) which she opposes to the method of 'induction' (i.e. agreement with the appearances, or *non-contestation*).[64] In a later publication, however, Asmis seems to subsume both types of scientific inference under the general heading of *non-contestation*, which makes her position come very close to Gisela Striker's (see p. 24 f. above).[65] Allen suggests that for Epicurus proofs by *contestation of the contradictory* would have been a special case of the more common proof by agreement with the appearances.[66] Asmis' and Allen's view may perhaps be summarized as follows: Epicurus claims that in matters of the non-evident truth is established by *non-contestation*, by which he means agreement with the appearances. This agreement with the appearances may be singular, as with the fundamental physical theories, or plural, as with most astronomical and meteorological phenomena. To establish a theory's singular agreement with the appearances Epicurus uses a special method that may be described as *contestation of the contradictory hypothesis*, which he seems to have considered a special kind of *non-contestation*. Sextus' report would then be inaccurate insofar as it simply identifies this *contestation of the contradictory hypothesis* with *non-contestation*, of which it is only a special kind.

Having reviewed a number of modern interpretations concerning the truth of multiple explanations it is now time to summarize the results. The first approach, defended by Gisela Striker, assuming two different kinds of *non-contestation*, one of which establishes the truth of single theories, while the other only establishes the possibility of each one of a number of alternative explanations, introduces a distinction that seems unwarranted by Epicurus' unqualified claim that *non-contestation* establishes truth.[67] A second approach, proposed by Jim Hankinson, claiming that *non-contestation* does not apply to, and therefore does not establish the truth of, individual alternative explanations, but applies only to the complete disjunction of alternatives, is refuted by the evidence. This leaves us only the third approach, which maintains that Epicurus really claimed the truth of all alternative explanations. As we saw above, two varieties of this approach can be distinguished. The first variety, defended

---

64 Asmis (1984) 211 *et passim*.
65 Asmis (1999) 289.
66 Allen (2001) 200.
67 See n. 36 on p. 14 above.

by David Sedley, rejects not only Sextus' account of *non-contestation* as contestation of the contradictory hypothesis, but also minimizes the importance of Epicurus' frequent use of this very method (even though he does not call it 'non-contestation') to prove the most fundamental physical theories. At the same time it seems to ignore Epicurus' explicit distinction of *two* methods of inquiry, instead suggesting that truth is ultimately always established by agreement with the appearances. The second variety, which may be attributed to Elizabeth Asmis and James Allen, deals more cautiously with the available evidence. It provides a plausible way in which contestation of the contradictory hypothesis may at once be distinguished from, and yet subsumed under, *non-contestation*. On this interpretation Sextus' report of *non-contestation* can be retained as long as we realise that it only applies to one special kind of *non-contestation*.

In sum, on the basis of the evidence we have thus far examined the conclusion that Epicurus really held all alternative explanations to be *true* seems inevitable.

2.3.2.3 Some Reservations

Those who maintain the truth of all alternative explanations (see p. 25 f. above), generally explain this *truth* by means of the *principle of plenitude*, referring to Lucretius' testimony in *DRN* V 526–533 (item F on p. 21 above), which is usually taken to mean that, although for each individual event only one explanation can be true (though we do not know which one), with respect to the general type of event that is being explained every alternative explanation that is *not contested* by the appearances is *true*.[68] I have a number of reservations about this claim.

My first reservation concerns the meaning of the word 'true'. Although the principle of plenitude provides a way in which each one of a number of alternative explanations may be called *true*, it must be observed that this *truth* is something very different from the universal and ubiquitous truth attaching to those theories which exhibit a singular agreement with the appearances, and—despite his unqualified claim that *non-contestation* establishes *truth*— Epicurus does seem to acknowledge the difference. Why else would he, in the introduction of his *Letter to Pythocles* (2 [86]; see § 2.3.2.1.5 on p. 20 above), oppose plural to singular agreement with the appearances? In order to serve as the foundation of a systematic physical theory some tenets not only allow but

---

[68] Asmis (1984) 322, 324–325; id. (1999) 289; Sedley (1982) 270 with n. 72; Long & Sedley (1987) vol. 1, 95–96; Allen (2001) 197–198.

actually require not just *agreement* but *singular agreement* with the appearances, so as to be pronounced universally true, and this singular agreement with the appearances can only be established by contestation of the contradictory hypothesis. It is almost as if Epicurus were saying that all explanations are true, but some (viz. singular explanations) are more true than others.[69] In this respect Gisela Striker's division of Epicurean *non-contestation* into a stricter kind which establishes truth and a weaker kind which establishes only possibility (first option above) is actually a good approximation of Epicurus' use of *non-contestation*.

My second reservation is about the general application of the principle of plenitude to all cases of multiple explanations. It seems to be universally agreed upon that Lucretius' account of the principle may be generalised in this way, but in fact his account is appended to, and only explicitly refers to, the alternative explanations of the motions of the stars. Yet, even if we allow that his words have a broader application, it seems legitimate to investigate the scope of their applicability. In order to be able to apply the *principle of plenitude* each phenomenon under investigation needs to be viewed as an instance of a general type of events. In the case of meteorological phenomena this general type can be easily envisaged even without reference to the infinity of worlds. Thunder, for instance, is accounted for by Lucretius with nine different explanations.[70] Only one of these will be true with respect to one particular thunderclap (although we do not know which one), but all of them are true with respect to thunder in general. In the case of astronomical phenomena, which are often concerned with unique objects, it is harder to accept them as instances of some general type. Only if we are prepared to accept 'sun' and 'moon' as generic terms for objects of which there may be only one in this world but infinitely many in the universe at large,[71] can we claim that each possible explanation is also *true*. Yet, what are we to do with multiple explanations for exceptional local phenomena such as eruptions of the Etna, the summer flooding of the Nile, the anomalous daily temperature fluctuation of the spring near the shrine of Hammon, etcetera, which Lucretius discusses in the second part of book VI (see also p. 109 ff. below)? Are we to suppose that in the universe at large there are infinitely many Niles flowing down from infinitely many Aethiopian mountain-ranges each overflowing in summer to irrigate infinitely many Egypts? That would, in fact, be a logical outcome of the assumption of an

---

69 On Epicurus' application of different degrees of exactitude to different fields of physics, see Bénatouïl (2003) 46–47.
70 See n. 13 above and table 3.8 on p. 143 below.
71 This actually appears to be Lucretius' point of view in *DRN* II 1084–1086.

infinity of worlds. Yet, if Lucretius had wished us to think of a general type of event, he would not have emphasized all these inessential particulars, but he would have spoken about rivers that overflow in summer generally, of which the Nile only presents the most notable example.[72] It is clear, then, that here and in similar passages Lucretius was not thinking of general types of events, but of particular and in some cases even unique[73] local phenomena, to which the 'principle of plenitude' does not apply. In these cases each of the alternative explanations can at best be called possible, not true.[74]

There is another reason why in some of these cases the alternative explanations cannot all be called true. Whereas astronomical and meteorological phenomena cannot be physically approached, and therefore rightly belong to the class of *non-evident* things (ἄδηλα) which are typically tested by *contestation* and *non-contestation* (see § 2.3.2.1.1 on p. 13 above), some of the exceptional local phenomena described in *DRN* VI do not necessarily defy closer observation, and so may seem open to testing by *attestation* and *non-attestation* (see § 2.3.2.1.1 on p. 13 and § 2.3.2.1.2, first paragraph, on p. 15 above). One could simply go to Egypt and observe whether the annual flooding of the Nile[75] is somehow correlated to the onset of the etesian winds (715–723), or to the formation of sandbanks in the mouths of the river (724–728), or to the onset of seasonal rains upstream (729–734), or to the melting of snow in the Aethiopian mountains (735–737).[76] So, rather than being *true* for *not being contested by the appearances*, these causes could be said to be still *waiting* (προσμένοντα) to be *attested* or *not* by closer observation (see § 2.3.2.1.1 on p. 13 above), and so be *neither true nor false*. In this respect the example Lucretius has chosen to illustrate

---

72 Seneca, *N.Q.* III 26, 1, informs us that, according to Theophrastus, a number of rivers in Pontus exhibited the same behaviour.

73 In *DRN* VI 712–713 Lucretius calls the Nile *unique* ('unicus') for overflowing in summer (but see n. 72 above).

74 It must be noted that for phenomena of this class Lucretius most often provides only one explanation, sometimes plus a subsidiary one. The summer flooding of the Nile, with four explanations, is exceptional in this respect too: see APPENDIX 2 on p. 272 below. For the general scarcity of alternative explanations for phenomena of this class see the second paragraph of § 3.3.3 on p. 125 below.

75 Or, more precisely, one might have done so before the completion of the Aswan Dam in 1970, which effectively cancelled the Nile's annual flooding.

76 In fact, we have reports of several expeditions going up the Nile in classical times in order to find its sources and establish the causes of its annual flooding: one or two sent out by Alexander the Great, one by Ptolemy II, and one by the emperor Nero. None of these, however, seems to have settled the problem definitively. For an overview see e.g. Gambetti (2015).

the method of multiple explanations in book VI is very appropriate after all. In VI 703–711 (§ 2.3.2.1.7 on p. 22 above), immediately preceding the account of the Nile flood, Lucretius compares the use of multiple explanations to the procedure one should adopt when viewing a dead body from afar: since no cause of death can be excluded all causes should be accepted as possible. Yet, just like the Nile, a dead body *can* be examined at closer range, and so (to a certain extent) reveal the causes of this particular death. In cases like these, then, we may not be justified in calling every alternative explanation *true*.

In sum: all alternative explanations are true, (1) *insofar as* they concern *non-evident* phenomena which are subject to *contestation* and *non-contestation*, and (2) insofar as these phenomena are conceived of as instances of a *general type of event*, and even then we have to subscribe to a very meagre conception of *truth*, which common parlance would rather refer to as *possibility*. Gisela Striker's interpretation (see p. 24f. above) turns out to be not so bad after all.

I would like to add one final observation concerning *the principle of plenitude*. This principle, on which the truth of all alternative explanations rests, itself depends on the assumed existence of a plurality of worlds, which in turn depends on the assumed existence of an infinite number of atoms in infinite space. However, as we shall see in the Chapter Four (see p. 209 below), Lucretius' argument for an infinite number of atoms may not be as strong as he would have wished.

2.3.2.4   Conclusion

Although the conclusion that Epicurus considered all alternative explanations true follows logically from his use of *non-contestation* to support individual alternative explanations and his claim that *non-contestation* establishes truth, there turns out to be much that detracts from this conclusion. Epicurus' efforts to provide a more certain basis than mere *non-contestation* (i.e. agreement with appearances) for his fundamental physical theories show that he did not set much value on the *truth* of multiple explanations, and Lucretius' failure to generalize the *principle of plenitude* to all instances of multiple explanations clearly shows the limitations of identifying *possibility* with *truth*. It need not surprise us therefore that Epicurus himself in the *Letter to Pythocles* most often presents the alternative explanations as simply *possible*, and never explicitly calls them 'true'.[77]

---

[77] See APPENDIX 1 on p. 269 below. Similar observations in Bénatouïl (2003) 44 and Masi (2015) 44.

2.3.3 Contestation or Disagreement with Appearances

The Epicureans' criterion for rejecting theories is pretty straightforward: a theory is false if it disagrees with, or is *contested* by, the appearances (see § 2.3.2.1.2 on p. 15 above). In practice Epicurus and Lucretius only rarely reject theories. Epicurus does, however, repeatedly mention one kind of explanations that should never be admitted in physical enquiry and especially astronomy and meteorology, viz. those explanations which attribute these phenomena to the involvement of the gods. Such an involvement, Epicurus holds, would be in conflict with the blessed nature of the gods and therefore must be rejected.[78] Several examples of such theories are mentioned by Lucretius, who in VI 379–422 argues against the popular view that thunderbolts are Jupiter's work, in VI 753–754 against the myth that crows avoid the Athenian Acropolis because of Pallas Athena's wrath, and in VI 762–766 against the belief that Avernian places are the gates to the Underworld. Yet, even explanations which do not rely on divine interference may sometimes be rejected. One example of this is, again, provided by Lucretius. In *DRN* VI 848–878 he discusses the curious behaviour of the spring near the shrine of Hammon, whose water is cold during the day and hot at night. Before embarking on his own account of the matter, he first describes and rejects a theory brought forward by 'people' (homines), who claim that the sun heats the spring from below during its nocturnal passage under the earth. Lucretius rejects this theory on the ground that, if the sun were able to affect the spring from below through the vast body of the earth, it would affect the spring even more when shining down on it unimpeded by the earth. But then the spring would have to be even hotter during the day, which is not observed to happen. Although the explanation is free from religious superstition, it is contested by appearances and therefore must be rejected.

2.3.4 Incompatibility with Explanations of Other Phenomena

While for the Epicureans incompatibility with the appearances was a sufficient and necessary ground for rejecting an explanation, incompatibility with explanations of other phenomena was not. In an article written in 1978, Abraham Wasserstein faults the Epicureans for this.[79] If the Epicureans had been truly committed to science, he says, they should have paid attention to the fact

---

[78] Epic. *RS* 1; *Hdt.* 76–77; 81; *Pyth.* 14 [97]; 33 [113]; 36 [115–116]. Cf. Lucr. II 1090–1104; V 156–234; 1183–1240; VI 50–79; 379–422; Cic. *ND* I, 52.

[79] Wasserstein (1978) 490–494. See also Bénatouïl (2003) 41.

that many of their theories are interdependent so that elimination of one theory may bring along the elimination of another. For instance, the theory that the sun is extinguished at night[80] is incompatible with the theory that the moon receives its light from the sun,[81] and hence with every explanation of the moon's phases and eclipses that presupposes that the moon shines with reflected light.[82] Although Wasserstein makes a very pertinent observation here, he overstates his point. It is true that in the *Letter to Pythocles* and in the astronomical and meteorological sections of the *DRN* phenomena are generally presented in isolation,[83] but from an Epicurean-epistemological perspective it makes no difference for a theory's possibility if it logically depends on another possible theory. It would only matter if this second theory were for some reason eliminated: in that case every theory that depends on it would be cancelled as well. However, since according to Epicurus all alternative explanations not contested by the appearances are *objectively* possible (see next subsection) and hence 'true', no possible theory will ever be eliminated and neither will such theories as depend on it. Wasserstein contrasts the Epicurean method with 'real astronomy', where consistency with all the other parts of the system is a prerequisite for a theory's acceptability. He fails to note, however, that the existence of such a unified system is in itself just another of those "empty assumptions and arbitrary principles" Epicurus warns us against, urging us to follow "the lead of the appearances" instead.[84] Hence Epicurus' strong opposition to "the partisans of the foolish notions of astronomy," who groundlessly reject any theory that does not fit their mathematical paradigm.[85] So, rather than call Epicurean meteorology unscientific I would say it represents a different type of science, viz. one which privileges observation over theory.

---

80 Epic. *Pyth.* 7 [92] 1st explanation; Lucr. v 650–653, 660–662.
81 Epic. *Pyth.* 11 [94–95] 2nd explanation; Lucr. v 705.
82 Lucr. v 705–714 and 762–767.
83 A salient example is Epicurus' and Lucretius' separate treatment of thunder (*Pyth.* 18 [100]; *DRN* VI 96–159), lightning (*Pyth.* 19 [101–102]; *DRN* VI 160–218) and thunderbolts (*Pyth.* 21 [103–104]; *DRN* VI 219–422), as if they were three independent phenomena, rather than symptoms of a single phenomenon. On the other hand in *DRN* V explanations of lunar phases (705–750) and eclipses (762–770) are explicitly distinguished according to whether they assume that the moon shines with its own or with reflected light.
84 *Pyth.* 2 [86]: "Οὐ γὰρ κατὰ ἀξιώματα κενὰ καὶ νομοθεσίας φυσιολογητέον, ἀλλ' ὡς τὰ φαινόμενα ἐκκαλεῖται." (translation Bailey, modified).
85 *Pyth.* 33 [113]: "τῶν τὴν ματαίαν ἀστρολογίαν ἐζηλωκότων" (translation Bailey, modified): see also p. 56 ff. below.

Nevertheless, Wasserstein's article does reveal a dangerous flaw in Epicurean meteorology. This flaw does not, however, consist in the use of multiple explanations, but, on the contrary, in Epicurus' failure to implement this method rigorously. It is the presence of *single* explanations in Epicurean meteorology that threatens to undermine multiple explanations, when every accepted alternative theory may yet be rejected due to incompatibility with certain single explanations. With 'single explanations' I mean, of course, those which exclude any alternative view (see on p. 11 above), such as the claim that the sun and the other heavenly bodies are the size they appear to be, or the view— nowhere stated explicitly but generally assumed to be implied—that the earth is flat. Both these views and their relation to each other, to multiple explanations and to mainstream astronomy will be investigated in Chapter Four below.

2.3.5   Non-Contestation and Analogy

Until now I have tried to evade the question of **how** *non-contestation* establishes the *possibility* (or *truth*) of multiple explanations. Above I have argued for the existence of two kinds of truth; now it will be necessary to distinguish two kinds of possibility as well. If *non-contestation* is the absence of *contestation*, and if *contestation* of a theory consists in tracing a fatal incompatibility with the appearances (as Sextus explains), then *non-contestation* might be interpreted as the failure to trace such an incompatibility. In that case we would have established the theory's *subjective possibility*: the theory is possible as far as our knowledge goes; it cannot be excluded that at some later point in time new information may force us to reconsider and reject the theory. However, Epicurus' equation of *possibility* with *truth* clearly shows that he had in mind something more fundamental than that, viz. *objective possibility*, a possibility beyond the limitations of our knowledge, residing in the structure of the universe itself.[86] However, in order to establish such an objective possibility we would need not just the absence of *contestation*, but the certainty that the theory will *never be contested* by the appearances. How can such a certainty be obtained?

In the astronomical and meteorological accounts of Epicurus and Lucretius an important role is assigned to analogy. The astronomical and meteorological sections of Lucretius' *DRN* abound in specific analogies (as does the rest of

---

86   Allen (2001) 197–198; Bénatouïl (2003) 42–44; Verde (2013) 134–135. See also Jürss (1994) 240.

his work).⁸⁷ Almost every single explanation is illustrated by a specific analogy from everyday experience. Although, as a poet, Lucretius knows how to exploit these analogies poetically, the fact that many of his particular analogies are identical to those known from other, non-poetical, works on these subjects (like the *Syriac meteorology*⁸⁸ (see p. 70 below) and Seneca's *Naturales Quaestiones*⁸⁹) suggests that their primary role was scientific. Lucretius does not tell us what this role is, but we may perhaps learn more from Epicurus.

Although the number of specific analogies in the *Letter to Pythocles* is very limited⁹⁰ (probably due to its being a summary),⁹¹ Epicurus does provide some useful theoretical remarks about the use of analogy in general. According to Epicurus (*Hdt.* 80), "we must carefully consider in how many ways a similar phenomenon is produced here with us, when we reason about the causes of the phenomena above as well as everything non-evident", and (*Pyth.* 2 [87]) "signs of what happens in the sky can be obtained from some of the appearances here with us: for we can observe how they come to pass, though we cannot observe the appearances in the sky: for they may be produced in several ways".⁹² Even more explicit is *Pyth.* 10 [94], on the phases of the moon, which, according to Epicurus, may be accounted for "in all the ways in which appearances here with us, too, invite us to explanations of this appearance".⁹³

In these passages analogy seems to be presented as a *heuristic device*: its purpose is to *provide signs*, or to *invite us* to consider certain explanations.⁹⁴ We might be tempted at this point to ascribe to Epicurus a scientific method consisting of two neatly distinguished stages, with analogy *providing* hypothe-

---

87 On Lucretius' use of analogies see e.g. Schrijvers (1978) and Garani (2007).
88 Many of the parallels are noted in Daiber (1992) 272–282. The degree of correspondence is variously assessed: while Kidd (1992), 301, observes 'close parallels including the illustrative analogies' between Lucretius and the *Syriac meteorology*, Garani (2007), 97, instead notes 'the remarkable lack of correspondence between Theophrastean [this is a reference to the *Syriac meteorology*] and Lucretian analogies.'
89 See e.g. Bailey's commentary on Lucretius book VI.
90 In *Pyth.* 6 [91] the sun is compared to terrestrial fires (οὕτω γὰρ καὶ τὰ παρ' ἡμῖν πυρὰ ἐξ ἀποστήματος θεωρούμενα κατὰ τὴν αἴσθησιν θεωρεῖται), and in 18 [100] (1) thunder-production due to the wind whirling about in a hollow cloud is compared to a similar effect occurring in vessels (καθάπερ ἐν τοῖς ἡμετέροις ἀγγείοις).
91 On the character and structure of Epicurus' *Letter to Pythocles* see p. 92 ff. below.
92 Translations by Bailey, with modifications.
93 Translation by Bailey, with modifications. A further example is found in *Pyth.* 19 [101].
94 Allen (2001) 196–197.

ses, and non-contestation *proving* them.⁹⁵ However, this is not the whole story. Sometimes analogy appears to be used not merely as a heuristic device, but as a *proof* in its own right.⁹⁶ This can be seen e.g. in *Pyth.* 11 [95], where two alternative theories about the light of the moon are backed up in the following way:

Καὶ γὰρ παρ' ἡμῖν θεωρεῖται πολλὰ μὲν ἐξ ἑαυτῶν ἔχοντα, πολλὰ δὲ ἀφ' ἑτέρων. Καὶ οὐθὲν ἐμποδοστατεῖ τῶν ἐν τοῖς μετεώροις φαινομένων, ἐάν τις τοῦ πλεοναχοῦ τρόπου ἀεὶ μνήμην ἔχῃ καὶ τὰς ἀκολούθους αὐτοῖς ὑποθέσεις ἅμα καὶ αἰτίας συνθεωρῇ καὶ μὴ ἀναβλέπων εἰς τὰ ἀνακόλουθα ταῦτ' ὀγκοῖ ματαίως καὶ καταρρέπῃ ἄλλοτε ἄλλως ἐπὶ τὸν μοναχὸν τρόπον.	For here with us, too, we see many things having light from themselves, and many having it from something else. And nothing in the appearances in the sky impedes this, if one always remembers the method of manifold causes and investigates hypotheses and explanations consistent with them, and does not look to inconsistent notions and emphasize them without cause and so fall back in different ways on different occasions on the method of the single cause.⁹⁷

The analogy with what happens 'here with us' is clearly presented as a *proof*. It is true that immediately afterwards Epicurus invokes *non-contestation* ('nothing impedes') as well, but significantly he restricts its use to 'appearances *in the sky*' only, thereby suggesting that the 'appearances *here with us*' have already been covered by the analogy. Apparently then, *analogy* with appearances *here with us* implies *agreement* (συμφωνία) with appearances *here with us*. If it is subsequently found that none of the appearances *in the sky* contests either, i.e. if the explanations are not at variance with the original (celestial) object of inquiry, the explanations must be accepted as objectively possible. Other examples of this probative, as opposed to heuristic, use of analogy are found in *Pyth.* 6 [91], on the size of the sun ("for so too fires on earth ...") and *Pyth.* 15 [98], on the length of nights and days ("as we observe occurs with some things on earth, with which we must be in harmony (συμφωνία) in speaking of celestial phenomena").

---

95   Allen (2001) 197.
96   Allen (2001) 197.
97   Translation by Bailey, modified, with my emphasis.

According to Epicurus, then, analogy performs two functions, a heuristic and a probative one.[98] The first function logically precedes the second. If an explanation is needed, it must first be found, and then be verified. One and the same analogy may perform both functions, but not simultaneously. Once an explanation has been found, the analogy has performed its heuristic function, and can no longer serve in that capacity. Of course one may still report the specific analogy that led to the discovery of a certain theory, but such a report can no longer be called heuristic but at best historical and anecdotal. The actual heuristic use of analogy is therefore rather limited. It can never be linked with specific theories, which, after all, have been found already. Epicurus does seem to realise this. In those passages which deal with analogy in its heuristic capacity (*Hdt.* 80; *Pyth.* 2 [87] and 10 [94]), he never refers to specific explanations.[99] We might of course still view Epicurus' and Lucretius' lists of alternative explanations as the *outcome* of an extensive heuristic use of analogy on Epicurus' part, but even that isn't exactly true. As will be demonstrated below (see §2.4 on p. 58ff.), almost all alternative explanations offered by Epicurus and Lucretius appear to derive from earlier thinkers. Not *analogy*, it turns out, but *doxography* seems to have been Epicurus' favourite heuristic device.

In explaining the many specific analogies in books V and VI of the *DRN* we may therefore disregard their *heuristic* function. The main purpose of these analogies is *probative*: they *prove* the objective possibility of an explanation.[100]

2.3.6 *Degrees of Probability and Personal Preferences*

Above we concluded that, at least with respect to a particular event, each one of a number of alternative explanations can at best be called possible. Yet, the question remains whether they are all *equally* possible. The only explicit statement on this subject is found in the Epicurean inscription of Diogenes of Oenoanda, fr.13 III 2–13, in the middle of a discussion of astronomical phenomena:

---

98 See Allen (2001) 195ff.
99 In *Pyth.* 10 [94] "καὶ κατὰ πάντας καθ' οὓς" ... κτλ. ["*and* in all ways in which" ... etc.] does not refer back to the three explicit explanations already given, but to other explanations that may at some point in the future be added to the list.
100 This does not mean that they may not perform other functions too; many of the analogies in Lucretius seem to have an illustrative function as well: they help the reader form a mental picture by providing a conceptual model. Besides they often provide Lucretius with an excellent excuse to show off his poetic genius.

Τὸν ζητοῦντά τι περὶ τῶν ἀδήλων, ἂν βλέπῃ τοὺς τοῦ δυνατοῦ τρόπους πλείονας, περὶ τοῦδέ τινος μόνου τολμηρὸν καταποφαίνεσθαι· μάντεως γὰρ μᾶλλόν ἐστιν τὸ τοιοῦτον ἢ ἀ⟨ν⟩δρὸς σοφοῦ. Τὸ μέντοι λέγειν πάντας μὲν ἐνδεχομένους, πιθανώτερον δ' εἶναι τόνδε τοῦδε ὀρθῶς ἔχει.	If one is investigating things that are non-evident, and if one sees that several explanations are possible, it is reckless to make a dogmatic pronouncement concerning any single one; such a procedure is characteristic of a seer rather than a wise man. It is correct, however, to say that, while all explanations are possible, *this one is more plausible than that.*[101]

The first part of this statement corresponds exactly to what we already know about Epicurus' method.[102] The last sentence, however, is not paralleled in any of Epicurus' surviving works, nor in Lucretius'.[103] Besides, there is something self-contradictory about Diogenes' words, for, after denouncing as 'seers' those who opt for a single explanation, he himself seems to be singling out one explanation under the guise of plausibility. It would have been interesting to know on what grounds Diogenes would have us consider one explanation more plausible than another, but unfortunately he either failed to inform the reader, or the relevant part of the inscription is lost. As neither Epicurus nor Lucretius have left us any explicit theoretical considerations about the admissibility or inadmissibility of applying different degrees of probability, we cannot know for certain how Diogenes' remark relates to Epicurean orthodoxy. However, even in the absence of such theoretical considerations, if Epicurus and Lucretius can be shown in particular cases to have assigned different degrees of probability or to have expressed a preference for one explanation over another, Diogenes' remark may not be as alien to Epicurean orthodoxy as it might appear.

In the *Letter to Pythocles* Epicurus nowhere expresses a preference for any particular explanation. We do, however, have one testimony which might be interpreted as attributing to Epicurus just such a personal preference. In the *Naturales Quaestiones*, VI 20, 7,[104] Seneca, having just reported a whole list of alternative explanations of earthquakes as brought forward by Epicurus, concludes with the following words:

---

101 Translation by Smith, slightly modified, with my emphasis.
102 For parallels in Epicurus and Lucretius, see Smith (1993), 455, n. 8.
103 Jürss (1994) 240 n. 32; Algra (2001) n. 28. See now also the discussion in Verde (2013) 135–137.
104 See also Verde (2013) 138–139.

Nullam tamen illi {sc. Epicuro} placet causam motus esse *maiorem* quam spiritum.	No cause of an earthquake, however, Epicurus deems to be *greater* than wind.

One's interpretation of these lines depends strongly on the meaning one wishes to attribute to *maiorem*, '*greater*'. One possible meaning in this context would indeed be '*more likely*'. Epicurus might have said that, although there are many possible causes, those involving wind are the *most likely*. However, as with most meteorological occurrences, an earthquake is not a single, recurrent, phenomenon, but every earthquake stands alone. It is perfectly possible, therefore, that one earthquake is brought about by cause A, while another is caused by B. Under these circumstances '*more likely*' becomes equivalent to '*more frequent*', which is another possible interpretation of *maiorem*. So, Epicurus might have said that, although earthquakes can be, and are, produced by many different things, they are *most often* produced by wind. Yet, there is another possible interpretation. Different causes may have different effects.[105] *Maiorem* in this context may also mean '*more powerful*'. Epicurus may have meant that, although earthquakes can be, and are, produced by many different things, the *strongest* ones are produced by wind. That this is in fact the correct interpretation is borne out by the way in which Seneca continues (VI 21, 1):

Nobis quoque placet hunc spiritum esse qui possit tanta conari, quo nihil est in rerum natura potentius, nihil acrius, sine quo ne illa quidem quae vehementissima sunt valent.	We (i.e. the Stoics) too believe that it is this wind, which can attempt so much, which is mightier and fiercer than anything in nature, without which not even those things which are strongest have power.

So, according to Seneca, Epicurus held that wind is the *most powerful* cause of earthquakes. Like Seneca himself, Epicurus may have been brought to this view by Aristotle.[106] Be that as it may, Seneca's testimony cannot serve to confirm

---

105   See p. 12 above.
106   Arist. *Mete.* II 8, 365b29–366a5: "Our next step should therefore be to consider what substance has the greatest motive power. This must necessarily be the substance whose natural motion is most violent. The substance most violent in action must be that which has the greatest velocity, as its velocity makes its impact most forcible. The farthest mover must be the most penetrating, that is the finest. If, therefore, the natural constitution of wind is of this kind, it must be the substance whose motive power is the greatest. For even fire when conjoined with wind is blown to flame and moves quickly. So the cause of

that Epicurus himself admitted different degrees of probability, or that he ever voiced personal preferences for any one of the alternative explanations. So much for Epicurus himself.

As for Lucretius: there is one instance in the *DRN* where Lucretius might seem to express a preference for a particular explanation. At V 621–622 he introduces his first explanation of the yearly and monthly motion of the sun and the moon with the following words:

Nam fieri vel *cum primis* id posse videtur, Democriti quod sancta viri sententia ponit: ...	For, *in the first place* it seems that this may be the case, what the sacred opinion of the man Democritus states: ...

There is some ambiguity in the words 'cum primis'. 'Cum primis' or 'cumprimis' literally means: *with* or *among the first*. This can be understood in two ways:

1. Most often 'cum primis' or 'cumprimis' (like the synonymous 'in primis' or 'imprimis') is used to indicate that what is said is so *in the highest degree*, or *particularly*.[107] On this interpretation 'vel' is best understood as an intensifying particle[108] with 'cum primis': '*among the very first*'. In the present case this would mean that Democritus' theory seems to be possible in the highest degree. This interpretation underlies the translations of e.g. Rouse & Smith ('For among the *most likely* causes is that ...') and Leonard ('Yet *chief in likelihood* seemeth the doctrine ...').
2. Occasionally 'cum primis' or 'cumprimis' is used to indicate *the first item in a series*.[109] (Its synonym 'in primis' or 'imprimis' is actually used in this sense quite often).[110] Interpreted in this way the expression may be rendered as *for a start* or *to begin with*. 'Vel' may again be an intensifier with 'cum primis',[111] or

---

earth tremors is neither water nor earth but wind, which causes them when the external exhalation flows inwards" (transl. Lee).

107   Lewis & Short 'primus' II B & '1. cum' II D; *OLD* 'cum¹' 6e & 'primus' 15c (cf. *OLD* 'imprimis' 1).
108   Lewis & Short 'vel' II B 1&2; *OLD* 'vel' 5c. Cf. Lucr. VI 1237 "vel in primis".
109   *Not* in *OLD* and Lewis & Short, but see Plaut. *Truc.* 660–661: "eradicarest certum *cumprimis* patrem,/ *post id locorum* matrem." and Apul. *Flor.* 16, 36: "*cum primis* commemoravit inter nos iura amicitiae {...}; *tunc postea* vota omnia mea {...} recognovit." *TLL* 'cumprimis' (vol. IV, p. 1380) does not register different meanings of the expression.
110   *OLD* 'imprimis' 2.
111   See n. 108.

it may be used to imply "that other instances might be mentioned at will".[112] Used in this sense 'vel' may be rendered as 'for instance' or 'for example'. This second interpretation is followed by e.g. Bailey ("For, *first and foremost*, it is clear that it may come to pass ...") and Ernout ("*Tout d'abord* il est possible semble-t-il, que les choses se passent ..."). See also Bailey's comment at V 621: "This makes it clear that Lucr. intends to expound *the first of a series* of alternative causes {...}".

Which of these two interpretations is the right one? The first interpretation attributes to Lucretius a preference for Democritus' view. Such a preference, however, seems to be unmotivated. The only way in which the present account differs essentially from other explanations in the astronomical and meteorological sections of *DRN*, is the explicit attribution to Democritus, whom Lucretius clearly admires. Yet, admiration for Democritus does not necessarily imply a preference for his theories: in III 370–373 another theory of Democritus, introduced with the same admiring words (III 371 = V 622), is flatly rejected! There is no reason, therefore, why in this case Democritus' view should be considered to be *among the most likely causes*. Besides, such a preference for a single theory is also quite unprecedented in the *DRN*. Books V and VI contain scores of problems for which several alternative explanations are offered. Why should Lucretius in this one case express his preference, and nowhere else? Finally, such a preference seems also to be unwarranted by Lucretius' own methodological remarks. Less than 100 lines earlier Lucretius stressed that out of several possible causes "one {...} is necessarily the case here, / {...} but which of them it is, / is not for them to lay down who proceed step by step." (V 531–533).[113] If Lucretius had thought it permissible to assign different degrees of probability to the alternative explanations *this* would have been the place to say so. But he did not. For these reasons I think this interpretation should be rejected. The second interpretation, although based on a less frequent use of the expression 'cum primis', provides a good alternative, which avoids all of the above problems.

This brings us back to Diogenes of Oenoanda's assertion that "it is correct {...} to say that, while all explanations are possible, *this one is more plausible than that*." Not only is the claim itself without precedent in earlier Epicurean writings, but now our search for *applications* of this principle has yielded nothing either. It seems safe to assume, therefore, that Diogenes' assertion is a later innovation.

---

112 Lewis & Short 'vel' II C. Cf. *OLD* 'vel' 4 a&b.
113 For a fuller quotation, see above, p. 21.

We do not know what the reason for this innovation was, nor how Diogenes himself applied it, but perhaps one example of its application can be unearthed from the ruins of his inscription.

Diogenes' claim is part of a fragment (fr.13) that begins with a promise to deal with the risings and settings of the sun, a problem for which Epicurus (*Pyth.* 7 [92]: see p. 19 above) and Lucretius (v 650–679) had proposed two possible explanations. Unfortunately Diogenes' fragment breaks off before he can deal with this specific problem. The same problem, however, is also discussed in another fragment (fr.66), where Diogenes criticises certain adversaries for "dismissing the unanimous opinion of all men, both laymen and philosophers, that the heavenly bodies pursue their courses round the earth both above and below …" (tr. Smith). It is clear that Diogenes himself shares this 'unanimous opinion of all men', silently passing by Epicurus' alternative explanation according to which the heavenly bodies are extinguished at night.

It is possible that with other astronomical problems too Diogenes preferred to follow the generally accepted view, and he may have found this appeal to greater and lesser plausibility a convenient way to express these preferences without explicitly rejecting Epicurus' alternative explanations,[114] thus reconciling Epicurean orthodoxy with the accepted astronomical views of his time.[115]

2.3.7 *Lucretius' Supposed Preference for the Theories of Mathematical Astronomy*

Although in the astronomical sections of their works Epicurus and Lucretius do not express preferences for individual alternative explanations, there are some passages in the *De rerum natura* where Lucretius is believed to betray at least an *implicit* preference for a certain class of explanations. If true, this observation would contradict our earlier conclusion that Lucretius, like Epicurus, was impartial to the individual alternative explanations. It will be necessary therefore to deal with this claim as well.

On p. 58 of his commentary Bailey writes: "in the astronomical passages he [i.e. Lucretius] frequently places the right explanation first, as though he had a personal preference for it". The point is repeated on p. 1394, in the introduction to Lucretius' astronomical section of v 509–770: "It should, however, be noticed that Lucr. usually places the true explanation first, as though he really preferred

---

114 With respect to meteorological phenomena Diogenes seems to have had no qualms about offering several alternative explanations: see fr.14 on the causes of hail, and fr.98.8–11 on the causes of earthquakes.

115 For other possible explanations of Diogenes' statement, see Verde (2013) 137.

it." Out of context, Bailey's observation seems a bit trivial, for: who wouldn't prefer the *right* and *true* explanation? In order to understand what Bailey really means with these remarks, we must have a closer look at his comments on the individual sections of Lucretius' astronomical passage.

On p. 1439, commenting on Lucretius' discussion of the phases of the moon in V 705–750, Bailey writes: "This [i.e. the first] view {...} is clearly the view of the astronomers to whom Lucr. refers as his authorities in 713–714, and again as the *astrologi* in 728. {...} Lucr. therefore included it, and probably by placing it first meant to suggest that he believed it to be the right explanation ...", and on pp. 1446–1447, commenting on Lucretius' discussion of solar and lunar eclipses in V 751–770: "This [i.e. the first] theory, which was no doubt that of the 'astronomers' and is in fact the true explanation {...}, is to be compared with the first theories put forward in 682–695 and 705–714. {...} Once again Lucr. by putting this theory first appears to give it the preference ..."

Apparently then, when Bailey speaks of the 'right' and the 'true' explanation he means the explanation of the mathematical astronomers, *which happens to be the true explanation*. But Lucretius couldn't have known that. That is precisely the point of his offering multiple explanations: that one cannot know for certain which is the right one.[116] What he could have known, and did know, is that certain explanations came from the stock of the (mathematical) astronomers or 'astrologi', as he calls them. We may therefore rephrase Bailey's observation as follows: "In the astronomical passages Lucretius frequently places the explanation *of the mathematical astronomers* first, as though he had a personal preference for it."

Now there are two sides to this observation: (a) the observed fact, and (b) Bailey's interpretation of the fact. Let us first turn to the observed fact: "in the astronomical passages Lucretius frequently places the explanation of the mathematical astronomers first."

In the astronomical section Lucretius covers the following eleven subjects:

1. Motions of the stars (509–533)
2. Immobility of the earth (534–563)
3. Size of the sun, moon and stars (564–591)
4. Source of the sun's light and heat (592–613)
5. Turnings of the sun, moon and planets (614–649)
6. Causes of nightfall (650–655)
7. Causes of dawn (656–679)

---

116 See e.g. *DRN* V 526–533 (see text on p. 21 above).

8. Varying lengths of day and night (680–704)
9. Phases of the moon (705–750)
10. Solar eclipses (751–761)
11. Lunar eclipses (762–770)

Two of these subjects (2 and 3) do not, apparently, admit of more than one explanation, and the explanation given in each case is certainly not that of the mathematical astronomers. Lucretius explains the immobility of the earth on the assumption that it floats on a cushion of air, a view not easily reconciled with the spherical earth of mathematical astronomy (see Chapter Four and esp. §4.3.6 below), and the heavenly bodies he claims to be the size they appear to be, which is usually interpreted as being very small, whereas the mathematical astronomers, for all their different estimates, at least agreed that the sun and the stars are much larger than the earth, and only the moon somewhat smaller.[117] That leaves us with nine cases where we can test Bailey's observation.

The first of these, about the (daily?) motions of the stars, is problematic. Lucretius offers *five* possible explanations in all, which—however—fall into *two* main divisions: either (a) the whole sphere of the sky revolves, carrying the heavenly bodies along, or (b) the sky stands still, while the heavenly bodies move independently. The first option was—in fact—the view favoured by the mathematical astronomers.[118] Lucretius, however, goes on to subdivide these main divisions, offering two possible physical explanations for the first option, and three for the second, in a way that goes beyond the constraints of mathematical astronomy, which only concerned itself with the mathematical, i.e. quantitative and geometrical, aspects of astronomy.[119] It remains unclear, therefore, whether we should consider this passage a case in point. Bailey himself does not seem to have viewed it as such, for in his commentary to this passage he makes no reference to his own observation.

The next subject where a plurality of explanations comes into play is item 4, on the source of the sun's light and heat. Lucretius offers three explanations, none of which can be related to the mathematical astronomers. In fact the

---

117 Cf. Cleomedes II 1–3 (esp. II 3.68 ff.). See also Heath (1913) 328–350, esp. the tables on p. 332 and p. 350.
118 Evans (1998), p. 75 with nn. 3 and 4.
119 See e.g. Arist. *Ph.* II 2, 193b22 ff.; Stoics apud D.L. VII 132; Posidonius F18 E–K (= Geminus apud Alexandrum apud Simpl. *In Arist. Phys.* 291.21–292.21); Sen. *Ep.* 88.25–28; Strabo 1.1.20 & 2.5.2; etc. For modern views on the matter see e.g. Bowen-Todd (2004), 6 & 193–199, and Evans (1998), 217–219.

ancient astronomers have left us no view on this subject at all, which falls outside the scope of their competence, i.e. the quantitative and geometrical aspects of astronomy (see above).

The following subject (5), the turnings of the sun, moon and planets, falls well within the competence of the mathematical astronomers and we know what their solution to this problem was. According to the astronomers, the sun, the moon and the planets exhibit a slower, secondary, east-ward motion on top of the daily, west-ward, revolution of the fixed stars. This secondary motion they all perform more or less along the same circular path, the so-called zodiac. The zodiac is not parallel to the equator but inclined to it by an angle of about 23.5°. This fact was referred to in antiquity as *the obliquity of the zodiac*—ἡ λόξωσις τοῦ ζῳδιακοῦ (sc. κύκλου) / *obliquitas signiferi* (sc. *orbis* / *circulus*). In Epicurus' own treatment of the subject, in the *Letter to Pythocles* 9 [93], this explanation is the first to be mentioned.[120] Lucretius too shows himself to be aware of the existence of this explanation, for, in a later passage (8), when discussing the related problem of the varying lengths of day and night, he clearly alludes to it.[121] In the present passage, however, he does not mention it. So, far from placing the mathematical astronomers' view first, Lucretius chooses to ignore it. Instead, as we have seen, he starts with an explanation explicitly ascribed to Democritus.

The first instance where Lucretius *does* include the view of the mathematical astronomers among a number of possible explanations is subject 6, on the causes of nightfall. Night ensues, he says, either because the sun, upon reaching the westernmost point of its orbit, is extinguished, or because the sun, upon reaching this point, passes out of sight below the plane of the earth. Once again Lucretius does *not* conform to Bailey's observation: the view of the mathematical astronomers is presented second, not first.

The next subject (7), on the causes of dawn, is the mirror image of the previous subject. Again Lucretius gives us two explanations, which correspond *chiastically* to the two possible causes of nightfall: either the same sun, having reached the easternmost limit of its orbit, emerges again above the plane of the earth, or a new sun is born from small fires which collect in the eastern sky each morning. This time, at last, the view of the mathematical astronomers is presented first.

---

120   Τροπὰς ἡλίου καὶ σελήνης ἐνδέχεται μὲν γίνεσθαι κατὰ λόξωσιν οὐρανοῦ οὕτω τοῖς χρόνοις κατηναγκασμένου ...
121   *DRN* v 691–693: ... propter *signiferi* posituram totius *orbis*, / annua sol in quo concludit tempora serpens, / *obliquo* terras et caelum lumine lustrans ...

TABLE 2.1 *Place of the mathematical astronomers' explanations in the astronomical section of* DRN *V*

Passage		Subject	Number of explanations	View of the mathematical astronomers
509–533	1.	Motions of the stars	5	–
534–563	2.	Immobility of the earth	1	–
564–591	3.	Size of the sun, moon and stars	1	–
592–613	4.	Source of the sun's light and heat	3	–
614–649	5.	Turnings of the sun, moon and planets	2	–
650–655	6.	Causes of nightfall	2	2
656–679	7.	Causes of dawn	2	1
680–704	8.	Varying lengths of day and night	3	1
705–750	9.	Phases of the moon	4	1
751–761	10.	Solar eclipses	3	1
762–770	11.	Lunar eclipses	3	1

The rest of the subjects (8–11) also follow this pattern, thereby conforming to Bailey's observation. Mathematical astronomers explained (8) the seasonal variation of the length of nights and days with reference to the sun's position in the slanting zodiacal belt, (9) the phases of the moon with reference to the relative positions of the sun and the moon—assuming that the latter shines with light reflected from the former—, (10) solar eclipses as the moon blocking the sun from our view, and (11) lunar eclipses as the moon falling into the earth's conical shadow and so being deprived of the sun's light. In Lucretius' account each of these theories is the first of a number of alternative explanations. Table 2.1 sums up our findings.

Only 5 out of 11 cases seem to conform to Bailey's thesis. If, however, we confine ourselves to those subjects where the view of the mathematical astronomers is included at all, the ratio becomes 5 to 6, which seems significant enough. We may therefore safely conclude that the explanations of the mathematical astronomers *were* somehow privileged, although we do not yet know why or in what way.

In only two of these cases the mathematical astronomers are explicitly mentioned or implied in a meaningful way. In lines 694–695, at the end of his first explanation of (8) the variation of day-length, Lucretius speaks of those "who have mapped the places of the sky, / all adorned with stars properly arranged",[122] which is a clear reference to the mathematical astronomers, who typically demonstrated their theories by means of celestial globes and planetaria.[123] In lines 727–728, at the end of the *third* explanation of (9) the phases of the moon Lucretius speaks of the "art of the astronomers" (*astrologorum artem*), by which he appears to be referring back to the *first* explanation of that section. Apparently Lucretius assumes his reader to be familiar enough with contemporary astronomy to recognize this reference. In the same way the reader may be assumed to recognize the other, unidentified, references to the mathematical astronomers as well.

According to Bailey, the privileged position of the mathematical astronomers' explanations indicates that Lucretius himself preferred these over the other views, believing them to be the right ones. To Bailey this is so self-evident that he does not even defend this assertion. In fact, his claim is quite unfounded. Not only is such a preference, as we have seen, hard to reconcile with Lucretius' insistence that all explanations offered have an equal claim to the truth (526–533), but it actually fails to take into account certain clues provided by Lucretius himself in several of the relevant passages.

The first of these clues is at 713–714, where Lucretius concludes his first explanation of the phases of the moon (i.e. the explanation of the mathematical astronomers) with the following words:

ut faciunt, lunam qui **fingunt** esse pilai consimilem cursusque viam sub sole tenere.	as they hold, who **imagine** the moon to be like a ball and to keep the path of her course below the sun.

Throughout the *DRN* the verb 'fingere' is used to stress the *unfoundedness* and even *falsehood* of theories,[124] which are subsequently rejected. It would be most surprising if Lucretius would now use this same verb to refer to a theory which, in Bailey's words, "he believed [...] to be the right explanation."

---

122 *DRN* V 694–695: ... qui loca caeli / omnia dispositis signis ornata locarunt.
123 Plato, *Ti.* 40c–d, claims that the planetary motions can be properly demonstrated by means of a visible model only, and Epicurus, *Pyth.* 9 [93] (cf. *On nature* XI fr.38 Arr. with Sedley (1976), 32, 37–39), seems to associate the use of such models with mathematical astronomy. On the use of visible models in astronomy see Cornford (1937) 74–76; and Evans (1998) 78–84.
124 See e.g. I 371, I 842, I 847, I 1083, II 175 and V 908.

Another clue, which will require a bit more work, but may also help us to find the real reason why Lucretius gives priority to the explanations of the mathematical astronomers, is found at the end of the astronomical passage (lines 751–770), where Lucretius discusses the causes of (10) solar and (11) lunar eclipses. In his discussions of other astronomical phenomena Lucretius simply enumerates his multiple explanations, saying something like: 'phenomenon x may be caused by A, or B, or C, etc.', where A most often represents the view endorsed by mathematical astronomy. However, in his discussions of solar and lunar eclipses Lucretius employs a slightly different structure. This is Lucretius' account of solar eclipses (753–761):

A   nam cur luna queat terram secludere solis lumine et a terris altum caput obstruere ei, obiciens caecum radiis ardentibus orbem,

For why should the moon be able to shut off the earth from the sun's light and obstruct the sun's high source from the earth, by interposing her dark orb to his burning rays,

B   tempore eodem aliut facere id non posse putetur
corpus, quod cassum labatur lumine semper?

and not at the same time some other body, which always glides with unseen light, be thought able to achieve this?

C   solque suos etiam dimittere languidus ignis tempore cur certo nequeat recreareque lumen,
cum loca praeteriit flammis infesta per auras,
quae faciunt ignis interstingui atque perire?

and why could not the sun at a certain time from weariness dismiss his fires and then again renew his light, when he has passed the regions harmful to his flames, which make his fires go out and die?

And this is how Lucretius explains lunar eclipses (762–770):

A   et cur terra queat lunam spoliare vicissim lumine et oppressum solem super ipsa tenere,
menstrua dum rigidas coni perlabitur umbras,

And why should the earth in turn be able to rob the moon of light, and keep the sun oppressed, being herself above, while in its monthly course the moon glides through the rigid shadows of the cone,

B tempore eodem aliud nequeat succurrere   and not at the same time some other
      lunae   body be able to pass beneath the
   corpus vel supra solis perlabier orbem,   moon or glide above the solar orb, to
   quod radios inter rumpat lumenque   interrupt his rays and flood of light?
      profusum?

C et tamen ipsa suo si fulget luna nitore,   and if, after all, the moon shines of
   cur nequeat certa mundi languescere parte,   herself with her own light, why could
   dum loca luminibus propriis inimica per   she not grow faint in a certain part of
      exit?   heaven, while passing through regions
                                                hostile to her own light?

For each of the two phenomena Lucretius offers three possible explanations, but, instead of simply enumerating them, as he usually does, he now marshals them into the format of a rhetorical question. The structure is the same in both cases: 'Why should the (solar/lunar) eclipse be caused by 'A', and not by 'B' or 'C'?' The implied answer is, of course, that there is *no good reason* to prefer A over B and C. (So much for Bailey's interpretation.) Yet, the way the question is put also suggests something else. Lucretius seems to be particularly worried that someone might consider 'A', i.e. the mathematical astronomers' view, the only possible explanation. That someone might single out 'B' or 'C' in this manner does not seem to occur to him. Why is that? Is it intrinsically worse to accept the view of the astronomers rather than any of the other theories? That is *not* what Lucretius is saying. Throughout the astronomical passage he has insisted upon the *equal* plausibility of each alternative explanation: no explanation is better or worse than any other. So, why then does Lucretius single out the explanation of the mathematical astronomers?

An account with a somewhat similar rhetorical structure is found in an astronomical passage of Epicurus' *Letter to Pythocles*. Here, in chapter 32 [112] Epicurus writes:

   Τινὰ ⟨ἄστρα⟩ στρέφεται αὐτοῦ, ὃ συμβαίνει   Some stars revolve in their place,
                                                which comes to pass

A οὐ μόνον τῷ τὸ μέρος τοῦτο τοῦ κόσμου   not only because this part of the
   ἑστάναι, περὶ ὃ τὸ λοιπὸν στρέφεται,   world is stationary and round it the
   καθάπερ τινές φασιν,   rest revolves, as some say,

B ἀλλὰ καὶ τῷ δίνην ἀέρος ἔγκυκλον αὐτοῖς περιεστάναι, ἢ κωλυτικὴ γίνεται τοῦ περιπολεῖν ὡς καὶ τὰ ἄλλα,

but also because a whirl of air is formed in a ring round it, which prevents their moving about as do the other stars,

C ἢ καὶ διὰ τὸ ἑξῆς μὲν αὐτοῖς ὕλην ἐπιτηδείαν μὴ εἶναι, ἐν δὲ τούτῳ τῷ τόπῳ ἐν ᾧ κείμενα θεωρεῖται,

or else it is because there is not a succession of appropriate fuel for them, but only in this place in which they are seen fixed,

καὶ κατ' ἄλλους δὲ πλείονας τρόπους τοῦτο δυνατὸν συντελεῖσθαι, ἐάν τις δύνηται τὸ σύμφωνον τοῖς φαινομένοις συλλογίζεσθαι.

and there are many other ways in which this may be brought about, if one is able to infer what is in agreement with the appearances.[125]

This passage is concerned with the problem (not discussed by Lucretius) why some stars (e.g. those of Ursa Major and Minor) never set but revolve in their place. The first option, 'A', corresponds to the view of the mathematical astronomers. Although Epicurus—like Lucretius—is normally perfectly happy to present his alternative explanations in the form of an uncomplicated disjunction ('either A or B or C'),[126] this time he has chosen a slightly more complex formulation: 'not only A, but also B and C'. This formulation implies that Epicurus in reality only contemplates the possibility that someone might say 'A', *not* that someone might say 'B' or 'C'. This is confirmed by the words "as some say" ("καθάπερ τινές φασιν"), which Epicurus adds to explanation A. Apparently, explanation A had some actual support, which B and C, as far as we are told, had not.

Something similar seems to be the case with Lucretius' account of eclipses. In order to confirm this I will try to establish the extent of the contemporary support for view A, and the lack of such support for the alternative views, B and C. I will start with Lucretius' account of *lunar* eclipses (V 762–770). A useful piece of information is provided by Aëtius, who in his chapter on the phases and eclipses of the moon (II 29),[127] also reports a view that can be identified

---

125 Translation Bailey, slightly modified.
126 See n. 56 on p. 24 above, and the text thereto. For the various connectors used by Epicurus to articulate his lists of alternative explanations, see APPENDIX 1 on p. 269 ff. below.
127 In Bakker (2013) I argue that Aëtius II 29 'On the eclipse of the moon' is a conflation of two separate chapters, one on the phases of the moon and one on its eclipse.

with Lucretius' first explanation of lunar eclipses (together with an account of the phases):

Θαλῆς Ἀναξαγόρας Πλάτων Ἀριστοτέλης οἱ Στωικοὶ τοῖς μαθηματικοῖς συμφώνως τὰς μὲν μηνιαίους ἀποκρύψεις συνοδεύουσαν αὐτὴν ἡλίῳ καὶ περιλαμπομένην ποιεῖσθαι, τὰς δ' ἐκλείψεις εἰς τὸ σκίασμα τῆς γῆς ἐμπίπτουσαν, μεταξὺ μὲν ἀμφοτέρων τῶν ἀστέρων γενομένης, μᾶλλον δὲ τῆς σελήνης ἀντιφραττομένης.	Thales, Anaxagoras, Plato, Aristotle and the Stoics (declare) in agreement with the mathematical astronomers that it (the moon) produces the monthly concealments by travelling together with the sun and being illuminated by it, and the eclipses by descending into the shadows of the earth which interposes itself between the two heavenly bodies, or rather when the moon is obstructed (by the earth).[128]

According to this report, Lucretius' first explanation, which Bailey attributed to the astronomers, was also accepted by Thales, Anaxagoras, Plato, Aristotle and the Stoics. Thales and Anaxagoras, who had been long dead and left no schools to continue their thought, are irrelevant for the present purpose, but Plato and Aristotle, whose teachings were still followed in Lucretius' time, and the Stoics, who had become the most influential philosophical sect of the period, are very relevant. From Aëtius' report, which is confirmed by many other sources,[129] it appears that Lucretius' first explanation was the view, not just of the mathematical astronomers, but of every major school of philosophy still in existence in Lucretius' day, Epicureanism excepted. The second and third explanations, on the other hand, do not seem to have been entertained by anyone later than Anaxagoras, who believed that lunar eclipses were also

---

128 I have basically followed the reconstruction offered by Mansfeld & Runia (2009a), 613–623, although I have opted for Stobaeus' *lectio difficilior* "τοῖς μαθηματικοῖς συμφώνως" ('in agreement with the mathematical astronomers') instead of Pseudo-Plutarch's "οἱ μαθηματικοὶ συμφώνως" ('and the mathematicians in agreement'), a variation which Mansfeld & Runia do not comment upon. The translation is freely adapted from Mansfeld & Runia (2009a).

129 The attribution of this theory to Plato cannot be verified from his own works, but that he accepted the view that the moon is illuminated by the sun is clear from *Resp.* X 617a (with Heath (1913) 158); cf. *Cra.* 409a–b. Aristotle refers to the theory in *Cael.* II 14, 297b24–31, *Mete.* II 8, 367b20–22; *Metaph.* VIII 4, 1044b9–15; *An. Post.* I, 31, 87b39–a2; II 2, 90a15–18; *et passim*. For the Stoics see e.g. *SVF* I 119, 120; II 676, 678 and Cleomedes II 6.

caused by interposition of other, unseen, heavenly bodies beside the earth, and Xenophanes, who used to ascribe all such phenomena to extinction and rekindling.[130]

Much the same can be said about Lucretius' treatment of solar eclipses (V 753–561): his first explanation can again be attributed, not just to the mathematical astronomers, but to Aristotle and the Stoics and probably Plato too,[131] while the second can only be attributed to certain unnamed people (ἔνιοι), mentioned in Aëtius II 24, and the third to no-one later than Xenophanes.[132]

In sum, Epicurus in the passage just quoted and Lucretius in his account of solar and lunar eclipses both start with the view of the astronomers, because that was the view that most people, including the other major philosophical schools, believed to be uniquely true. To this view Epicurus and Lucretius oppose other views that may not be current and popular, but which they consider equally possible.

It seems reasonable to suppose that the same pattern applies also to the other cases in the astronomical section of the *DRN* where Lucretius starts with the view of the astronomers:

(7) The view that dawn is caused by the same sun re-emerging from below the horizon can safely be attributed to Plato, Aristotle and the Stoics,[133] all of whom conceived of the heavenly bodies as permanent entities, and we even have the explicit statement of a later Epicurean, Diogenes of Oenoanda (fr.66), that this was 'the *unanimous* opinion of all men, both laymen and philosophers'. On the other hand, the alternative view that the sun is rekindled every morning can at best be assigned to Xenophanes, and perhaps Heraclitus and Metrodorus of Chios as well, but to no one later.[134]

---

130   For attempts to identify Lucretius' theories with those of his predecessors see the various commentaries.
131   Plato, *Ti.* 40c–d, rightly attributes (solar and lunar?) eclipses to the interposition of another heavenly body (see also Cornford (1937) 135–136); for Aristotle see *Div. Somn.* 1 462b28–29; for the Stoics see *SVF* I 119; II 650 etc.
132   For attempts to identify Lucretius' theories with those of his predecessors see the various commentaries.
133   For the Stoics see also Ar. Did. fr.32 (= *SVF* II 683) and Cleom. II 1.426–466.
134   Xenophanes A32, A33, A38, A40; Heraclitus B6; Metrodorus of Chios A4 D–K. On the other hand, the theory that the sun is quenched and rekindled is explicitly rejected by the Peripatetic Eudemus apud Theon phil. *Expos.* 199.21–22; and the Stoic Cleomedes II 1.426–

(8) The theory that the annual variations in day-length are caused by the obliquity of the zodiac was at least maintained by the later Stoic Cleomedes,[135] while the theory of the obliquity of the zodiac as such is attributable to Plato, Aristotle and the Stoics in general.[136] The two alternative views, on the other hand, do not seem to have been held by anyone at all.[137]

(9) The section on the phases of the moon presents a slightly different story. Lucretius offers four alternative explanations. The first explanation—according to which the moon is illuminated by the sun, and the phases result from the changing relative positions of the two bodies—is easily recognized as the view of the mathematical astronomers.[138] Lucretius does not, at this point, explicitly identify them, apart from the vague reference to those "who imagine the moon to be like a ball".[139] Yet, the use of present tense and plural already suggests that the theory did at least have some advocates in Lucretius' day. In this particular case, however, the astronomers' view seems to have met with a more serious challenge: at the end of his *third* explanation—according to which the moon is a sphere, one half of which is fiery, and which by revolving around its own axis produces the phases—Lucretius writes (727–730):

466; further by Ptol. *Alm.* I 3, 11.24–12.18 (= I 1, 12 Heiberg); and Theon math. *Comm. in Ptol. synt.* 340–341 (Rome).

135 Cleom. I 3.76–4.17. See also Gem. 6,29 ff., Vitruv. IX 3, 1–3 and Plin. *N.H.* II 17, 81.

136 Ascription to Plato and Aristotle in Aët. II 23.5, and to the Stoics in *SVF* I 542, II 650.3 & 651.5 (= D.L. VII 144.3 & 155.9). For Plato see also *Ti.* 36b–d & 38e–39a (with Heath (1913) 159–160; or Cornford (1937) 72 ff.). For Aristotle see *Gen. Corr.* I 10.336a32–b24, 337a8 and *Metaph.* XII 5.1071a16 & 8.1073b17 ff. The theory is also described by Aratus 525–544 and Vitr. IX 1,3.

137 See the various commentaries ad loc.

138 Aëtius II 29 (see p. 51 above) ascribes the theory to the mathematical astronomers, Plato, Aristotle and the Stoics. The attribution of this theory to Plato cannot be verified from his own works, but he certainly accepted the view that the moon is illuminated by the sun (see n. 129 above). Aristotle alludes to the astronomical theory of the lunar phases in *Cael.* II 11, 291b18–21 and *An. post.* I 13, 78b4–11. For the Stoics cp. D.L. VII 145 (= *SVF* II 650). Later Stoics (perhaps from Posidonius onwards: see Bowen-Todd (2004) 138 n. 8, 141 n. 19) held a slightly different theory, maintaining that the moon, on the side where it is touched by the rays of the sun, responds by emitting its *own* light: see Cleom. II 4, 21–32. On this later Stoic view see also p. 54 (following n. 140) below.

139 *DRN* V 713–714: lunam qui fingunt esse pilai / consimilem ...

ut Babylonica Chaldaeum doctrina refutans	as the Babylonian doctrine of the Chaldeans, refuting the science of the astronomers, strives to uphold against them; just as if that which each of them fights for could not be, or as if there were less reason to embrace this than that.
astrologorum artem contra convincere tendit,	
proinde quasi id fieri nequeat quod pugnat uterque	
aut minus hoc illo sit cur amplectier ausis.	

The Babylonian theory is presented differently from the other two alternative explanations. Whereas the second and fourth explanations are, as we have come to expect, mere museum pieces, the view of the Chaldeans still seems to have been able to muster some real support among Lucretius' contemporaries and later. This is confirmed by several other sources. In Vitruvius' *De architectura* IX 2, written several decades after Lucretius' death, two different theories of the phases of the moon are presented. The first, which corresponds to Lucretius' third explanation, is attributed to the *Chaldean* Berosus, and the second, which corresponds to Lucretius' first explanation, is attributed to the *mathematician* Aristarchus of Samos. It is significant that Vitruvius does not choose between the two. The same impartial attitude towards these two explanations we also encounter in Apuleius' *De deo Socratis* 1.14–30 and Augustine's *Enarratio in Psalmos 10, 3*. According to Augustine, both theories are probable, but it is humanly impossible to know which one is true. A different approach is found in the work of the Stoic Cleomedes, who may have lived some time around 200 AD.[140] In II 4.1, Cleomedes discusses no less than *three* different theories concerning the phases of the moon. The first is again that of Berosus, the second the traditional view of the astronomers and the Peripatetics,[141] and the third a Stoic modification of the former, according to which the moon, on the side where it is touched by the rays of the sun, responds by emitting its own light. As a Stoic, Cleomedes of course opts for the third alternative, but what is significant here is that he feels compelled to refute not only the theory of traditional astronomy, but also that of Berosus, as if *both* theories were *equally* relevant. We may assume therefore that Berosus' theory was widely regarded as a reasonable alternative to the view of the astronomers, and one which could not be discarded as easily as other alternative theories. Lucretius shows himself to be aware of this. He chooses, however, to stick to his usual pattern,

---

140 For Cleomedes' dates, see Bowen-Todd (2004) 2–4.
141 See n. 138 above.

TABLE 2.2 *Place of the astronomers' explanations in the astronomical passages of the* Letter to Pythocles

Chapter	Subject	Number of explanations	View of the mathematical astronomers
7 [92]	Risings and settings	2	2
8 [92–93]	Motions of the stars	3	1
9 [93]	Turnings of the sun and moon	4	1
10 [94]	Phases of the moon	3+	–
11 [94–95]	Origin of the moon's light	2	2
12 [95–96]	Face in the moon	2+	–
13 [96–97]	Solar and lunar eclipses	2	2
14 [97]	Regularity of periods	1	–
15 [98]	Varying lengths of day and night[142]	2	2
32 [112]	Stars turning in their place	3+	1
33 [112–113]	Planets	2	1
34 [114]	Stars lagging behind	3	2

starting with the theory of the astronomers, and leaving the competing view of the Chaldeans for later.

We have now established that Lucretius, when choosing to include the explanation of the mathematical astronomers, usually mentions it first. By doing so, Lucretius demonstrates both his awareness of the predominant position of their theories, and his determination to combat this predominance by pointing out that other solutions, whether newly invented or long-forgotten or still in vogue (like the Chaldean theory of lunar phases), are just as plausible.

In the discussion above we have also examined one passage in Epicurus' *Letter to Pythocles*, viz. ch. 32 [112], which—like Lucretius, and for probably similar reasons—starts with the view of the astronomers. It would be interesting to see whether in Epicurus' *Letter* this is a sustained practice, like in Lucretius' astronomical passages, or just an isolated case.

In table 2.2 I have set out the astronomical subjects of Epicurus' *Letter to Pythocles* (excluding chapters 31 [111] on comets and 35 [114–115] on shooting

---

142 Sections 16–31 are devoted to subjects that traditionally belonged to meteorology and are therefore irrelevant for the present subject.

stars, which, in antiquity, were not generally considered astronomical), each with the number of explanations given,[143] and the place of the astronomers' explanation, if included.

The total number of astronomical subjects in the *Letter* is 13. In 9 cases we find the view of the astronomers included among a number of possible explanations.[144] In only 4 of these cases the view of the astronomers is presented first. The general pattern Bailey detected in the astronomical passages of the *DRN* does not seem to apply to the corresponding portion of Epicurus' *Letter to Pythocles*.

Yet, Lucretius' critical attitude towards the views of the astronomers is not unlike Epicurus'. The *Letter to Pythocles* contains two explicit references to the (mathematical) astronomers (both in chapters where their view *is* presented first): in chapter 9 [93] Epicurus says:

Πάντα γὰρ τὰ τοιαῦτα καὶ τὰ τούτοις συγγενῆ οὐθενὶ τῶν ἐναργημάτων διαφωνεῖ, ἐάν τις ἀεὶ ἐπὶ τῶν τοιούτων μερῶν ἐχόμενος τοῦ δυνατοῦ εἰς τὸ σύμφωνον τοῖς φαινομένοις ἕκαστον τούτων δύνηται ἐπάγειν, μὴ φοβούμενος τὰς ἀνδραποδώδεις *ἀστρολόγων* τεχνιτείας.	For all these and kindred explanations are not at variance with any clear-seen facts, if one always clings in such departments of inquiry to the possible and can refer each point to what is in agreement with the appearances without fearing the slavish artifices of the *astronomers*.[145]

And in chapter 33 [113]:

Τὸ δὲ μίαν αἰτίαν τούτων ἀποδιδόναι, πλεοναχῶς τῶν φαινομένων ἐκκαλουμένων, μανικὸν καὶ οὐ καθηκόντως πραττόμενον ὑπὸ τῶν τὴν ματαίαν *ἀστρολογίαν* ἐζηλωκότων καὶ εἰς τὸ κενὸν αἰτίας τινῶν ἀποδιδόντων, ὅταν τὴν θείαν φύσιν μηθαμῇ λειτουργιῶν ἀπολύωσι.	But to assign a single cause for these phenomena, when the appearances demand several explanations, is madness, and is quite wrongly practised by persons who are partisans of the foolish notions of *astronomy*, and who give futile explanations of the causes of certain

---

143 A '+'-sign after the number indicates that Epicurus explicitly tells us that there may be still more explanations.

144 For identification of the individual explanations see the commentaries to the *Letter to Pythocles*.

145 Translation Bailey, slightly modified, with my Italics.

phenomena, whenever they do not by
any means free the divine nature
from the burden of responsibilities.[146]

In his article 'Epicurus and the mathematicians of Cyzicus',[147] David Sedley argues that such references should be viewed in the light of Epicurus' rather personal feud with the Eudoxan school of mathematics and astronomy at Cyzicus.[148] Although I agree that such personal animosity may certainly have added to the vehemence of Epicurus' attacks, I think that here these attacks must be seen as having a broader application. In *Pyth*. 33 [113] Epicurus speaks of τῶν τὴν ματαίαν ἀστρο¬λογίαν ἐζηλωκότων: 'the partisans of the foolish notions of astronomy'. This is certainly an odd way to refer to just the astronomers, let alone such a specific group of astronomers. In fact the expression applies just as well, if not more, to all those who, while not being astronomers themselves, passionately embraced their findings, like e.g. Plato and Aristotle and their followers, and later the Stoics. At any rate, the second part of Epicurus' criticism, that 'they do not by any means free the divine nature from the burden of responsibilities', applies more naturally to these philosophers, who made the gods responsible for the heavenly motions, than to the mathematical astronomers.[149]

In *Pyth*. 33 [113] the 'partisans of the foolish notions of astronomy' are attacked, among other things, for assuming a single explanation, when the appearances call for several. The same criticism occurs throughout the *Letter to Pythocles*,[150] but only here the target is specified.[151] Yet, it is very likely that in the other instances too Epicurus was thinking in particular about these devotees of astronomy. Perhaps Anaxagoras in his time had been a proponent

---

146 Translation Bailey, slightly modified, with my Italics.
147 Sedley (1976) 26–43; see esp. p. 43 above.
148 See, however, Podalak (2010), 45–55, who is sceptical about the existence of such a school.
149 In Cic. *ND* I 30–39 the Epicurean spokesman Velleius explicitly criticises, among others, Plato (30), Aristotle (33), Theophrastus (35) and the early Stoics (36–39) for assigning divinity to the heavens and the heavenly bodies. For Plato's views see e.g. *Ti*. 40a–d; *Resp*. VI 508a; *Leg*. VII 821b–c, X 899a–b, XII 950d; *Epin*. 981e, 983a–b, 984d (cf. Barnes (1989) 41); for Aristotle's see e.g. *Metaph*. XII 8, 1074a38–b14, *Cael*. I 3.270b6–25, 9.278b14–16, II 1.284a12–14 & 284b3–5, 3.286a10–13, 12.292b32–293a1; for the Stoics' see e.g. Cic. *ND* I 36–39, II 39, 42, 44, 54, 80. On the connection between mathematical astronomy and the deification of the stars, see also Jürss (1994) 245–246 and Algra (2001) n. 39.
150 *Pyth*. 2 [87], 10 [94], 15 [98], 33 [113], 34 [114]. See also *Hdt*. 80.5–6.
151 On Epicurus' polemic, for both ethical and epistemological reasons, against the proponents of single explanations in astronomy, see e.g. Jürss (1994) 240, 245, Bénatouïl (2003) 28–35, and Verde (2013) 131.

of single causes, or Empedocles, or Democritus, but that was a long time ago. In Epicurus' days the only advocates of single explanations to be reckoned with were the astronomers and their partisans. In this respect Epicurus seems to have had the same reasons as Lucretius afterwards for attacking their views, and Lucretius turns out the be firmly rooted in Epicurus' track.

Bailey's observation that Lucretius in his astronomical passage usually presents the views of the astronomers first, appears to be basically correct. However, Bailey's interpretation of the fact is wrong. Far from actually *preferring* the views of the astronomers, as Bailey supposed, Lucretius, like Epicurus before him, singles them out as the principal representatives of the *wrong attitude* towards the explanation of the non-evident. While the appearances call for several explanations and do not permit us to choose between them, the astronomers and their followers idly opt for a single explanation.

2.4 Multiple Explanations and Doxography

Although, as Epicurus claimed, the appearances themselves invite us to adopt certain explanations,[152] many of Epicurus' and Lucretius' alternative explanations were actually derived from earlier thinkers. For instance, to take our stock example, the theory that the sun is extinguished at sunset and rekindled at sunrise, which Epicurus and Lucretius consider a viable alternative for the common view that it passes unaltered below the earth, can be confidently identified with the view espoused by Xenophanes.[153] Although the sources of individual alternative explanations are generally left unspecified, Lucretius does occasionally identify them: Democritus (V 621–622), the Chaldeans (V 727) and the 'astrologi' (V 728).[154] We have also seen that in his astronomical section Lucretius often consciously (although mostly without explicit reference) starts his lists of alternative explanations with the views of the mathematical astronomers. In addition, the commentators of Epicurus and Lucretius have traced the ultimate sources of many more of the alternative explanations. It appears that almost every one of their alternative explanations has been borrowed, sometimes with minor modifications, from one or other of their predecessors.

---

152 Epic. *Pyth.* 10 [94]. Cf. also 2 [86], 18 [100] and 33 [113].
153 See n. 134 above.
154 As noted by Runia (1997a) 95.

Epicurus' dependence on earlier theories was already recognised in antiquity. In doxographical reports Epicurus' opinion, if included, is usually mentioned last,[155] and expressed in terms which relate it to the preceding views. For instance, in Aët. II 13.15 (on the substance of the stars) we read:

Ἐπίκουρος οὐδὲν ἀπογινώσκει τούτων ἐχόμενος τοῦ ἐνδεχομένου.[156]	Epicurus rejects none of these (explanations), clinging to what is possible.

And in Aët. II 22.4 (on the shape of the sun):

Ἐπίκουρος ἐνδέχεσθαι τὰ προειρημένα πάντα.	Epicurus holds all the aforementioned (explanations) to be possible.

A similar report is found in Seneca's *Naturales Quaestiones* VI 20.5 (on earthquakes):

Omnes istas esse posse causas Epicurus ait pluresque alias temptat.	Epicurus says that all these causes may apply and he tries his hand at several more.

Judging from these testimonies one gets the impression that Epicurus himself must have had before him some doxographical work, very much like Aëtius' *Placita*, where he would have found all the relevant theories on each topic neatly listed side by side, which he could have simply copied out, striking the names of the original authors, expressing his consent with all of them, and sometimes adding a few of his own. That Epicurus might have followed such a procedure was first suggested by Diels and Usener, and has been generally accepted since.[157] For chronological reasons Aëtius' *Placita*—the only virtually

---

155 Runia (1992) 135, n. 76.
156 For the formula "clinging to what is possible", cp. Epicurus himself in *Pyth.* 9 [93]: πάντα γὰρ τὰ τοιαῦτα καὶ τὰ τούτοις συγγενῆ οὐθενὶ τῶν ἐναργημάτων διαφωνεῖ, ἐάν τις ἀεὶ ἐπὶ τῶν τοιούτων μερῶν ἐχόμενος τοῦ δυνατοῦ εἰς τὸ σύμφωνον τοῖς φαινομένοις ἕκαστον τούτων δύνηται ἐπάγειν. ["For all these and kindred explanations are not at variance with any clear-seen facts, if in such departments of inquiry one always *clings to what is possible* and can refer each point to what is in agreement with the appearances." (tr. Bailey, slightly modified)]
157 Diels (1879) 225: "Epicuri epistula ad Pythoclem {...} tanquam ex doxographis nominibus

complete doxographical work that has come down to us—cannot itself have been Epicurus' source (as is obvious from the inclusion of Epicurus' own name in Aëtius' work), but it is believed that Aëtius' work derives from earlier works of a similar nature, having a similar structure and lay-out, which may have been used by Epicurus.[158]

There are two important arguments in favour of this theory. In the first place it is clear that Lucretius' representation of earlier views sometimes depends on doxographical reports, rather than autopsy of the original works. This has been demonstrated convincingly by Wolfgang Rösler (1973) with respect to *DRN* I 635–920, where Lucretius successively deals with the views of Heraclitus, Empedocles and Anaxagoras concerning the ultimate constituents of reality.[159] As Rösler pointed out, certain misrepresentations, generalisations and choice of terminology, such as the designation of the Anaxagorean first principles as *homoeomeria*,[160] betray Lucretius' dependence on a doxographical tradition that goes back to Aristotle and Theophrastus.[161] This does not prove, of course, that Epicurus' and Lucretius' accounts of astronomical and meteorological theories, in the *Letter to Pythocles* and *DRN* V 509–770 and VI, depend on doxographical reports as well, but it certainly adds to the likelihood of this hypothesis.

philosophorum omissis raptim corrasa est {...}." Usener (1887) xl–xli: "Elegisse autem Epicurum perquisitis omnium physiologorum libris quis credat? Quem etsi Democriti et Democriteorum, Anaxagorae et Archelai opiniones facile concedemus ipsum ex illorum libris novisse, reliquorum ut cognosceret rationes consentaneum est librum ei ad manum fuisse, quo conpositas et conparatas nullo negotio inveniret, hoc est Theophrasti φυσικῶν δόξας." See also Ernout-Robin (1925–1928) III 201–202; and Runia (1997a).

158 In Usener's and Diels' wake this earlier work, from which Aëtius' *Placita* is supposed to be ultimately derived, is often identified with Theophrastus' Φυσικαὶ δόξαι ('Physical Opinions'), a work of which only a single fragment remains (see Runia (1992) 117). However, following the important studies on these subjects by Jaap Mansfeld and David Runia it seems more prudent now to simply state that Aëtius' *Placita* is based on, and influenced by, several works of Theophrastus as well as Aristotle. See e.g. Mansfeld (1989) esp. 338–342; Mansfeld (1992b) and Mansfeld (2005).

159 An expanded version of the argument is offered by Montarese (2012) 11–56.

160 Lucr. *DRN* I 830 & 835; on the provenance of this term see Rösler (1973), 67–68 with nn. thereto, Sedley (1998a), 124–125, and Montarese (2012) 36–42.

161 Whether or not these doxographical data were mediated by a Greek Epicurean source is still open to debate: Rösler (1973) 71–72 and Montarese (2012) 56–57. Sedley (1998a) 123–126 & 145–146 argues that books XIV and XV of Epicurus' *On nature* were Lucretius' direct source for this passage, a view that is strongly rejected by Montarese (2012) 58–146.

Another argument in favour of an ultimately doxographical origin of these passages is the close correspondence in the order of subjects that can be observed in—on the one hand—the meteorological sections of Lucretius' *DRN* and Epicurus' *Letter to Pythocles*, and—on the other hand—book III of Aëtius' doxography.[162] The question of the precise relations between these three texts is, however, complicated by the existence of a further parallel, a meteorological treatise ascribed to Theophrastus and preserved in Syriac and Arabic, which exhibits more or less the same order of subjects. We will examine this treatise and the nature of the complications involved more closely on p. 70 ff. below, while the correspondences in the order and scope of subjects of these four works, as well as a number of other texts, will be investigated more thoroughly in Chapter Three.

It is remarkable, however, that the observed correspondence in the sequence of subjects does not extend to the astronomical sections of these works: here Epicurus, Lucretius and Aëtius differ considerably from each other both in the range and the order of subjects (see also p. 160 below). Nor does the correspondence extend to the inclusion and order of individual explanations per meteorological subject: in this respect too Epicurus, Lucretius, Aëtius and, in addition, the *Syriac meteorology* show but little agreement.[163]

One aspect of Epicurean multiple explanations does not seem to be accounted for by the assumption of a doxographical origin. As we saw above, analogy plays an important role in the validation of individual explanations. Accordingly, in the *DRN* many explanations of astronomical and especially meteorological phenomena are supported by references to similar appearances here with us. For instance, in *DRN* VI 121–131 one possible cause of thunder—wind being trapped in a hollow cloud and then violently bursting forth—is compared to the explosion of an air-filled bladder. Since both Epicurus' *Letter to Pythocles* and the corresponding books of Aëtius' *Placita* (book II, on cosmology and astronomy, and book III, on meteorology and terrestrial phenomena)

---

162 As observed by Reitzenstein (1924) 34–35; Runia (1997a) 97; Sedley (1998a) 158. See also table 3.4 on p. 128 below, where the order of subjects of Lucretius VI, Epicurus *Pyth.*, Aët. III and the *Syriac meteorology* is compared.

163 As observed by Sedley (1998a) 182. For instance, Lucretius' 'habit' of mentioning the views of the mathematical astronomers first (see §2.3.7 on p. 42 ff. above) is quite unlike the way Aëtius structures his chapters according to *diaeresis* and *diaphonia*. Aëtius' method is becoming increasingly clear thanks to the works of David Runia and Jaap Mansfeld: see esp. Runia (1989) and (1992), and Mansfeld & Runia (2009a) part 1, pp. 3–16 *et passim*, and part 2, *passim*.

provide very few specific analogies, one might be tempted to ascribe the addition of such analogies to Lucretius himself rather than—through Epicurus' mediation—a doxographical source. As we have seen, however, the virtual absence of specific analogies from Epicurus' *Letter to Pythocles* may well be due to its being a summary of a more extensive work.[164] In fact, many of the analogies provided by Lucretius are old, much older even than Epicurus—the comparison of thunder with an exploding bladder, for instance, is found in Aristophanes' *Clouds* (lines 403–407)—and it seems unlikely that the theories and the accompanying analogies should have reached Lucretius by different roads. It is also noteworthy that the *Syriac meteorology*, which I mentioned above, offers many of the same analogies as Lucretius, including the exploding bladder (1.18–20).

In sum, it is very likely that the meteorological and astronomical portions of Lucretius' *DRN* ultimately derive—probably for the most part through Epicurus' works—from a doxographical source, which in the sequence of its subjects may have resembled Aëtius, but which, in contrast to Aëtius, combined the unattributed *doxai* with explanatory analogies. What place should be assigned to the *Syriac meteorology* in this transmission is as yet unclear and depends largely on its authorship (see pp. 70 ff. and 156 ff. below).

2.5 The Sources of the Method of Multiple Explanations

2.5.1 *Introduction*
Epicurus and Lucretius may have derived most of their alternative explanations from doxography, but the result is something new and different. Stripped of their name-labels, theories devised by earlier thinkers have been transformed into truly alternative explanations endorsed by Epicurus and Lucretius themselves.[165] So, even though doxography may explain where the individual explanations came from, it does not explain how Epicurus and Lucretius came to use them as they did.

In this section I want to examine a number of other texts and authors who also sometimes resorted to the use of multiple explanations, and find out if and to what extent they resemble and may have influenced Epicurus. The

---

164 See p. 35 with nn. 90 and 91 above.
165 This is not always appreciated by the commentators: see Ernout-Robin (1925–1928) III 202: "c'est en somme une doxographie, mais sans nom propre."

examination will include Democritus, Aristotle, Theophrastus, and the *Syriac meteorology* commonly ascribed to Theophrastus. In order to be able to make a comparison, it will be useful first to indicate some general characteristics of the method as employed by Epicurus and Lucretius. The following features I consider typical of Epicurean multiple explanations:

1. They are applied to *non-evident* matters, such as the nature and causes of astronomical and meteorological phenomena.
2. In those fields of physical enquiry where multiple explanations are used (astronomy and meteorology) they are used *systematically*.
3. Lists of multiple explanations may consist of up to *eight* or *nine* explanations.[166]
4. Many explanations, especially in meteorology, are supported with analogies from everyday experience.
5. Most alternative explanations can be related to the views of earlier thinkers.

Armed with these five distinctive features we may now proceed to investigate possible parallels to Epicurus' and Lucretius' method of multiple explanations.

2.5.2 *Democritus*

Already in antiquity Epicurean physics was often viewed as a modernised version of the teachings of Democritus.[167] It is not unreasonable therefore to start our investigation into the origins of Epicurus' method of multiple explanations with Democritus.

A very promising testimony in this respect is provided by Seneca. In *NQ* VI 20.1 (Democritus fr.A98 D–K), having just presented an overview of ancient theories on earthquakes, Seneca continues thus:

Veniamus nunc ad eos qui *omnia* ista quae rettuli in causa esse dixerunt aut ex his *plura*. Democritus *plura* putat. Ait enim motum (i) aliquando spiritu

Let us now come to those who said that *all* these causes, which I recounted, are responsible or *several* of these. Democritus thinks *several*.

---

166   Epic. *Pyth.* 19 [101–102] offers *eight* causes of lightning, and *DRN* VI 96–159 *nine* causes of thunder (for the structure of this section see table 3.8 on p. 143 below).

167   Cf. Cic. *N.D.* I 73 "quid est in physicis Epicuri non a Democrito?" See also Cic. *Acad. Post.* I 6; *De fin.* I 17–18, 21; II 102; IV 13; and *N.D.* I 93 & 120; Diog. Laërt. X 4; Plut. *Adv. Col.* 3, 1108e–f.

fieri, (ii) aliquando aqua, (iii) aliquando utroque, et id hoc modo prosequitur: ...	For he says that an earthquake (i) sometimes happens because of wind, (ii) sometimes of water, (iii) sometimes of both, and he pursues this in the following manner: ...

Seneca goes on to describe each of these three explanations in some detail—first explanation (ii) and then (iii) and (i)—after which he continues (20.5), picking up the reference at the beginning of the chapter to 'those who said that all these causes ... are responsible':

Omnes istas esse posse causas Epicurus ait pluresque alias temptat ...	Epicurus says that *all* these causes may apply and he tries his hand at several more ...

According to Seneca, then, both Democritus and Epicurus explained earthquakes with a number of alternative theories, but whereas Epicurus accepted all available theories, Democritus was more selective. This explicit contrast seems to suggest that Seneca knew something about Democritus' method and how this differed from Epicurus'.[168] It is quite probable, however, that Democritus' selectivity is only apparent. Democritus wrote at a time when many of the theories later described by Seneca had not yet been devised and the major doxographical works reporting them not yet been written.[169] I am inclined to think, therefore, that Seneca had no positive information about Democritus' methodology at all, but simply inferred so much from the three alternative explanations he found attributed to Democritus, which he himself then contrasted to the larger number of explanations offered by Epicurus.

The amount of detail with which Seneca is able to report Democritus' three explanations may suggest that he had a very good source for them, but in fact the wording of the text indicates that he may have filled in much of the detail himself. At the beginning of section 2, for instance, he writes: 'And now, just as

---

168 On Seneca's contrasting of Epicurus' and Democritus' methods in this passage see now also Verde (2013) 137–138.

169 Jaap Mansfeld points out to me that the fifth century BC already knew some doxographical overviews, such as Hippias' presentation of related views and Herodotus' account of the Nile flood, although these overviews fall far short of the doxographical works and passages of later times. See also Mansfeld & Runia (2009a) 154 ff.

we spoke of wind, we must also speak of water',[170] as if he were describing his own theory, instead of someone else's.

Seneca's report on Democritus also strangely deviates from the account offered by Aristotle in *Mete.* II 7, 365b1–6 (Dem. fr.A97 D–K). According to Aristotle, Democritus held that earthquakes occur both (a) when the earth is full with water and its cavities overflow, and (b) when the earth is dried up and its cavities draw water from elsewhere. Aristotle's first explanation may perhaps be identified with Seneca's second, but Aristotle's second explanation has nothing to do with either Seneca's first or his third theory.

In sum, it is quite possible that Democritus offered more than one explanation for earthquakes, but in view of the discrepancy between Aristotle's and Seneca's accounts we cannot be certain which explanations these were. Yet, neither Aristotle nor Seneca tells us why Democritus resorted to multiple explanations. For all we know Democritus may have simply offered his two or three explanations by way of a hypothesis, without any strong epistemological motives. There is no indication, moreover, that Democritus extended this use of multiple explanations to other phenomena as well. On the contrary, although there are many ancient reports concerning Democritus' views on specific astronomical, meteorological and terrestrial problems, none of these (beside those on earthquakes) attribute to Democritus anything other than single explanations. There is no good reason therefore to consider Democritus a major source of inspiration for Epicurus' method of multiple explanations.[171]

2.5.3 *Aristotle*

A more promising candidate in this respect is Aristotle.[172] Several cases of multiple explanations are found in his works, especially in the *Meteorology*. In I 3, 341a12–31, for instance, Aristotle gives *two* reasons why the sun, though not fiery in itself, produces heat on earth; in I 4, 341b36–342a13, he offers *two* explanations for the production of shooting stars and the like; and finally, in I 7, 344a5–b4, he gives *two* accounts of the production of comets, corresponding to two different *types* of this phenomenon. This last subject is introduced with the following lines (344a5–8):

---

170   Sen. *NQ* VI 20,2: "Etiam nunc, quomodo de spiritu dicebamus, de aqua quoque dicendum est."
171   *Pace* Asmis (1984) 328–329.
172   Asmis (1984) 329–330, Mansfeld (1994) 33, n. 18.

Ἐπεὶ δὲ περὶ τῶν ἀφανῶν τῇ αἰσθήσει νομίζομεν ἱκανῶς ἀποδεδεῖχθαι κατὰ τὸν λόγον, ἐὰν εἰς τὸ δυνατὸν ἀναγάγωμεν, ἔκ τε τῶν νῦν φαινομένων ὑπολάβοι τις ἂν ὧδε περὶ τούτων μάλιστα συμβαίνειν·	For we consider that we have given a sufficiently rational explanation of things non-evident to sense-perception if we have referred them to what is possible; and, on the basis of the present appearances, one may assume that they are best accounted for as follows.[173]

Here, just like Epicurus afterwards, Aristotle applies multiple explanations to things *non-evident*, inferring the *possibility* of each explanation on the basis of the *appearances*.[174] It must be observed, though, that while Aristotle is here thinking of the appearances of the object of inquiry itself, Epicurus usually refers to the appearances of analogous phenomena here with us. Sometimes, however, Aristotle too supports his alternative explanations by reference to appropriate *analogies with phenomena here with us*, as we can observe in his first account of the sun's heat-production (341a23–27):

Τὸ δὲ μᾶλλον γίγνεσθαι ἅμα τῷ ἡλίῳ αὐτῷ τὴν θερμότητα εὔλογον, λαμβάνοντας τὸ ὅμοιον ἐκ τῶν παρ' ἡμῖν γιγνομένων· καὶ γὰρ ἐνταῦθα τῶν βίᾳ φερομένων ὁ πλησιάζων ἀὴρ μάλιστα γίγνεται θερμός.	That the heat is increased by the presence of the sun is easily enough explained by considering *analogies with phenomena here with us*: for here too the air in the neighbourhood of a projectile becomes hottest.[175]

Yet, in spite of these similarities there is still a huge gap between Epicurus' and Aristotle's approaches to multiple explanations. In the first place, Aristotle only very rarely resorts to multiple explanations: in the entire body of the *Meteorology*, only three clear cases are found. Most often Aristotle is perfectly happy to give just *one* explanation. Secondly, the number of alternatives offered is much smaller: in the *Meteorology* Aristotle in each case offers no more than

---

173 My translation.
174 Cf. Epic. *Pyth.* 9 [93]: πάντα γὰρ τὰ τοιαῦτα καὶ τὰ τούτοις συγγενῆ οὐθενὶ τῶν ἐναργημάτων διαφωνεῖ, ἐάν τις ἀεὶ ἐπὶ τῶν τοιούτων μερῶν ἐχόμενος τοῦ δυνατοῦ εἰς τὸ σύμφωνον τοῖς φαινομένοις ἕκαστον τούτων δύνηται ἐπάγειν ... ["For all these and kindred explanations are not at variance with any clear-seen facts, if one always clings in such departments of inquiry to *what is possible* and can *refer* each point to what is in agreement with the appearances ..." (tr. Bailey, slightly modified)]
175 Translation by Lee, modified.

two explanations, whereas Epicurus and Lucretius may offer up to *eight* or even *nine* possible causes. Thirdly, Aristotle only occasionally uses analogies to support an explanation, whereas for Lucretius and Epicurus analogy with everyday appearances is essential for accepting an explanation. Finally, Aristotle's multiple explanations do not seem to relate to earlier views: his accounts of the sun's heat-production, of shooting stars and of comets proceed from Aristotle's own physical theory. When, on the other hand, Aristotle *does* engage with older theories, he usually rejects them, and substitutes them with a *single* theory of his own making.[176]

2.5.4 *Theophrastus*

With Theophrastus, Aristotle's successor as head of the Lyceum, the use of multiple explanations becomes much more prominent. Many instances are found in Theophrastus' minor treatises *De ventis*, *De lapidibus* and *De igne*, and many more in his botanical writings, especially his *De causis plantarum*.[177] Yet, even in these works multiple explanations, though by no means rare, are still the exception. When offering multiple explanations, Theophrastus most often gives two, but occasionally more; the maximum number I have found is five.[178]

Sometimes the explanations offered are explicitly linked to analogous occurrences within our sensory experience, as can be seen in *De igne* 1, 4–11:

Ἔτι δὲ αἱ γενέσεις αὐτοῦ {sc. τοῦ πυρὸς} αἱ πλεῖσται [καὶ] οἷον μετὰ βίας, καὶ γὰρ ἡ πληγὴ τῶν στερεῶν ὥσπερ λίθων, καὶ ⟨αἱ⟩ τρίψει καὶ πιλήσει καθάπερ τῶν πυρείων ⟨καὶ πάντων⟩ ὅσα ἔχει φοράς, ὥσπερ τῶν πυρουμένων καὶ τηκομένων (ἐκ δ' αὐτοῦ τοῦ ἀέρος καὶ τοῖς νέφεσι συστροφαὶ καὶ θλίψεις· βίαιοι γὰρ δὴ αἱ φοραί, δι' ὧν δὴ οἱ πρηστῆρες καὶ κεραυνοὶ γίνονται), καὶ ὅσους δὴ τρόπους ἄλλους τεθεωρήκαμεν εἴθ' ὑπὲρ γῆς γινομένων εἴτ' ἐπὶ γῆς εἴθ' ὑπὸ γῆς. Αἱ γὰρ πολλαὶ δόξειαν ἂν αὐτῶν μετὰ βίας.	Moreover, most forms of generation of fire take place by force, as it were; for instance, that caused by the striking of solids like stones, and those caused by friction and compression, as in firesticks and in all those substances which are in process, such as those which are ignited and fused (in fact, it is from air that the clouds undergo their concentrations and compressions, for of course the motions by which firewinds (*prēstēres*) and

---

176 So Taub (2003) 94: "In the Meteorology, Aristotle does not usually accept the views of his predecessors, even when they are those of 'the majority or the wise'."
177 Steinmetz (1964) 32–33, 46, 82, 88, 91, 103, 122–123, 132, 139; Eichholz (1965) 6; Wöhrle (1985) 145–148; Vallance (1988) 34–35; Daiber (1992) 285; Gottschalk (1998) 287.
178 *CP* I 17.5.

thunderbolts are generated are forcible), and whatever other ways we have observed, whether above the earth, on it, or beneath it. Most of these appear to come about by force.[179]

Theophrastus only rarely comments on his motives for accepting several explanations; the only clear instance I have found is *De lapidibus* 3, 1–3:

Ἡ δὲ πῆξις τοῖς μὲν ἀπὸ θερμοῦ τοῖς δ' ἀπὸ ψυχροῦ γίνεται. κωλύει γὰρ ἴσως οὐδὲν ἔνια γένη λίθων ὑφ' ἑκατέρων συνίστασθαι τούτων.	This solidification is due in some cases to heat and in others to cold, for there may be nothing to prevent certain kinds of stone from being formed either by heat or by cold.[180]

It is interesting to note that—apart from the modest ἴσως ('perhaps')[181]—Theophrastus' "κωλύει γὰρ οὐδὲν" ('for nothing prevents') is very similar to the formulas Epicurus later uses to signal the validity of his alternative explanations, like "οὐδὲν γὰρ ἀντιμαρτυρεῖ" ('for nothing contests') and "οὐθὲν ἐμποδοστατεῖ" ('nothing stands in the way').[182]

It must also be observed that, when Theophrastus offers multiple explanations, these hardly ever relate to earlier views, and when he does adduce older theories it is usually to reject them and replace them with a theory of his own.[183]

Until now we may have been comparing apples and oranges: Epicurus and Lucretius typically apply multiple explanations to astronomical and meteorological problems, and the few instances of multiple explanations in Aristotle's works also occur in his *Meteorology*. It would be interesting therefore to see to what extent Theophrastus used multiple explanations in *his* meteorological treatise. The *Syriac meteorology*, which is commonly, but in my view prematurely, ascribed to Theophrastus, will be dealt with in the next subsection. Here I will confine myself to Greek and Latin testimonies of Theophrastus' meteoro-

---

179 Translation by Coutant, slightly modified.
180 Translation by Eichholz.
181 According to Baltussen (1998), 171, this use of ἴσως was a mannerism of Theophrastus', indicating firm conviction rather than uncertainty.
182 See p. 19 n. 48 above.
183 See e.g. Theophr. *De igne* 52–56; *HP* III 1, 4–5; III 2, 2. Cp. what was said about Aristotle in n. 176 above and text thereto.

logical views. A very interesting text in this respect is fr.211B FHS&G, preserved by Olympiodorus *In Arist. Mete.* I 9, 346b30 (p. 80.30–81.1 Stüve):

Ἰστέον δέ, ὅτι ὁ μὲν Ἀριστοτέλης αἴτιον λέγει τῆς εἰς ὕδωρ μεταβολῆς τὴν ψύξιν μόνον· Θεόφραστος δὲ οὐ μόνον τὴν ψύξιν αἰτίαν φησὶ τῆς τοῦ ὕδατος γενέσεως, ἀλλὰ καὶ τὴν πίλησιν. ἰδοὺ γὰρ ἐν Αἰθιοπίᾳ μὴ οὔσης ψύξεως ὅμως ὑετὸς κατάγεται διὰ τὴν πίλησιν· φησὶ γὰρ ὄρη εἶναι ἐκεῖσε ὑψηλότατα, εἰς ἃ τὰ νέφη προσπταίουσι, καὶ εἶθ' οὕτως ὑετὸς καταρρήγνυται διὰ τὴν γινομένην πίλησιν. ἀλλὰ μὴν καὶ ἐπὶ τῶν λεβήτων ὑγρότης, φησίν, ἀντικαταρρεῖ, ἔτι δὲ καὶ ἐπὶ τῶν θόλων τῶν λουτρῶν μὴ παρούσης ψύξεως, διὰ τὴν πίλησιν δηλονότι τούτου γινομένου.	One should known that Aristotle says that cooling alone causes the transformation into water. Theophrastus, however, says that not only cooling causes the generation of water, but also condensation. Note that in Aethiopia where there is no cooling, rain nevertheless pours down due to condensation. For he says there are very high mountains there, against which the clouds collide and that subsequently rain pours down because of the ensuing condensation. Yet also in the case of cauldrons, says he, moisture runs down, and also in the case of the vaults of baths, where there is no cooling, this obviously occurs due to condensation.

According to this report,[184] Theophrastus accepted two different causes of rain-production: cooling and condensation. The first view is authorized by Aristotle, the second supported by one example (rain in Aethiopia) and two *analogies* (cauldrons and the vaults of baths). Both views can be related to earlier theories: the first view is Aristotle's (*Mete.* I 9, 346b30–31), as Olympiodorus informs us, and the second corresponds to the views of several earlier thinkers.[185] For rain production, then, Theophrastus accepted two explanations, both deriving from earlier thinkers.

There is one other meteorological problem for which Theophrastus' view is explicitly reported. In *NQ* VI 13.1 (fr.195 FHS&G), Seneca ascribes to Theophras-

---

184 Cf. Proclus *In Plat. Tim.* 22E (= Theophr. fr.211A FHS&G); Galen *In Hippocr. Aer.* 8.6 (= fr.211C ibid.), and Theophrastus *De ventis* 5.1–5.
185 Anaximenes A17 D–K (= Aët. III 4.1), Xenophanes A46 D–K (= Aët. III 4.4), Hippocrates *Aer.* 8.7 (II 34 Littré) and Democritus *apud* Diod. I 39.3 (cf. fr.A99 D–K = Aët. IV 1.4) all ascribe rain formation to condensation of clouds; this condensation being due, according to Hippocrates, to compression by contrary winds and other clouds, and, according to Democritus, to compression against high mountains.

tus and Aristotle the *single* view that earthquakes come about through exhalations rising from the earth and then, for lack of place, turning back on themselves. If this report is correct, Theophrastus did not use multiple explanations to account for every meteorological problem.

Our findings may be summarized as follows. In his meteorological treatise (to judge from Greek and Latin testimonies), as in his works on fire, stones, winds and plants, Theophrastus frequently uses multiple explanations, although single explanations remain the norm. When he does give several explanations, the number of alternatives rarely exceeds two, although cases with up to five can be found. Alternative explanations are occasionally supported with analogies from everyday experience, but more often they are not. They may be derived from the views of earlier thinkers, but in general they are not. In short, most aspects of Epicurean multiple explanations can be found in Theophrastus, but on a far more modest scale and never systematically.

2.5.5 *The* Syriac Meteorology

In the previous subsection I have deliberately left out of account a treatise which most scholars now agree in identifying with (a part or a summary of) Theophrastus' lost meteorological treatise, the *Metarsiology*.[186] I have done so because I think there is still reasonable doubt about this identification. Since, however, this treatise furnishes the closest existing parallel to Epicurus' and Lucretius' use of multiple explanations, it must here be dealt with.

The treatise is preserved in three versions: two mutually independent Arabic translations of a Syriac original, and a single, badly mutilated, copy of this Syriac original.[187] In both Arabic versions the work is ascribed to *Theophrastus* (the corresponding section of the Syriac manuscript is lost). Together, the three versions seem to provide a good basis for a reconstruction of (the contents of) the Syriac original.[188] The treatise covers a range of meteorological phenomena,[189] explaining most of them with *a number of alternative explanations*.

---

186 E.g. Daiber (1992) 282 ff.; Mansfeld (1992a) 314–316; id. (1994) 30; Sedley (1998a) 158, 179; Sharples (1998) 17, 144; Taub (2003) 116.

187 The Syriac version was first edited, with German translation and commentary, by Wagner & Steinmetz (1964). The Arabic version of Bar Bahlūl was first edited, with a German translation and commentary, by Bergsträßer (1918). The second Arabic version, probably made by Ibn Al-Khammār, was edited, with an English translation and commentary by Daiber (1992), who also offered improved editions of the other two versions, unfortunately without translation.

188 *Not* necessarily of the *Greek* original, as Daiber (1992) claims on pp. 219 & 282–283.

189 See the table of contents on p. 91 below.

This pervasive use of multiple explanations in a treatise purportedly written by Theophrastus caused some embarrassment to the earlier commentators. They could not believe that Theophrastus would have employed a method so radically different (as it appeared to them) from Aristotle's and Theophrastus' known works.[190] In fact, as Gotthelf Bergsträßer observed,[191] if the treatise had not been explicitly ascribed to Theophrastus, no one would have guessed that it was his. In view of its offering multiple explanations for most of the phenomena, and in view of the close parallels with *DRN* VI, both in the order of subjects (on which see pp. 127 ff. below) and in the treatment of individual subjects, attribution to Epicurus or his school would have seemed obvious. For this reason Bergsträßer and Boll, and later Reitzenstein and a few others, felt compelled at least to consider the possibility of an Epicurean origin.[192] Unfortunately, they did not follow up on this hypothesis, but simply dismissed it in favour of another hypothesis, to the effect that the treatise's Greek original had been "a doxography or at least a discussion with a strong doxographical character [...], in which the excerptor had deleted the names of the inventors of each individual theory as being of no consequence, and rather blurred the traces of the author's own position."[193] In this way they were able to uphold Theophrastus' authorship, without saddling him with a method of multiple explanations that seemed alien to him. The attribution to Theophrastus has never been seriously doubted again,[194] even though the views about the treatise's real character have radically changed. Since Daiber's publication of the text with translation and commentary in 1992, it is generally accepted that the treatise really offers multiple alternative explanations, and not, as was previously believed, a (crit-

---

190 See e.g. Strohm (1937) 411: "daß die verworrene Folge von sieben Erklärungen des Donners und vier Gründen des Blitzes, die in einer an Epikurs Probabilismus gemahnenden Reihe nebeneinanderstehen, in dieser Form Theophrast fremd ist, liegt auf der Hand."
191 Bergsträßer (1918) 28.
192 Bergsträßer (1918) 28; Boll in his epilogue to Bergsträßer (1918) 30; Reitzenstein (1924) 7–11; Drossaert Lulofs (1955) 438. The question is wisely left open by Robin in Ernout-Robin (1925–1928) III 200–201 and 249.
193 Bergsträßer (1918) 28. See also Reitzenstein (1924) 8 ff., Strohm (1937) 411; Wagner-Steinmetz (1964) 14, 34; Steinmetz (1964) 55.
194 Bergsträßer (1918), 28, also advances the possibility of a late compendium, and so does Gottschalk (1965), 759–760, who identifies some passages as deriving from Strato rather than Theophrastus. Kidd (1992), 294, also leaves open the possibility of a late compendium, and Van Raalte (2003) believes that the so-called Theological Excursus cannot derive from Theophrastus. Doubts about the correctness of the attribution to Theophrastus also in Ernout-Robin (1925–1928) III 200–201.

ical) doxography.¹⁹⁵ Yet, if this is true (as I believe it is), the grounds on which Bergsträßer and Reitzenstein were able to reject the possibility of an Epicurean rather than a Theophrastean origin of the treatise, seem to have been removed: Theophrastus' authorship can no longer be taken for granted. I will therefore refer to the treatise simply as the '*Syriac meteorology*' and from this neutral position try to establish how the treatise relates to Theophrastus on the one hand and Epicurus and Lucretius on the other.

Although, as we saw above, Theophrastus was not as averse to multiple explanations as the earliest commentators of the treatise believed, there is still a huge gap between the use of multiple explanations in the *Syriac meteorology* and in Theophrastus' undisputed writings:

1. While in Theophrastus' undisputed writings single explanations still seem to be the rule, in the *Syriac meteorology* most problems are accounted for by a number of explanations.¹⁹⁶
2. While in Theophrastus' undisputed writings the number of alternative explanations rarely exceeds *two*, the *Syriac meteorology* frequently offers *three* or more, the maximum number being *seven* (as in the case of thunder).¹⁹⁷
3. While in Theophrastus' undisputed writings alternative explanations are only rarely supported with analogies, the *Syriac meteorology* abounds with them.¹⁹⁸
4. While in Theophrastus' undisputed writings alternative explanations only rarely derive from earlier, Presocratic, theories, in the *Syriac meteorology* the vast majority of theories can be identified with specific views of earlier thinkers.¹⁹⁹

In all these respects the *Syriac meteorology* is much closer to the meteorological sections of Epicurus' *Letter to Pythocles* and Lucretius' *De rerum natura* than to Theophrastus' undisputed works.

---

195 Daiber (1992) 285; Kidd (1992) 303–304; Mansfeld (1992a) 325; Mansfeld (1994) 33; Sedley (1998a) 181; Sharples (1998) 17 with n. 58; Taub (2003) 116–117.
196 See APPENDIX 3 on p. 274 below: 26 cases of multiple explanations vs. 22 with single explanations.
197 See APPENDIX 3 on p. 274 below: 9 cases with three or more explanations vs. 17 with just two explanations.
198 See Daiber (1992) 284, 285, 288 ('illustrative examples' or 'illustrative experiments') and esp. Taub (2003) 117–120.
199 Daiber (1992) 287–288, 290; Taub (2003) 117.

It has been observed that in the *Syriac meteorology* different explanations sometimes apply to different types of a certain phenomenon, each exemplified by a different analogy.[200] It has also been suggested that in this respect the use of multiple explanations in the *Syriac meteorology* differs from the way they are used by Epicurus and Lucretius, who typically conceived of their several explanations as equally possible alternative causes of a single undifferentiated phenomenon.[201] However, this supposed contrast between the *Syriac meteorology* and Epicurus / Lucretius is based on a misrepresentation of the latter: as we have seen above (p. 12), in his meteorology Lucretius too sometimes differentiates phenomena by type, and Epicurus probably did so as well.

As far as the application of multiple explanations to physical problems is concerned, the Syriac meteorology closely resembles the meteorological sections of Epicurus and Lucretius, but differs markedly from other, undisputed, writings of Theophrastus. From this point of view, then, an Epicurean origin of the treatise seems more obvious than attribution to Theophrastus. The question of the treatise's origin and its relations with Epicurus and Lucretius will be considered more thoroughly in Chapter Three (p. 145 ff.) below.

2.5.6   *Conclusions about the Origins of the Method*
It is now time to formulate some conclusions concerning the possible origins of Epicurus' method of multiple explanations.

However great Democritus' influence on Epicurean physics may have been, there is little reason to assume that he was a major source of inspiration for Epicurean multiple explanations as well. Even if Democritus gave two or three alternative explanations of earthquakes, there is nothing to suggest that this was motivated by any strongly felt epistemological considerations, or that the use of multiple explanations was extended to other physical problems as well.

The first clear instances of multiple explanations are found in the (undisputed) writings of Aristotle and Theophrastus. Although in these works multiple explanations are not applied systematically, when they are applied, they seem to be confined to a certain class of phenomena. Aristotle remarks that *non-evident* things need to be accounted for with reference to what seems *possible* on the basis of *the present appearances*—a procedure which, apparently, will sometimes result in the acceptance of multiple explanations—, and Theophrastus ascertains the validity of his alternative explanations from the

---

200   Steinmetz (1964) 322; Daiber (1992) 279 (*ad* 13.22–32), 285, 288; Kidd (1992) 299–304; Sharples (1998) xv; Taub (2003) 117, 130–131; Garani (2007) 97.
201   Kidd (1992) 303–304; Sharples (1998) xv; Taub (2003) 130–131; Garani (2007) 97.

observation that *nothing prevents* any one of them. In both these respects Aristotle's and Theophrastus' use of multiple explanations resembles Epicurus'. On the other hand, Aristotle and Theophrastus (in his undisputed works) rarely derive their alternative explanations from doxography, and when they do deal with earlier views they usually refute them.

It is very likely that Epicurus' use of multiple explanations was influenced by Aristotle and Theophrastus. Yet, the precise extent of their influence remains uncertain as long as Theophrastus' authorship of the *Syriac meteorology* is not firmly established. If, as is commonly believed, the *Syriac meteorology* is either the whole or a part or a summary of Theophrastus' lost *Metarsiology*, then Theophrastus' influence on Epicurus must indeed be deemed very great, to the extent that Epicurus must have derived his method of multiple explanations and its application to meteorological problems almost entirely from Theophrastus. If, on the other hand, the *Syriac meteorology* is not related to Theophrastus' treatise, but instead somehow derives from one of Epicurus' own works, then—although Theophrastus and Aristotle may still have exerted some influence—the method of multiple explanations must be considered Epicurus' own contribution to philosophy.

2.6 Conclusions

In this chapter various aspects of the Epicurean method of multiple explanations, as applied in Epicurus' *Letter to Pythocles* and Lucretius' *DRN*, have been explored. In the first part of the chapter I have dealt with certain epistemological aspects of the method of multiple explanations. Although it would seem that Epicurus really affirmed the truth of each one of a number of alternative explanations (as long as these were not contested by the appearances), it turns out that this 'truth' is fundamentally different from the singular truth which attaches to the basic tenets of Epicurean physics. Whereas these are true always and everywhere, alternative explanations are only true in view of the *principle of plenitude*: given the infinity of time and space everything that is possible must be true sometimes and somewhere. In addition, I argued that Diogenes of Oenoanda's claim that some explanations are more plausible than others is a departure from Epicurus and Lucretius for whom all alternative explanations have the same truth-value. I also examined Bailey's assertion that Lucretius' habit of mentioning the views of the mathematical astronomers first betrays a preference for their views. I argued that such a preference goes against the principles of Epicurean multiple explanations, which Lucretius himself subscribes to, and that instead this habit should be seen as serving a polemical purpose in line with these Epicurean principles: against those people (i.e. most of his contemporaries) who believe the explanations of the mathematical astronomers to be uniquely true, Lucretius produces a number of alternative explanations which, on the evidence of the senses, are just as likely and should therefore be accepted on a par with those others. Although this devaluation of mathematical astronomy contrasts sharply with the attitude adopted by, amongst others, the Peripatetics, in other respects Epicurus and Lucretius are clearly indebted to the Peripatos. It is well known, for instance, that most, if not all, of the alternative explanations which the Epicureans mobilize against the theories of mathematical astronomy, were borrowed from doxography, a genre of writings rooted in the works of Aristotle and Theophrastus. However, not just individual alternative explanations, but the very method of multiple explanations may owe something to Aristotle and Theophrastus. Indeed, both their works present occasional instances of multiple explanations as well as epistemological justifications for their use, which seem to prefigure Epicurus' method of multiple explanations. Theophrastus' influence on Epicurus might turn out to be even greater if the *Syriac meteorology* could be proved to be Theophrastus'. However, comparison of the use of multiple explanations in Theophrastus' undisputed works and the *Syriac meteorology* indicates that Theophrastus' authorship of the latter work is still far from certain. The question of the *Syriac meteorology*'s authorship will be further explored in the next chapter.

CHAPTER 3

Range and Order of Subjects in Ancient Meteorology

3.1 Introduction

Both Epicurus' *Letter to Pythocles* and the sixth book of Lucretius' *DRN* are sometimes styled *meteorological* treatises.[1] This description puts them in a league with a number of other ancient meteorological writings, first an foremost Aristotle's *Meteorology*, the work which gave this branch of physical inquiry its name. Other notable examples are Seneca's *Naturales Quaestiones* ('Natural Questions') and the *Syriac meteorology* commonly ascribed to Theophrastus.[2] Although each of these works can be and has been called a meteorology, it is remarkable how ill-defined the word meteorology actually is, and how variable its subject matter.

Before I go on, it will be therefore be expedient to try and offer something of a definition of meteorology, to state what it is, and especially how it relates to astronomy. The English word 'meteorology' derives from the Greek μετεωρολογία, which is the study of τὰ μετέωρα, 'lofty things'. Before Aristotle the word μετέωρος and its derivatives appear to have been used indiscriminately to refer to *astronomical* as well as *atmospherical* phenomena.[3] This does not necessarily mean that philosophers before Aristotle did not somehow distinguish between these two fields of physical inquiry.[4] That at least Democritus did, is suggested by the presence of two separate titles, Αἰτίαι οὐράνιαι ('celestial causes') and Αἰτίαι ἀέριοι ('atmospherical causes'), in the catalogue of his writings in Diog. Laërt. IX 47. Yet, it is to Aristotle that we owe the first clear delimitation of the two fields with respect to each other, and the restriction of the terms μετέωρος and μετεωρολογία to *sublunary* phenomena only. This division is followed by most subsequent meteorologists.

---

1 See e.g. Ernout-Robin III 199–200; Taub (2003) 127–137; O'Keefe (2000), §2 'Sources'. See also p. 1 above.
2 See p. 70 ff. above.
3 Capelle (1912a) 421–441; id. (1935), col. 316.
4 As Capelle seems to think: see Capelle (1912a) 425, 427, 447–448; id. (1913) 322; id. (1935), col. 316, lines 15–16, 22–28.

A curious exception is Epicurus, who in the *Letter to Pythocles* deals with astronomical as well as atmospherical phenomena under the single heading of μετέωρα, 'lofty things', as if ignoring Aristotle's contribution. Nonetheless, as we shall see, the *Letter to Pythocles* shows so many similarities in scope and even structure to Aristotle's *Meteorology*, that Epicurus must have been acquainted with crucial aspects of Aristotle's views, either directly or through Aristotle's pupil Theophrastus. Epicurus' departure from this tradition must therefore have been deliberate. It is tempting to connect this with Epicurus' rejection of mathematical astronomy, to which, among others, Aristotle was committed (see p. 56 ff. above). By reassigning astronomical phenomena to meteorology, Epicurus would then be effectively cancelling the epistemological *Sonderposition* of these phenomena.

Interestingly, in antiquity Aristotle's limitation of meteorology to the sublunary sphere was more influential than the actual terms he chose to refer to this field of inquiry: just two generations later Epicurus could still use the word μετέωρος to refer to both meteorological and astronomical phenomena, and much later still the famous Stoic scholar Posidonius too seems to have used the word and its derivatives indiscriminately to denote both astronomical and meteorological matters.[5] Perhaps because of this continuing ambiguity Aristotle's associate and successor Theophrastus found it necessary to introduce another word altogether (μετάρσιος and μεταρσιολογία) to refer to atmospherical phenomena.[6]

Apart from Epicurus, then, most ancient meteorologists clearly distinguished meteorology from astronomy, although the precise location of the demarcation line may vary. But there are other points of divergence as well. In addition to atmospherical phenomena a meteorological treatise may or may not cover certain geological and hydrological phenomena. However, although the precise boundaries of the subject matter may vary from one meteorological treatise to the next, they are by no means arbitrary, but depend on certain underlying assumptions and traditions, which may be brought to light by a thorough comparison of the extant works.

---

5 Posidonius is reported to have defined 'cosmos' in a work called Μετεωρολογικὴ στοιχείωσις (D.L. VII 138 = fr.14 E–K), to have discussed the substance of the sun in his Περὶ μετεώρων (D.L. VII 144 = fr.17 E–K), and contrasted the physical and mathematical approaches to astronomy in his Μετεωρολογικά (Simpl. *In Arist. Phys.* 291.21–292.21 = fr.18 E–K). On the other hand he is said to have discussed the rainbow in a work called Μετεωρολογικὴ (D.L. VII 152 = fr.15 E–K). See also Capelle (1913) 337 ff.
6 See Capelle (1913) 333–336.

In section 3.2 of this chapter I propose to carry out such a comparison, in order to elucidate the various traditions of ancient meteorology and the position of Lucretius and Epicurus therein. Special attention will be given to the second part of Lucretius' book VI, which is largely devoted to exceptional local phenomena or *mirabilia*. Although such phenomena are sometimes mentioned in other meteorologies as well, they belong more properly to *paradoxography*. Section 3.3 will therefore deal with this genre as well, in order to define the precise relations and the division of labour between the two genres, and Lucretius' position vis-à-vis the two.

Another matter is the order in which the various meteorological subjects are treated. It has often been observed that the order of subjects in Lucretius' book VI closely resembles those of the *Syriac meteorology* and Aëtius' book III, and to a lesser degree Epicurus' *Letter to Pythocles*, but the precise extent of these similarities and the exceptions to them have never been thoroughly assessed. In section 3.4 I will therefore delve into this matter as well, with a view to assessing the relations between these four works. In this context it will also be necessary to return to the question of the identity and authorship of the *Syriac meteorology*, which is generally—but in my view prematurely—identified with Theophrastus' *Metarsiologica*, but could in fact be a later work based largely on Epicurus' own meteorology: to this question section 3.5 will be devoted.

3.2   Range, Delimitation and Subdivisions of Meteorology

3.2.1   *Introduction*

In this section I will compare a number of writings dealing exclusively or for the most part with meteorology. By meteorology I mean the study of atmospherical phenomena as well as such phenomena as were often associated with them, such as the Milky Way, comets, shooting stars, earthquakes and terrestrial waters. The comparison in this section will be confined to such matters as the range of subjects of meteorology, its delimitation from astronomy and its major subdivisions. I will not, in this section, go into the treatment of individual subjects or the theoretical background of each work,[7] unless these throw some light upon the reasons for including a certain subject in meteorology or in one of its major subdivisions. From this point of view I am only interested in those

---

7   For individual subjects see the commentaries to the relevant passages. For ancient meteorology in general see Taub (2003).

works or testimonies that present a reasonably complete and coherent account of meteorology, and I will pass by the meteorological views, however interesting they may be, of e.g. Posidonius or Arrian, which are known to us from scattered references only.[8] The writings that meet the above criteria are the following[9] (the page numbers indicate the beginning of my brief introduction of each work):

- Aristotle *Meteorology*, books I–III   p. 79
- [Aristotle] *De mundo*, ch. 4   p. 82
- Aëtius *Placita*, book III (+ IV 1)   p. 84
- Pliny *Naturalis Historia*, book II, §§ 89–248   p. 87
- Seneca *Naturales Quaestiones*   p. 89
- Stoics apud Diog. Laërt. VII 151–154   p. 89
- The *Syriac meteorology*   p. 91
- Epicurus *Letter to Pythocles*, 16 ff. [98 ff.][10]   p. 92
- Lucretius *De rerum natura*, book VI   p. 95

3.2.2 The Texts

3.2.2.1 Aristotle *Meteorology* I–III

In the opening chapter of his *Meteorology*[11] Aristotle defines the province of meteorology as 'everything which happens naturally, but with a regularity less than that of the first element of material things, and which takes place in the region which borders most nearly on the movements of the stars'.[12] Further down he says: 'The whole region around the earth, then, is composed of these bodies {i.e. earth, water, air and fire}, and it is the conditions which affect them

---

8 Posidonius' meteorological fragments are collected in Edelstein & Kidd (1972) (frs.11, 15, 121 and 129–138; see also frs.214–229 on tides and hydrology and frs.12 and 230–232 on seismology) and Arrian's in Roos & Wirth (1967/8), vol. 2, pp. 186–195.

9 All these works but one (viz. Aëtius III) are discussed in Taub (2003), and all but one (viz. Pliny *NH* II) feature in the appendix 'On the Order of Presentation of Meteorological Phenomena' in Kidd (1992) 305–306.

10 Passages in the *Letter to Pythocles* will be referred to by the chapter number of Bollack & Laks (1978), with the traditional numbering of Meibom's edition of Diogenes Laërtius in square brackets.

11 For a general account of Aristotle's *Meteorology* see Taub (2003) 77–115.

12 Arist. *Mete.* I 1, 338b1–3, ὅσα συμβαίνει κατὰ φύσιν μέν, ἀτακτοτέραν μέντοι τῆς τοῦ πρώτου στοιχείου τῶν σωμάτων, περὶ τὸν γειτνιῶντα μάλιστα τόπον τῇ φορᾷ τῇ τῶν ἄστρων. Transl. Lee (1952), slightly modified. According to Lee (1952), xii note *a*, this refers to the entire sublunary region, according to Capelle (1912b), 516–517, to its fiery upper part only.

which, we have said, are the subject of our inquiry'.[13] While thus delimiting meteorology from astronomy, at the same time Aristotle extends its subject matter to include not just atmospherical, but also fiery, watery and even earthly phenomena. Astronomy, which is the subject of his *De caelo*, deals with the orderly and eternal movements of the heavens and the stars, which are made up of the 'first element' (a.k.a. *aether*); meteorology on the other hand studies the less orderly phenomena of the region around the earth, which is occupied by the four classic elements, earth, water, air and fire. The boundary between the two regions is marked by the orbit of the moon.

Yet, Aristotle's definition is not extremely precise and leaves open many questions. It is instructive therefore to have a closer look at the actual range of phenomena covered by Aristotle's *Meteorology*. I will leave book IV out of account, which, as is generally agreed, was not part of the original work.[14] In books I–III, then, the following subjects are discussed (I have italicised those subjects we would nowadays no longer call meteorological):[15]

A. PHENOMENA OF THE FIERY UPPER ATMOSPHERE
    Shooting stars   I 4
    Other luminary phenomena   I 5
    Comets   I 6–7
    The Milky Way   I 8

B. PRODUCTS OF MOIST EXHALATION
    Mist, clouds and rain   I 9
    Dew and hoar-frost   I 10
    Snow   I 11
    Hail   I 12

C. PRODUCTS OF DRY EXHALATION (PLUS TERRESTRIAL WATERS)
    Winds I   I 13
    { *Springs and rivers*   I 13
      Climatic and coastal change   I 14   } TERRESTRIAL WATERS
      The sea   II 1–3
    Winds II   II 4–6

---

13   Arist. *Mete.* I 2, 339a19–21, ὁ δὴ περὶ τὴν γῆν ὅλος κόσμος ἐκ τούτων συνέστηκε τῶν σωμάτων· περὶ οὗ τὰ συμβαίνοντα πάθη φαμὲν εἶναι ληπτέον. Transl. Lee (1952), slightly modified.

14   As was observed already by Alexander of Aphrodisias (2nd cent. AD) *In Arist. Mete.* 4.1, 179.1–5 Hayduck. See also Taub (2003) 206 n. 24 and Wilson (2013) 8–9.

15   For the organization of *Meteorology* I–III see e.g. Capelle (1912b) or Louis (1982) XXVIII–XXXIV.

	Earthquakes	II 7–8
	Thunder and lightning	II 9
	Thunderbolts and whirlwinds	III 1
D.	NON-SUBSTANTIAL PHENOMENA	
	Rainbows I	III 2
	Haloes	III 2–3
	Rainbows II	III 4–5
	Rods and mock suns	III 6
E.	PHENOMENA OF THE EARTH	
	Minerals and metals	*III 6*

As the table shows, Aristotle's *Meteorology* covers more than just atmospherical phenomena. In the first place, it includes a number of subjects which we would nowadays call *astronomical*, like the Milky Way, comets and shooting stars, probably owing to their—real or apparent—irregularity, which to Aristotle seemed incompatible with the supposed orderly character of the supralunary world.[16] Although not *meteorological* in our sense of the word, these phenomena are at any rate μετέωρα, i.e. 'lofty things'.

This is not true of some other phenomena included in Aristotle's *Meteorology*, viz. rivers and the sea, earthquakes, and minerals and metals. Minerals and metals seem to be included because they too, just like some of the phenomena above the earth, are products of the two exhalations.[17] Earthquakes are more closely connected with the μετέωρα proper. They are treated right after winds, because they are themselves caused by subterranean winds.[18] Not so clear are Aristotle's motives for including rivers and the sea. As the table shows, the entire passage on rivers and the sea is inserted in the section on winds. This

---

16   Already in antiquity Aristotle was criticised for assigning the Milky Way to meteorology. See e.g. Olymp. *In Arist. Mete.* [CAG 12.2] 10.33; 66.17–20; 75.24–76.5 (citing Ammonius); Philop. *In Arist. Mete.* [CAG 14.1] 113.33–118.26 (citing Damascius 116.36 ff.). It may therefore not be a coincidence that the Milky Way is absent from almost all subsequent writings on meteorology (Aëtius excepted). Based on Theophr. fr.166 FHS&G, Steinmetz (1964) 167–168 concludes that Aristotle's view was already rejected by Theophrastus; for a more critical attitude concerning the evidence see Sharples (1985) 584–585 and id. (1998) 108–111.

17   Arist. *Mete.* III 6, 378a13 ff.

18   Arist. *Mete.* II 7, 365a14–15 (quoted on p. 105 below). Cf. ibid. II 8, 366a3–5, οὐκ ἂν οὖν ὕδωρ οὐδὲ γῆ αἴτιον εἴη, ἀλλὰ πνεῦμα τῆς κινήσεως {sc. τῆς γῆς}, ὅταν εἴσω τύχῃ ῥυὲν τὸ ἔξω ἀναθυμιώμενον.—"So the cause of an earthquake is likely to be neither water nor earth, but wind, when the external exhalation happens to flow inwards" (my translation).

move seems to have been prompted by the observed analogy between wind and flowing water. Such analogies, however, do not in themselves justify the classification of rivers and the sea as meteorological phenomena. A more plausible reason for their inclusion in a work on meteorology would be that rivers and the sea, being sustained by rain and melting snow, are an integral part of the hydrological cycle,[19] and thus intimately related to the subject of meteorology, but Aristotle does not adduce this justification. Whatever Aristotle's reasons were, from then on rivers and the sea were frequently included in works on meteorology.

Aristotle's *Meteorology* does *not* deal with volcanoes as a separate subject, although it occasionally refers to volcanic phenomena as by-products of earthquakes.[20] Neither does it concern itself with problems pertaining to the earth as a whole, like its shape, position and stability, which had already been treated in Aristotle's *De caelo*.[21]

3.2.2.2 [Aristotle] *De mundo* 4

Practically the same subject matter is also discussed in chapter 4 of the *De mundo*.[22] Although the work is clearly rooted in Aristotelian philosophy, most scholars reject Aristotle's authorship and date the work to some time between 50 BC and 150 AD.[23] The subject of chapter 4 is introduced as 'the most notable

---

19   On the hydrological cycle see Arist. *Mete*. I 9, 346b22–347a8; I 13, 349b3–8; II 2, 354b28–34 and II 3, 356b22–357a2.
20   See e.g. Arist. *Mete*. II 8, 367a1–11 on the eruption of the Aeolian island of Hiera.
21   Arist. *Cael*. II 13–14. The occasional references to such matters in the *Meteorology* (I 3, 339b7–9 on the relatively small size of the earth; I 9, 346b24 on the earth being at rest; II 5, 362a33–b33 on the five terrestrial zones) merely serve as a background for real meteorological problems.
22   See Taub (2003) 161–168.
23   See Furley (1955) 337–341. Mainly on linguistic and stylistic grounds G. Reale en A.P. Bos believe that the work's ascription to Aristotle is correct (Reale (1974); Reale & Bos (1995); etc.), while J. Barnes (1977) and D.M. Schenkeveld (1991), though excluding Aristotle's authorship, argue for a much earlier date than has hitherto been accepted. Their arguments have failed to convince the majority of scholars: see esp. J. Mansfeld (1991), 541–543, and (1992c). Although the question of the work's authorship is no concern of the present work, it may yet contribute to an answer: it will be shown (see table 3.1 below) that the meteorological portion of the *De mundo* differs considerably from Aristotle's *Meteorology* in its subdivision of the subject matter, its omission of the Milky Way (see also n. 16 above), and its inclusion of tides, volcanoes and poisonous exhalations (on the inclusion of the two last subjects in meteorology see p. 119 below).

phenomena in and about the inhabited world (i.e. land and sea)'.[24] The structure of the chapter can be set out as follows:

GENERAL INTRODUCTION	394a7–8
A. PHENOMENA OF THE AIR	394a9–395b17
1. The two exhalations	394a9–19
2. Products of the moist exhalation	394a19–b6
(mist, dew, ice, hoar-frost, dew-frost, cloud, rain, snow, hail)	
3. Products of the dry exhalation	394b7–395a28
(winds, violent winds, incl.: thunder, lightning, thunderbolt, πρηστήρ, τυφών)	
4. Appearances vs. substantial {sc. luminary} phenomena	395a28–32
5. Appearances	395a32–b3
(rainbow, 'rod', halo)	
6. Substantial {sc. luminary} phenomena	395b3–17
(σέλας, shooting star, comet)[25]	
B. PHENOMENA IN THE EARTH	395b18–396a16
(hot springs, volcanoes, noxious exhalations, earthquakes)	
C. PHENOMENA IN THE SEA	396a17–27
(chasms, retreats and incursions of waves, submarine volcanoes, springs and rivers, trees growing in the sea (!),[26] currents, eddies, tides)	
GENERAL CONCLUSION	396a27–32

As the table shows the subject matter is basically divided into three parts: phenomena of the air, phenomena in the earth, and phenomena in the sea.[27]

---

24 394a7–8: Περὶ δὲ τῶν ἀξιολογωτάτων ἐν αὐτῇ {sc. γῇ καὶ θαλάττῃ, ἥντινα καλεῖν εἰώθαμεν οἰκουμένην} καὶ περὶ αὐτὴν παθῶν νῦν λέγωμεν.

25 In *De mundo* 2, 392a32–b5 these same phenomena are assigned to the layer of fire above the air, not to the air itself.

26 If this is a reference to the observation Antigonus *Hist. mir.* 132 ascribes to Megasthenes, the author of the *Indica*, this would be another argument against Aristotle's authorship of the *De mundo* (see n. 23 above).

27 In 395b17–18 we read: Τὰ μὲν τοίνυν ἀέρια τοιαῦτα. Ἐμπεριέχει δὲ καὶ ἡ γῆ πολλὰς ἐν αὐτῇ, and in 396a17 it says: Τὰ δὲ ἀνάλογον συμπίπτει [τούτοις] {sc. τοῖς ἐν γῇ πάθεσι} καὶ ἐν θαλάσσῃ ...

Unlike Aristotle's *Meteorology*, the *De mundo* seems to recognize volcanoes as phenomena to be studied in their own right, mentioning them (incl. the Etna) briefly but separately before a longish discussion of earthquakes. Like Aristotle's *Meteorology* the meteorological section of the *De mundo* excludes problems concerning the earth as a whole.

3.2.2.3 Aëtius *Placita* III

More or less the same subject matter is also dealt with in the third book of Aëtius' *Placita*.[28] This work, to be dated most likely to the first century AD, has not come down to us directly, but can be reconstructed to a large degree from two later works that largely derive from it: Pseudo-Plutarch's *Placita* and Stobaeus' *Eclogae Physicae*. Of these two Pseudo-Plutarch has most faithfully preserved the work's original division into books and chapters,[29] and it is this division which is generally followed and which we shall follow too.

Book III of the *Placita* is not a meteorology in the sense of the two works mentioned above. While Aristotle's *Meteorology* and the *De mundo* aim to give a coherent theory of the whole field of meteorology, Aëtius is concerned with presenting the various and often conflicting views brought forward by a variety of thinkers. Yet the subject matter of the book coincides largely with, and betrays a strong dependence on, Aristotelian meteorology in the range, subdivision and order of its subjects.[30] The book lacks a single general heading:

Finally, in the general conclusion (396a27–32), we read: Ὡς δὲ τὸ πᾶν εἰπεῖν, τῶν στοιχείων ἐγκεκραμένων ἀλλήλοις ἐν ἀέρι τε καὶ γῇ καὶ θαλάσσῃ κατὰ τὸ εἰκὸς αἱ τῶν παθῶν ὁμοιότητες συνίστανται, τοῖς μὲν ἐπὶ μέρους φθορὰς καὶ γενέσεις φέρουσαι, τὸ δὲ σύμπαν ἀνώλεθρόν τε καὶ ἀγένητον φυλάττουσαι.

By the way, in chapters 2 and 3 the same subject matter is organized differently. There, having first dealt with a number of cosmological and astronomical issues (391b9–392a32), the author moves on to the phenomena of the sublunary sphere. This falls apart into three regions: that of *fire* (392a32–b5), that of *air* (392b5–13), and that of *earth and sea* taken together (392b14–394 a6). To the fiery region are attributed such phenomena as comets and shooting stars. The air is the abode of clouds, rain, snow, frost, hail, winds, whirlwinds (τυφῶνες), thunder, lightning and thunderbolts. In the subsequent section on the earth and the sea, the author, rather than enumerate the corresponding physical phenomena, offers a picturesque geographical description of the terrestrial sphere, which need not concern us here.

28   On Aëtius' work in general see now Mansfeld & Runia (1997) and id. (2009a). On book III in particular see Mansfeld (2005).
29   Diels (1879) 61; Mansfeld & Runia (1997) 184–185.
30   On Aristotle as a source for Aëtius as to methodology and contents see Mansfeld

having dealt with cosmology and astronomy in the previous book, in book III Aëtius goes on to discuss, first, in chapters 1–8, what he calls τὰ μετάρσια (lofty phenomena),[31] and then, in chapters 9 and following, what he calls τὰ πρόσγεια (down-to-earth phenomena).[32] The latter section also includes a number of chapters, 9–14, dealing with the earth as a whole,[33] which are not part of the scope of Aristotle's *Meteorology* or chapter 4 of the *De mundo*. Their inclusion may have been prompted by Aëtius' wish to present his subjects in a rigorous top-down (or 'outside-in') order, based on the *location* of each cosmic part and each phenomenon rather than its *nature*. The structure of the book (as preserved by Pseudo-Plutarch) is as follows:

Τὰ μετάρσια
1. Milky Way
2. comets, shooting stars and the like
3. thunder, lightning, thunderbolts and whirlwinds (*prēstēres* and *typhōnes*)
4. clouds, rain, snow and hail
5. rainbow
6. rods and mock suns
7. winds
8. winter and summer

(1992b). On Aëtius III depending on, and deriving from, Arist. *Mete.* I–III see Mansfeld (2005). See also Mansfeld & Runia (2009a).

31 Aëtius III 0: Περιωδευκὼς ἐν τοῖς προτέροις ἐν ἐπιτομῇ τὸν περὶ τῶν οὐρανίων λόγον, σελήνη δ' αὐτῶν τὸ μεθόριον, τρέψομαι ἐν τῷ τρίτῳ πρὸς τὰ μετάρσια· ταῦτα δ' ἐστὶ τὰ ἀπὸ τοῦ κύκλου τῆς σελήνης καθήκοντα μέχρι πρὸς τὴν θέσιν τῆς γῆς, ἥντινα κέντρου τάξιν ἐπέχειν τῇ περιοχῇ τῆς σφαίρας νενομίκασιν. ἄρξομαι δ' ἐντεῦθεν.—"Having briefly traversed in the previous chapters the account of the *heavenly phenomena*, of which the moon is the border region, I shall in the third book turn to *lofty phenomena*. These are what is from the circle of the moon to where the earth is situated, which they are convinced occupies the position of the centre in relation to the circumference of the sphere. I shall begin from here." On this passage see now Mansfeld & Runia (2009a) 54, whose translation I have basically followed.

32 Aëtius III 8.2: Περιγεγραμμένων δέ μοι τῶν *μεταρσίων*, ἐφοδευθήσεται καὶ τὰ *πρόσγεια*.—"Now that the *lofty phenomena* have been described, the *down-to-earth phenomena* will be inspected as well." On this passage see also Mansfeld & Runia (2009a) 55.

33 On Aëtius' inclusion of these subjects with meteorology see p. 103 below.

Τὰ πρόσγεια
9. the earth (being unique and limited)
10. shape of the earth
11. position of the earth   PROBLEMS PERTAINING TO
12. inclination of the earth   THE EARTH AS A WHOLE
13. motion of the earth
14. division of the earth
15. earthquakes
16. the sea: its origin and bitterness
17. the sea: ebb and flood
18.* halo
IV 1.* the flooding of the Nile
(*Stob. 39* water properties*)

Two chapters appear to have been misplaced. Chapter 18 on the halo does *not* belong in the section on τὰ πρόσγεια, which it now concludes, but must have been part of the preceding section on τὰ μετάρσια. There is in fact quite some evidence to connect it more specifically with chapters 5 and 6, on the rainbow and 'rods and mock suns' respectively.[34] Also misplaced is the first chapter of book IV, which discusses the topic of the Nile flood. This subject was clearly meant to go with τὰ πρόσγεια in the second part of book III, where also the origin and salinity of the sea and the causes of ebb and flood are dealt with.[35] It is quite out of place in book IV, which is otherwise about the soul and its functions.

There is reason to believe that a further chapter existed, which is now missing. In Stobaeus' *Eclogae Physicae*, one of the two main sources for the reconstruction of Aëtius' text,[36] there is a chapter (39), titled Περὶ ὑδάτων ('On waters'),[37] which has no counterpart in Pseudo-Plutarch's *Placita*, the other main source for Aëtius and our principal guide as to the table of contents of

---

34 On the dislocation of Aët. III 18 see Diels (1879) 56, 60–61, Lachenaud (1993) 25, Mansfeld (2005) 26–27, 37 (n. 52) and 56 and Mansfeld & Runia (2009a) 44. Note that in Arist. *Mete.* III 2–6, [Arist.] *De mundo* 4, 395a32–b3, and Sen. *NQ* I 2–13, the subjects of the rainbow, the halo and rods and mock suns are also discussed successively (see also p. 134 below).

35 See Diels (1879) 56 and 61, Lachenaud (1993) 274.

36 On Stobaeus as a source for the reconstruction of Aëtius, see Mansfeld & Runia (1997) 196–271.

37 *Not* included in the list of possibly lost chapters in Mansfeld & Runia (1997) 186, but see the table printed ibid. 214–216.

Aëtius' work.[38] In its present state this chapter contains only one lemma reporting Aristotle's views on water properties, which Stobaeus probably derived not from Aëtius but from Arius Didymus (fragment 14a).[39] However, since the *subjects* covered by Stobaeus' *Eclogae Physicae* derive to a large extent from Aëtius[40] (even though Stobaeus often adds or substitutes lemmas from other sources), it is possible that this chapter's *title* and *subject* too derive from Aëtius. That the chapter in its present state contains no Aëtian material can be ascribed to the very selective transmission of that part of Stobaeus' work which corresponds to the second half of Aëtius book III.[41] Stobaeus' chapter 'On waters' follows immediately upon the chapter on tides and so has the same relative position as Aëtius' chapter on the flooding of the Nile, which only Pseudo-Plutarch has preserved. If, as I have suggested, Stobaeus' chapter 'On waters' derives from Aëtius, it will have immediately preceded, or followed on, the chapter on the Nile flood. It is worth noting at this point that in Seneca's *Naturales Quaestiones* (see below) a book dealing with (terrestrial) waters (III) is followed by a book on the summer flooding of the Nile (IVa).

3.2.2.4 Pliny *Naturalis Historia* II §§ 89–248

Roughly the same range of subjects, including (like Aëtius) a number of sections dealing with the earth as a whole,[42] is discussed in the second part of book II (from § 89 onwards) of Pliny's *Naturalis Historia*.[43] The first part of the book is devoted to cosmology and astronomy. Pliny divides astronomical and meteorological phenomena differently from Aristotle, Pseudo-Aristotle and Aëtius. Not only erratic celestial phenomena like comets and shooting stars, but also merely apparent phenomena like haloes, 'rods' and mock suns (but not the rainbow!) are classified by Pliny among the 'stars'. The text of book II (like every

---

38 On Pseudo-Plutarch as a source for the reconstruction of Aëtius, see Mansfeld & Runia (1997) 121–195.
39 Diels (1879) 854; see also Mansfeld & Runia (1997) 249 n. 167.
40 Mansfeld & Runia (1997) 216: "the topics covered by the book [i.e. Stobaeus' *Eclogae Physicae*] have been largely based on the subjects dealt with in the *Placita* [of Aëtius]. Only 7 or 8 of the 60 chapters find no equivalent in A[ëtius]."
41 Mansfeld & Runia (1997) 202–203: "When we further examine the epitomized chapters ¶31–60 in Book I, we soon observe that a very one-sided selection has taken place. [...] In various chapters that must have contained copious extracts from Aëtius just one or two lemmata containing the views of Plato and Aristotle are written out (¶32, 36, 38–39, 42–43, 45, 51–60)." Note the inclusion of ch. 39 'On waters' in the list.
42 On Pliny's inclusion of these subjects with meteorology see p. 103 below.
43 On Pliny's meteorology see Taub (2003) 179–187.

other book) abounds in repetitions, interruptions and all kinds of excursuses, which makes it hard to summarize its contents. The following overview (from § 89 onwards) is no more than an impression. For a more complete summary see Pliny's own table of contents in book I of the *Naturalis Historia*.

A. Cosmos/heavens and stars (i–xxxvii, §§ 1–101)
    xxii–xxxvii §§ 89–101 'Sudden stars' (incl. comets, shooting stars, but also haloes and other insubstantial luminary phenomena)

B. Atmospherical phenomena (xxxviii–lxii, §§ 102–153)[44]
    xxxviii   §§ 102–104   Nature of air
    xxxix–xli   §§ 105–110   Influence of astronomy on the weather and animals
    xlii   § 111   Rain, wind and clouds
    xliii   §§ 112–113   Storm-winds and thunderstorms
    xliv–l   §§ 114–134   Winds (incl. whirlwinds)
    li–lvi   §§ 135–146   Thunderbolts
    lvii–lix   §§ 147–150a Miraculous phenomena in and from the sky
    lx   §§ 150b–151 Rainbows
    lxi   § 152   Hail, snow, hoar-frost, mists, dew, clouds
    lxii   § 153   Particular local climates.

C. Earthly phenomena (lxiii–cxiii, §§ 154–248)[45]
    lxiii–lxxx   §§ 154–190   The earth as a whole (shape, position, seasons)
    lxxxi–lxxxvi §§ 191–200   Earthquakes
    lxxxvii–xciv §§ 201–206a Formation of new land
    xcv   §§ 206b–208 Products of the earth (incl. mines, gems, peculiar stones, medicinal springs, volcanoes, poisonous exhalations)
    xcvi   § 209   Vibrating lands and floating islands

---

44 At the beginning of §102, referring back to the previous sections, *including* those on comets, shooting stars, haloes and 'rods', Pliny writes: "Hactenus de *mundo* ipso *sideribus-que*. Nunc reliqua caeli memorabilia: namque et hoc caelum appellavere maiores quod alio nomine *aëra*."

45 The last words of § 153 are: "Haec sint dicta de *aëre*." § 154 starts with: "Sequitur *terra*."

xcvii–xcviii	§§ 210–211	Local earth marvels
xcix–civ	§§ 212–223	Tides (and other effects of the moon and sun)
cv	§ 224a	Depth of the sea
cvi	§§ 224b–234a	Miraculous waters
cvii–cix	§§ 234b–235	fiery phenomena
cx	§§ 236–238	volcanoes
cxi	§§ 239–241	marvels of fire
cxii–cxiii	§§ 242–248	size of the earth

In table 3.1 (p. 100) I will not include every one of the many subjects touched upon by Pliny, but only those which have a clear counterpart in one or more of the other texts.

3.2.2.5 Seneca *Naturales Quaestiones*

More or less the same range of subjects is also covered by Seneca's *Naturales Quaestiones*.[46] At the outset of the second book,[47] Seneca divides the study of natural phenomena into three parts: the *caelestia* (heavenly things), the *sublimia* ('lofty' things) and the *terrena* (earthly things). The term *caelestia* denotes the phenomena of the heavens and the heavenly bodies, i.e. cosmology and astronomy. The *sublimia* cover all phenomena occurring in the region between the heavens and the earth, i.e. atmospherical phenomena, but also earthquakes. Finally, among *terrena* are understood such subjects as waters, lands, trees, plants and 'everything contained in the ground'.[48] If we compare the range of subjects actually covered in the *Naturales Quaestiones*

---

46 See Taub (2003) 141–161. On the macro-structure of the *NQ* as compared to Seneca's own programmatic remarks at *NQ* II 1, 1–2 see Mansfeld & Runia (2009a) 46–48 and 119–121.

47 According to Carmen Codoñer Merino (1979), xii–xxi, and independently Hine (1981), 6–19, originally the eighth and last book.

48 Sen. *NQ* II 1, 1–2: "Omnis de uniuerso quaestio in *caelestia, sublimia, terrena* diuiditur. Prima pars naturam siderum scrutatur et magnitudinem et formam ignium quibus mundus includitur, solidumne sit caelum ac firmae concretaeque materiae an ex subtili tenuique nexum, agatur an agat, et infra sese sidera habeat an in contextu sui fixa, quemadmodum anni uices seruet, solem retro flectat, cetera deinceps his similia. | Secunda pars tractat inter caelum terramque uersantia. Hic sunt nubila, imbres, niues, ⟨uenti, terrae motus, fulgura⟩ et humanas motura tonitrua mentes; quaecumque aer facit patiturue, haec sublimia dicimus, quia editiora imis sunt. Tertia illa pars de aquis, terris, arbustis, satis quaerit et, ut iurisconsultorum uerbo utar, de omnibus quae solo continentur."

with Seneca's theoretical division of natural phenomena, it appears that the work is concerned almost exclusively with *sublimia* and *terrena*. There is only one exception: comets, which according to Seneca should be classed with the *caelestia*.[49] However, in holding this view he is, as he himself admits, dissenting from the accepted Stoic (and Aristotelian) view that comets are irregular and therefore necessarily sublunary phenomena.[50] By including the subject in a treatise otherwise devoted to atmospherical and earthly phenomena, Seneca is simply following the tradition. In the *Naturales Quaestiones* Seneca does not deal with questions concerning the earth as a whole, but he informs us that some of these (esp. those concerning the earth's position) should be classified among the *caelestia*,[51] while the rest belong with the *terrena*.

Two parts of the work appear to be lost: book IVa, on the Nile flood, lacks its final part, and IVb, presently on hail and snow, its beginning. It is likely that IVb originally included such subjects as clouds and rain as well.[52] The work as it has come down to us covers the following subjects:

I	Lights in the sky (both substantial an insubstantial)
II	Lightnings and thunders
III	Terrestrial waters (almost entirely excluding the sea)
IVa	Nile
IVb	Hail and snow
V	Winds, incl. whirlwinds
VI	Earthquakes
VII	Comets

3.2.2.6 Overview of Stoic Meteorology in Diogenes Laërtius VII 151–154

In chapters 151–154 of book VII of his *Lives of Eminent Philosophers*, Diogenes Laërtius offers an overview of the Stoic theories on 'things taking place in the

---

49 Sen. *NQ* VII 22 1.1–3: "Ego nostris {sc. Stoicis} non assentior. Non enim existimo cometen subitaneum ignem sed inter aeterna opera naturae." See also VII 4 and 21.1.

50 Comets are included unreservedly in Diogenes Laërtius' overview of Stoic meteorology (VII 151–154), in Aristotle's *Meteorology* (I 6–7), and in the meteorological sections of the *De mundo* (4, 395b8–9) and Aëtius (III 2).

51 Seneca, *NQ* II 1, 5: "ubi quaeretur quis terrae situs sit, qua parte mundi consederit, quomodo aduersus sidera caelumque posita sit, haec quaestio cedet superioribus et, ut ita dicam, meliorem condicionem sequetur."

52 Corcoran (1971/2), 'Introduction' xx; Hine (1981) 10, 29–30; Gross (1989) 185.

air' (VII 151.1: τῶν δ' ἐν ἀέρι γινομένων).[53] The overview comprises the following subjects (I have numbered them as they occur in the text):

1. Seasons
2. Winds
3. The rainbow
4. Comets
5. Shooting stars
6. Rain
7. Hoar-frost
8. Hail
9. Snow
10. Lightning
11. Thunder
12. Thunderbolts
13. Whirlwinds (*typhōnes* & *prēstēres*)
14. Earthquakes.

As table 3.1 (on p. 100 below) will show, this range of subjects corresponds almost exactly to that of Seneca's *sublimia* (if we include comets).

3.2.2.7 The *Syriac Meteorology*

In the previous chapter the *Syriac meteorology* has been introduced already (see p. 70 ff. above). This work, which is preserved in one Syriac and two Arabic versions, is now commonly believed to be either the complete text of, or an extract from, Theophrastus' lost two-book treatise Μεταρσιολογικά.[54] For reasons explained above I am not convinced of this identification, although the work is obviously Greek in origin. At the beginning of the treatise its subject matter is stated as 'lofty phenomena'.[55] It deals with the following subjects:[56]

1. Thunder
2. Lightning
3. Thunder without lightning
4. Lightning without thunder
5. Why lightning precedes thunder
6. Thunderbolts
7. Clouds
8. Rain
9. Snow
10. Hail
11. Dew
12. Hoar-frost

---

53 On D.L.'s account of Stoic meteorology see Taub (2003) 137. On the origins of D.L.'s account of Stoic philosophy in general (VII 38–160) see Mansfeld (1986); Mejer (1978) 5–7; and id. (1992) 3579–3582.
54 Cited by Diogenes Laërtius (V 44) in his list of works by Theophrastus.
55 Both Arabic versions have *al-āṯār al-'ulwīyah* 'lofty phenomena'—the title by which also Aristotle's *Meteorology* was known; in the Syriac version the title is missing.
56 I will follow Daiber's 1992 edition and translation unless otherwise specified.

13. Wind, incl. whirlwinds (*prēstēres*)
14a. Halo around the moon[57]
14b. Theological excursus
15. Earthquakes

There has been much speculation about whether or not the treatise as we have it is complete. It has been suggested that it might originally have included subjects like comets and shooting stars, the rainbow and a number of terrestrial phenomena other than earthquakes. In this section I will try to avoid such speculation; instead I will compare the text as we have it with other Graeco-Roman meteorologies and see where that may lead us.

3.2.2.8 Epicurus *Letter to Pythocles*

The *Letter to Pythocles* is one of Epicurus' three doctrinal letters quoted in full in the tenth book of Diogenes Laërtius' work on the lives and doctrines of the philosophers. In the introduction Epicurus claims that with this letter he is complying with Pythocles' request for a 'concise and well-outlined discussion of "lofty matters"' (περὶ τῶν μετεώρων σύντομον καὶ εὐπερίγραφον διαλογισμὸν). The *Letter*'s subject, "lofty matters" (τὰ μετέωρα), is nowhere clearly defined, but appears to cover not just meteorological, but also cosmological and astronomical matters. By combining these subjects under a single heading and subjecting them to a single scientific method, Epicurus seems to be intentionally distancing himself from Aristotle's sharp demarcation of astronomical from atmospherical phenomena.[58] Nevertheless, the distinction appears to play some role in the organization of the subject matter, as the table below will show. The chapter numbers used in the table are those of Bollack & Laks (1978), which I think provide more insight into the structure of the text (in addition I have split chapters 17 and 27, which each deal with two separate subjects, into an A and a B part). For the sake of reference the traditional numbering of Meibom's edition of Diogenes Laërtius is added in the second column.

1. Introduction   pp. 84–85
2. Method   pp. 85–88

---

[57] I have divided chapter 14 of the Arabic version by (presumably) Ibn Al-Khammār into 14a and 14b.
[58] See also Jürss (1994) 237–238.

3.	Definition of 'cosmos'	p. 88	
4.	Number and origin of cosmoi	pp. 89–90	
5.	Formation of the heavenly bodies	pp. 90–91	
6.	Size of the heavenly bodies	p. 91	
7.	Risings and settings	p. 92	
8.	Motions of the heavenly bodies	pp. 92–93	(COSMOLOGY &
9.	Turnings of the sun and moon	p. 93	ASTRONOMY)
10.	The phases of the moon	p. 94	
11.	The light of the moon	pp. 94–95	
12.	The face in the moon	pp. 95–96	
13.	Eclipses of the sun and moon	pp. 96–97	
14.	The heavenly bodies' regular periods	p. 97	
15.	The length of nights and days	p. 98	
16.	Weather signs	pp. 98–99	
17A.	Clouds	p. 99	
17B.	Rain	pp. 99–100	
18.	Thunder	p. 100	
19.	Lightning	pp. 101–102	
20.	Why lightning precedes thunder	pp. 102–103	
21.	Thunderbolts	pp. 103–104	
22.	Whirlwinds (*prēstēres*)	pp. 104–105	
23.	Earthquakes	pp. 105–106	(METEOROLOGY)
24.	(Subterranean) winds	p. 106	
25.	Hail	pp. 106–107	
26.	Snow	pp. 107–108	
27A.	Dew	p. 108	
27B.	Hoar-frost	p. 109	
28.	Ice	p. 109	
29.	The rainbow	pp. 109–110	
30.	The halo around the moon	pp. 110–111	
31.	Comets	p. 111	
32.	Revolution of the stars	p. 112	
33.	Planets	pp. 112–113	(ASTRONOMY)
34.	Lagging behind of certain stars	p. 114	
35.	Shooting stars	pp. 114–115	
36.	Weather signs from animals	pp. 115–116	} (METEOROLOGY)
37.	Conclusion	p. 116	

Many scholars have commented upon the *Letter*'s strange order of subjects.[59] At first the order is clear enough: chapters 1–2 are introductory, 3–5 deal with cosmological matters, 6–15 are astronomical, and 16–29 meteorological (including earthquakes). It is at this point that the confusion begins: some of the following chapters (32–34) deal with subjects that are undeniably *astronomical* again,[60] while three others (viz. 30 on haloes, 31 on comets and 35 on shooting stars) are concerned with phenomena which most ancient meteorologists considered *meteorological*, although Pliny classified them as *astronomical* (as table 3.1 on p. 100 below will show). Chapter 36 is again *meteorological* and seems to supplement what was said in chapter 16.

Although, as I argued above, Epicurus does not formally distinguish between astronomical and meteorological phenomena, both of which are subsumed under the general heading of "lofty matters" (μετέωρα), the division does play some role at the organizational level. For our present purpose it would be very useful if haloes (30), comets (31) and shooting stars (35) could also each be assigned to one of the two divisions. In fact, despite the confused order of its last portion, the *Letter* does provide some clues. First, it can hardly be a coincidence that Epicurus deals with rainbows and haloes in two successive chapters, when these same two phenomena were believed to be related and were discussed together not only in Aristotle's *Meteorology*, but also in the meteorological section of the *De mundo* and in Seneca's *Naturales quaestiones* (see the overviews above). It would seem therefore, that in this respect Epicurus followed the tradition inaugurated by Aristotle, which classified haloes together with rainbows as atmospherical phenomena. Second, the way in which the chapters on comets and shooting stars straddle three undeniably astronomical subjects seems to favour their classification as astronomical phenomena.[61] Not coincidentally, both phenomena were traditionally called ἀστέρες 'stars'.[62] In this respect, then, Epicurus seems to have departed from Aristotle, and instead to have followed the popular association of comets and shooting stars with (fixed)

---

59 Usener (1887) xxxviii–xxxix; Reitzenstein (1924) 36, 40–43; Bailey (1926) 275; Arrighetti (1973) 524 & 691 ff.; Arrighetti (1975) 39–42; Bollack & Laks (1978) 11–18; Sedley (1998a) 122–123 (n. 75), 157; Montarese (2012) 285–287. Podolak (2010) 41–45.
60 Podolak (2010) 44.
61 So Usener (1887) xxxviii.
62 Comets were known as κομῆται ἀστέρες 'long-haired stars' (so e.g. Arist. *Mete*. I 6, 343b5 and Epic. *Pyth*. 31 [111]) and shooting stars were called (δι-)ᾄττοντες ἀστέρες 'jumping stars' (so e.g. Arist. *Mete*. I 4, 341b34–35) or διαθέοντες ἀστέρες 'running stars' (so e.g. Arist. *Mete*. I 4, 341b2–3); Epicurus calls these οἱ λεγόμενοι ἀστέρες ἐκπίπτειν 'stars said to be falling' (*Pyth*. 35 [114]).

stars and planets. In the table above, the likely implicit subdivisions of the *Letter's* subject matter have been indicated behind curly brackets.⁶³

In the introduction to the *Letter* Epicurus refers to what he had written elsewhere (τὰ γὰρ ἐν ἄλλοις ἡμῖν γεγραμμένα) on these matters. We do not know where exactly Epicurus had dealt with these subjects before, but from fragments and citations we do know that at least some cosmological and astronomical problems were discussed in books XI and XII of his *On nature*. There is no evidence of any atmospherical phenomena being discussed anywhere in the *On nature*, although it is clear that Epicurus must have discussed at least some of them outside the *Letter to Pythocles* as well, witness e.g. the long Epicurean account of earthquakes in Seneca's *Naturales Quaestiones* VI 20.5–7, which cannot derive from the corresponding passage in the *Letter to Pythocles*. In his reconstruction of Epicurus' *On nature*, David Sedley suggests that meteorological phenomena may have been discussed in book XIII of this work, separated from the discussion of astronomical phenomena in books XI and XII by the interposition of a passage on other worlds and the origin of civilisation, which would have occupied the later part of book XII. If this reconstruction is correct, this would confirm that Epicurus did indeed subdivide his μετέωρα into astronomical and atmospherical phenomena. However, as Sedley himself admits, this part of his reconstruction is highly speculative, and it cannot be excluded that meteorological phenomena were dealt with in book XII immediately following the discussion of astronomical phenomena.⁶⁴ In short, the little evidence we have about Epicurus' *On nature* cannot help us to decide whether or not he separated astronomical from atmospherical phenomena in this work.

In the *Letter to Pythocles* Epicurus does not deal with subjects concerning the earth as a whole, but we know that at least one such subject, the earth's stability, was discussed at the end of book XI of his *On nature*, following on and preceding a number of astronomical subjects in books XI and XII.⁶⁵ This suggests that, as far as Epicurus did distinguish astronomical and meteorological phenomena, problems pertaining to the earth as a whole were classed with the former.

3.2.2.9 Lucretius *De rerum natura* VI
In book VI of the *DRN* Lucretius discusses a number of atmospherical and terrestrial phenomena roughly coinciding with the subjects of Aristotle's *Meteo-*

---

63   My subdivion of the *Letter's* subject matter is virtually identical to Podolak's (2010) 41–43.
64   Sedley (1998a) 122–123 with n. 76. Sedley's views on the matter are criticised by Montarese (2012) 285–286.
65   Sedley (1998a) 119–122.

rology, and other works of this genre. More precisely, the following subjects are dealt with (or, in the case of snow, wind, hail, hoar-frost and ice, merely enumerated and then passed over):

96–159	Thunder
160–218	Lightning
219–422	Thunderbolts
423–450	Whirlwinds (*prēstēres*)
451–494	Clouds
495–523	Rain
524–526	Rainbow
527–534	Snow, wind, hail, hoar-frost, ice
535–607	Earthquakes
608–638	Why the sea does not grow bigger
639–702	Etna
(703–711	Brief exposition of the principle of multiple explanations)
712–737	The summer flooding of the Nile
738–839	Avernian places
840–847	Water in wells colder in summer
848–878	The spring of Hammon
879–905	A cold spring which kindles tow
906–1089	The magnet
1090–1286	Diseases

In contrast to Epicurus in the *Letter to Pythocles*, Lucretius deals with astronomical and meteorological phenomena separately: the first are discussed in *DRN* V 509–770, while the latter form the subject matter of the whole of book VI. Yet, it may be surmised that Lucretius too subscribed to the fundamental unity of these two classes of natural phenomena, which in the introductions to books V an VI he refers to with one and the same expression (V 84–85 = VI 60–61): "rebus [...] illis / quae supera caput aetheriis cernuntur in oris."—"those things which are observed in the heavenly regions above our heads," in which, with a little phantasy, one may recognize Epicurus' μετέωρα.[66] Another indication for the essential unity of both classes of phenomena in Lucretius is a passage in book V (1189–1193), where Lucretius lists a number of phenomena, covering without any break both astronomical and atmospherical subjects:

---

66   See Robin in Ernout-Robin (1925–1928) III 14.

per caelum volvi quia nox et luna videtur, luna dies et nox et noctis signa severa noctivagaeque faces caeli flammaeque volantes, nubila sol imbres nix venti fulmina grando et rapidi fremitus et murmura magna minarum	because night and the moon are seen to revolve through the sky: the moon, day and night, and the stern stars of the night, and the sky's night-wandering torches and flying flames; clouds, sun, showers, snow, winds, thunderbolts, hail, and the rapid roarings and great threatening rumblings.

Among the phenomena mentioned in this list we also find shooting stars ('noctivagaeque faces caeli flammaeque volantes'). Since for practical purposes Lucretius does distinguish astronomical from atmospherical phenomena, it would be interesting to know to which subdivision he would have assigned comets and shooting stars. Unfortunately, neither the astronomical passage of v 509–770, nor the meteorological account of book vi mentions these phenomena. Yet, the evidence seems to be slightly in favour of Lucretius' assigning comets and shooting stars to astronomy. The range of atmospherical phenomena in book vi seems to be quite exhaustive (as the table below will show) and even those subjects, like snow, wind, hail, hoar-frost and ice, which Lucretius chooses not to discuss, he still feels obliged to mention. Had he felt that comets and shooting stars too belong to this class of phenomena, he would surely have mentioned them. The account of astronomical phenomena in book v, on the other hand, is rather selective, omitting even several phenomena that were covered in the astronomical sections of Epicurus' *Letter to Pythocles*. In his astronomical section, therefore, Lucretius could have omitted comets and shooting stars without explicitly saying so.

Problems pertaining to the earth *as a whole* are not discussed in Lucretius' book vi either. He does, however, discuss one such problem—the stability of the earth—in book v (534–563), right in the middle of his astronomical section, which suggests that he might have considered other problems concerning the earth as a whole as belonging in that class as well.

Book vi falls apart into two main divisions, the first dealing with atmospherical, and the second with terrestrial phenomena. There is some uncertainty about the exact place of the cut, especially with respect to the section on earthquakes (535–607). In lines 527–534 (i.e. just before the account of earthquakes) Lucretius invites the reader to find out for himself the causes of "the other things that grow above and are produced above" (527, tr. Rouse-Smith), such as snow, wind, hail, hoar-frost and ice. This seems to imply that the account of these "things that grow *above* and are produced *above*" is hereby concluded, and that all subsequent subjects, beginning with earthquakes, belong to another class of phenomena.[67] It is also possible, however, to place the dividing line *after* the subject of earthquakes. The next subject, the constant size of the sea, starts with the following words (608–609):

Principio mare mirantur non reddere maius naturam, …	In the first place people wonder why nature does not make the sea bigger, …

The word 'principio' ('in the first place') seems to suggest that Lucretius is now passing on to something new, viz. phenomena that inspire wonder—, of which the constant size of the sea presents the first instance.[68] Below we shall further explore this group of problems, both in relation to the preceding section in Lucretius, and to more or less corresponding sections in other meteorological works (see § 3.3 on p. 109 ff. below).

Comparison with other meteorological accounts may tell us, among other things, how each of the two proposed divisions relates to the traditional divisions of the subject, and to what extent Lucretius fits in this tradition.

3.2.3 *The Table*

Table 3.1 on page 100 ff. provides a synopsis of the subjects that are dealt with in each of the nine 'meteorologies'. Subjects that are *explicitly* excluded from meteorology by the respective authors have been shaded grey. Also indicated are the major subdivisions of the subject matter as applied in each of the nine texts. The order in which the texts and the subjects are presented is my own and is not at stake here.[69]

For the sake of brevity, passages of Pseudo-Aristotle's *De mundo* 4 are indicated by the last digit of the Bekker page only, the first two digits of the relevant pages being always 39. In Diogenes Laërtius' account of Stoic meteorology I have numbered the subjects as they occur in the text. Passages from Epicurus' *Letter to Pythocles* are indicated with the chapter numbers of the edition of Bollack & Laks (1978), to which I have made the minor adjustment of dividing chapters 17 and 27 each into an A and a B part.

---

67 Bailey (1947), 1567 and 1632, classifies all phenomena up to and including snow, wind, hail, hoar-frost and ice (i.e. lines 96–534) with 'atmospheric phenomena' and the rest (i.e. lines 535–1137, *including earthquakes*, with 'terrestrial phenomena'. Cf. Giussani (1896–1898), ad *DRN* VI 535–607.

68 So Giussani (1896–1898), ad *DRN* VI 608–638, and Bailey (1947), 1646–1647.

69 The *order* of subjects in a number of texts *is* dealt with in § 3.4 on p. 127 below.

3.2.4 *Some Observations*

3.2.4.1   Fiery Phenomena of the Upper Atmosphere

In *Meteor.* I 4–8 Aristotle sets one group of phenomena apart as belonging to the fiery upper part of the atmosphere. To this group he assigns comets, shooting stars, and the Milky Way. These three phenomena are variously treated in subsequent works on meteorology. The Milky Way seems to have been excluded from this group quite early in the tradition.[70] Except for Aëtius, all subsequent 'meteorologists' have omitted the subject. There seems to have been some doubt about the assignation of comets to this group as well, a doubt reported and shared by Seneca.[71] In this case, however, most meteorologists were happy to follow Aristotle's lead, discussing both shooting stars and comets under the general heading of atmospherical or lofty phenomena. The only explicit exception is Pliny, who classes comets and shooting stars, together with some other luminary phenomena, among astronomical matters.

The positions of the Syriac meteorologist, Epicurus and Lucretius are harder to ascertain. Comets and shooting stars are not discussed in the *Syriac meteorology* and Lucretius book VI, which suggests that they fell outside the scope of these works. It must be noted, however, that both subjects are also absent from Lucretius' astronomical passage in book V.

Epicurus does not formally distinguish between astronomical and atmospherical phenomena, both of which he calls 'lofty matters' (μετέωρα). Yet, the structure of his *Letter to Pythocles*, at least up to chapter 29, indicates that he accepted at least a practical division between the two groups of phenomena. Unfortunately, the confused order at the end of the letter, where comets and shooting stars are discussed together with a number of unmistakably astronomical phenomena makes it hard to decide to which group Epicurus would have assigned comets and shooting stars, although the placement of the chapters on these subjects suggests that Epicurus associated both phenomena with astronomy (see p. 94 above).

3.2.4.2   Non-Substantial Luminary Phenomena

Another sub-class of atmospherical phenomena distinguished by Aristotle is that of the non-substantial luminary phenomena.[72] The most notable of these are rainbows, 'rods', mock suns and haloes. Most subsequent meteorologists follow Aristotle and assign these to atmospherical or lofty phenomena. Again,

---

70 See n. 16 on p. 81 above.
71 See n. 49 on p. 90 above.
72 On the ultimate Aristotelian origin of the distinction between substantial and non-substantial phenomena see Mansfeld (2005).

TABLE 3.1 *Range of subjects and subdivisions in various ancient accounts of meteorology*

Subject	Work	Aristotle *Meteorology* I–III	[Aristotle] *De mundo* 4 (394a19–396a27)	Aëtius *Placita* III (+ IV 1)
		μετεωρολογικά	ἀέρια	τὰ μετάρσια
Milky Way		I 8		1
Comets		I 6–7	(5a32) & 5b8–9	2
Shooting stars		I 4	(5a32)	2
Other luminary phenomena		I 5	5b3–17	2
Rods and mock suns		III 6	5a35–36	6
Halo		III 2–3	5a36–b3	18
Rainbow		III 2 & 4–5	5a32–35	5
Thunder		II 9	5a11–14	3
Lightning		II 9	5a14–21	3
Thunderbolts		III 1	5a21–23	3
Whirlwinds (*prēstēres* etc.)		III 1	5a23–24	3
Mist		I 9	4a19–23	
Clouds		I 9	4a26–27	4
Rain		I 9	4a27–32	4
Snow		I 11	4a32–b1	4
Hail		I 12	4b1–5	4
Dew		I 10	4a23–24	
Hoar-frost		I 10	4a25–26	
Ice			4a25	
Winds		I 13 & II 4–6	4b7–5a10	7
Seasons				8
Weather signs				
			γῆ	τὰ πρόσγεια
Earthquakes		II 7–8	5b30–6a16	15

Pliny	Seneca	Stoics	*Syriac*	Epicurus	Lucretius
NH II	*NQ*	ap. Diog. Laërt.	*meteorology*	*Pyth.*	*DRN* VI
89–248		VII 151–154		16–36	96–end

sidera	sublimia	τὰ ἐν ἀέρι γινόμενα	lofty phenomena		
89–94	VII	4		31	
96a	I 1 & 14–15	5		35	
96b–101	?				
99a	I 9–13				
98a	I 2		14a	30	

aër

150b–151	I 3–8	3		29	524–526
112–113	II	11	1	18	96–159
112–113	II	10	2–5	19–20	160–203
112–113 & 135–146	II	12	6	21	204–422
131–134	V 13	13	13	22	423–450
152					
111 & 152			7	17A	451–494
111		6	8	17B	495–523
152	IVb	9	9	26	(527–534)
152	IVb	8	10	25	(527–534)
152			11	27A	
152		7	12	27B	(527–534)
				28	(527–534)
111 & 114–130	V	2	13	24 (?)	(527–534)
		1			
				16 & 36	

terra

191–200	VI	14	15	23	535–607

TABLE 3.1 Range of subjects and subdivisions in various ancient accounts of meteorology (cont.)

Subject	Work	Aristotle *Meteorology* I–III	[Aristotle] *De mundo* 4 (394a19–396a27)	Aëtius *Placita* III (+ IV 1)
PROBLEMS PERTAINING TO THE EARTH AS A WHOLE				9–14
Volcanoes (Etna)		(II 8, 367a1–11)	5b19–23	
Poisonous exhalations			5b26–30	
Springs and rivers		I 13	5b19 & 23–26	(*Stob.* 39?)
Miraculous waters		(II 3, 359a18–b22)		
Summer flooding of the Nile				IV 1
			θάλασσα	
Constant size of the sea		II 2, 355b20–32		
Coastal change		I 14	6a18–21	
Origin & salinity of the sea		II 1–3		16
Tides			6a25–27	17
Minerals and metals		III 6		
Magnets				
Diseases				

Pliny is the only explicit exception. He splits up the group, assigning mock suns (§ 99) and haloes (§ 98) to astronomy, and leaving only the rainbow (§§ 150–151) among atmospherical phenomena. In this case the *Syriac meteorology* seems to follow the majority view: among the otherwise meteorological phenomena it also includes the halo. It is strange, however, that the rainbow, which is the best known of this class of phenomena, should not have been included, especially since all the other meteorologies do include it. Lucretius' account in book VI only mentions the rainbow, but omits the halo. Epicurus, in the *Letter to Pythocles*, discusses both the rainbow and the halo. The fact that he discusses them in consecutive chapters suggests that he too, like most other meteorologists, considered these two phenomena to be related and hence to belong to the same class of phenomena, i.e. atmospherical phenomena.

Pliny	Seneca	Stoics	Syriac	Epicurus	Lucretius
NH II	*NQ*	ap. Diog. Laërt.	*meteorology*	*Pyth.*	*DRN* VI
89–248		VII 151–154		16–36	96–end
	terrena				
154–190 & 242–248					
207 & 236–238	(VI 4.1, etc.)				639–702
207–208	VI 28 & III 21				738–839
233–234a	III				840–847
224b–232	III 20 & 25–26				848–905
	IVa				712–737
166.9–11	III 4–5				608–638
201–206a					
222b					
212–220					
207					
					906–1089
					1090–1286

3.2.4.3 Problems Pertaining to the Earth as a Whole

Aëtius' section on τὰ πρόσγεια contains a number of chapters relating to the earth as a whole: (9) on the earth {being unique and limited}, (10) on the earth's shape, (11) on the earth's position, (12) on the earth's inclination, (13) on the earth's motion {or immobility}, and (14) on the earth's division {into five zones}.[73] The only other work to deal with such subjects within the scope of meteorological phenomena, is Pliny's *Naturalis Historia*. These subjects are absent from all the other works in the table and we even have explicit infor-

---

73 The words between {...} have been added to the Aëtian chapter titles in order to better specify the precise subject of each chapter.

mation that most of their authors considered such subjects cosmological and astronomical rather than meteorological: Aristotle discusses the shape, position and stability of the earth in his cosmological and astronomical work *De caelo* (II 13–14), and the same subjects, as well as the earth's size and division into zones, are dealt with in the cosmological and astronomical treatise of the Stoic Cleomedes (I 1 & 5–8); Lucretius discusses the stability and location of the earth in the astronomical section of *DRN* book V (534–563), and Epicurus dealt with the same subjects in book XI of his magnum opus *On nature* (fr.42 Arr.), amidst a number of cosmological and astronomical problems.[74] Seneca too, in the introduction to book II of the *Naturales Quaestiones*, tells us that certain problems concerning the earth, like its position, belong not to the *terrena* or *sublimia* but to the *caelestia*.[75] It would appear therefore that the inclusion of such problems among otherwise meteorological phenomena is an innovation by Aëtius and Pliny, probably inspired by their wish for a rigorous top-down presentation of natural phenomena.[76] In general, then, such problems did not belong to meteorology.[77]

3.2.4.4 Earthquakes

In the table, earthquakes are variously placed among *terrestrial* or *atmospherical* phenomena. In the *De mundo* 4, in Aëtius III and in Pliny II, earthquakes are dealt with under the general heading of *terrestrial* phenomena. The reason for this seems to be that in all three works precedence is given to the *location* of the phenomenon. Seneca and the Stoics, on the other hand, agree in classing earthquakes with *lofty* or *atmospherical* phenomena. Seneca provides us with the reason for this—perhaps—surprising move (*NQ* II 1, 3):

"Quomodo," inquis, "de terrarum motu quaestionem eo posuisti loco quo de tonitribus fulguribusque dicturus es?"	"Why," you ask, "have you put the study of earthquakes in the section where you will talk about thunder and lightning?"

---

[74] See Sedley (1998a) 119–121.
[75] See n. 51 above.
[76] On this top-down presentation of cosmological problems in Aëtius and other writers, and its consequences for the location of the sections dealing with the earth as a whole, see Mansfeld & Runia (2009a) 40–41 with n. 71, and 133–134. On the order of Pliny's cosmology see Kroll (1930) p. 2, and Hübner (2002).
[77] Similar observations and conclusions in Mansfeld (1992b) n. 124 and Mansfeld & Runia (2009a) 119–122.

Quia, cum motus spiritu fiat, spiritus autem aer sit agitatus, etiamsi subit terras, non ibi spectandus est; cogitetur in ea sede in qua illum natura disposuit.	Because, since an earthquake is caused by a blast, and a blast is air in motion, therefore, even if air goes down into the earth, it is not to be studied there; let it be considered in the region where nature has placed it.[78]

According to Seneca, who may be supposed here to speak on behalf of all Stoics, earthquakes, being caused by *air*, should be dealt with in connection with other phenomena of the *air*. In this respect the Stoics follow Aristotle's lead. Although Aristotle does not yet apply the neat bipartition of meteorological phenomena into those of the *earth* and those of the *air* (or 'lofty': μετάρσια / *sublimia*), such as we find with many of his successors, he does explicitly link the subject of earthquakes with that of winds (*Mete.* II 7, 365a14–15):

Περὶ δὲ σεισμοῦ καὶ κινήσεως γῆς μετὰ ταῦτα λεκτέον· ἡ γὰρ αἰτία τοῦ πάθους ἐχομένη τούτου τοῦ γένους ἐστίν.	About earthquakes and earth tremors we must speak after the previous subject (i.e. wind): for the cause of this phenomenon is akin to that of wind.[79]

For the *Syriac meteorology*, and the accounts of Epicurus and Lucretius the story is a bit different. In all three works earthquakes are accounted for by a number of alternative explanations, not just wind, which makes their link to atmospherical phenomena less obvious. Yet, it can hardly be a coincidence that, just as in the account of Stoic meteorology, so too in the *Syriac meteorology* and in the *Letter to Pythocles*, earthquakes are the only 'terrestrial' phenomenon to be discussed among a number of otherwise atmospherical phenomena. It would appear that even though the *Syriac meteorology* and Epicurus do not share the Stoics' and Aristotle's assumptions, they do follow the tradition that incorporates earthquakes among atmospherical phenomena. It seems reasonable to suppose that Lucretius, being a follower of Epicurus, and one, moreover, whose meteorological account closely matches the *Syriac meteorology*, belongs to this same tradition.

---

78   Transl. Corcoran (1971/2), slightly modified.
79   My translation.

3.2.4.5   Terrestrial Phenomena (Other Than Earthquakes)

In most of the meteorological accounts a number of terrestrial phenomena (other than earthquakes) are included. Lucretius, too, discusses a number of terrestrial phenomena. It is remarkable that the two closest parallels to Lucretius, viz. the *Syriac meteorology* and Epicurus' *Letter to Pythocles*, do not deal with this class of phenomena. It could be and has been argued, in view of the close similarity between Lucretius' book VI and the *Syriac meteorology*, that the latter must originally have dealt with such subjects too.[80] There is no reason to assume that the *Letter to Pythocles* is incomplete, but Epicurus might have dealt with terrestrial phenomena somewhere else, perhaps in his *On nature*. Yet, there is no direct evidence for this claim, and the only ones of Lucretius' 'terrestrial phenomena' that were certainly discussed by Epicurus—although we do not know in what context—are magnets and diseases.[81] It might therefore be claimed just as well that the section on 'terrestrial phenomena' in *DRN* VI was Lucretius' own invention. In § 3.3 below I will investigate how Lucretius' account of terrestrial phenomena in the second half of book VI relates to his account of atmospherical phenomena in the first half, and also how it relates to the discussion of similar matters in other meteorological works.

3.2.5   *Some Conclusions*

3.2.5.1   The *Syriac Meteorology*

The range of phenomena covered in the *Syriac meteorology* is smaller than in any of the other meteorologies. For this reason it has often been argued that the treatise must be an extract from a larger work in which more subjects were dealt with. Steinmetz, for instance, suggests that the original work would have included chapters on the rainbow, mock suns and volcanoes.[82] Many of the arguments for its completeness or incompleteness are based on its identification with Theophrastus' lost 2-volume *Metarsiologica*. Mansfeld, for instance, following a suggestion made by Daiber, believes that the chapter on earthquakes may originally have been the first chapter of Theophrastus' second book, which, in addition to this chapter, 'may have included treatment of other so-called terrestrial phenomena, e.g. "the advances and regressions of the sea and the extensions of the land"'.[83]

Now let us for a moment forget the ascription to Theophrastus and compare the work as we have it with other meteorologies. As we saw above (p. 104),

---

80   Steinmetz (1964) 216 with n. 3; Mansfeld (1992a) 315–317.
81   See p. 109 with nn. 86 and 87 below.
82   Steinmetz (1964) 216 with n. 3.
83   Mansfeld (1992a) 315–317.

the *Syriac meteorology*, just like D.L.'s account of Stoic meteorology and Epicurus' *Letter to Pythocles*, includes earthquakes, but excludes other terrestrial phenomena. Now, the reason for this practice in Stoic meteorology is clear: its subject matter is confined to 'things happening in the air' (τὰ ἐν ἀέρι γινόμενα); earthquakes are caused by moving air, and therefore must be classed with phenomena in the air. This motive, however, is not valid for Epicurus, for whom subterranean winds are only one of several explanations. The inclusion of earthquakes (*Pyth.* 23 [105–106]) among Epicurus' μετέωρα (lofty things) can therefore only be explained by his dependence on a tradition, whose principles he no longer subscribes to. The same explanation can be applied to the *Syriac meteorology*. The inclusion of earthquakes (ch. 15) in no way entails the inclusion of other terrestrial phenomena.

Another group of subjects that appears to be missing is that to which comets and shooting stars belong. From Aristotle onwards these two subjects appear to have been standard ingredients of meteorological treatises. We find them included under atmospherical / lofty phenomena in the *De mundo* 4, in Aëtius book III, and in the account of Stoic meteorology in Diogenes Laërtius. As we have seen above,[84] Seneca prefers to see comets as astronomical rather than meteorological phenomena, but in doing so he also testifies that comets were a traditional part of meteorology. Pliny is the only author who explicitly classes both comets and shooting stars under the general heading of astronomy. We do not know why the Syriac meteorologist omits both subjects, but he is not alone in doing so: they are also missing in Lucretius' meteorological survey in *DRN* VI. It is possible that both authors, like Pliny, assigned these subjects to astronomical rather than atmospherical phenomena (for Lucretius see p. 97 above). Both subjects *are* discussed in Epicurus' *Letter to Pythocles*. Here, the fact that they are discussed contiguously with a group of undeniably astronomical subjects may suggest that Epicurus too associated them with astronomical rather than meteorological phenomena (see p. 94 above). However this may be, the absence of comets and shooting stars from a meteorological treatise is not unparalleled, and there is no need to suppose that the Greek original of the *Syriac meteorology did* discuss these subjects.

This leaves us with only one more omission: the rainbow. The *Syriac meteorology* is the only meteorological account which does not include the rainbow.[85] This is the more striking as the treatise *does* discuss the halo, with which the rainbow was traditionally associated. If, for the sake of brevity, one of the

---

84 See n. 49 on p. 90 above and text thereto.
85 See Mansfeld (1992a) 315–316.

two is omitted, it is usually the halo: this is the case in Diogenes' account of Stoic meteorology, and Lucretius book VI. In the meteorological section of Epicurus' *Letter to Pythocles*, whose range of subjects is otherwise very close to that of the *Syriac meteorology*, both the rainbow and the halo are discussed. There seems to be some reason, then, to suppose that the *Syriac meteorology*, or its Greek source, may have contained a chapter on the rainbow, which was lost in the course of the transmission of the text.

It appears to me, therefore, that, as far the range of subjects is concerned, the *Syriac meteorology* may well be complete, except perhaps for the omission of the rainbow.

What does all this mean for the treatise's attribution to Theophrastus or, alternatively, to Epicurus? As the table shows, in its range of subjects it most closely resembles Diogenes Laërtius' account of Stoic meteorology, the latter part of Epicurus' *Letter to Pythocles*, and the first half of the sixth book of Lucretius' *DRN*. In fact, (except for the rainbow) its range of subjects is an almost perfect intersection of the *Letter to Pythocles* and *DRN* VI. It differs from most other ancient meteorologies, including Aristotle's *Meteorology* and Aëtius III, by its omission of comets and shooting stars. In this respect it is resembles Lucretius VI, and perhaps also Epicurus' *Letter*, if—as I argued above—comets and shooting stars are taken to be astronomical (see p. 94 above). On the other hand, its inclusion of earthquakes within a range of otherwise atmospherical phenomena suggests a dependence on a tradition in which the explanation of earthquakes was closely linked to that of wind. This tradition is most clearly exemplified by Seneca and the Stoics, and can be traced back to Aristotle. In this respect, however, the *Syriac meteorology* does not differ from Epicurus' *Letter to Pythocles*; both works stand in the same relationship to the (Aristotelian) tradition, and there is no reason why one should be closer to the origins of the tradition than the other.

3.2.5.2   Epicurus' *Letter to Pythocles*
Most that can be said about the *Letter* has been said already: it does not explicitly differentiate between astronomical and meteorological phenomena, yet in the organization of the *Letter* some kind of a division appears to be present. At the end of the *Letter* the order of subjects is a bit confused, and due to this confusion it is not entirely clear whether Epicurus considered comets and shooting stars astronomical or rather atmospherical. The evidence seems slightly in favour of the first option (see p. 94 above), and if this is true the range of truly atmospherical phenomena in the *Letter* would correspond almost exactly to that of the *Syriac meteorology* (except for the rainbow which is absent from the Syriac as we have it).

3.2.5.3 Lucretius' *DRN* Book VI

Lucretius' meteorological account differs from Epicurus' in two important respects. Firstly, whereas Epicurus does his best to obscure the difference between astronomical and meteorological phenomena by discussing them under the single heading of μετέωρα, Lucretius deals with both fields separately. The astronomical passage in book V 509–770 is firmly separated from the discussion of meteorological phenomena in book VI by the intervention of a very long section on the origins of life and civilisation, which occupies the second half of book V (771–1457). Secondly, the range of meteorological subjects discussed by Lucretius is considerably longer than that of either Epicurus' *Letter to Pythocles*, or the *Syriac meteorology*, which in many other respects appears the closest parallel to Lucretius' book VI. Whereas the *Letter to Pythocles* and the *Syriac meteorology* confine themselves to atmospherical phenomena and earthquakes, Lucretius proceeds to deal (608–1286) with a range of (other) terrestrial phenomena, such as the sea, the Etna, the Nile, poisonous exhalations, springs and wells, magnets and diseases, which are not represented in the *Letter to Pythocles* or in the *Syriac meteorology*.

It cannot be excluded, of course, that Epicurus dealt with such matters elsewhere: Galen credits him with an elaborate theory of magnetism,[86] and Diogenes Laërtius ascribes to him a work titled Περὶ νόσων δόξαι πρὸς Μίθρην, *Opinions on diseases, to Mithres*,[87] in which Epicurus may have dealt with the physical side of diseases. However, there is no evidence that he discussed any of the other subjects which Lucretius covers in *DRN* VI 608 ff. It is possible, therefore, that Epicurus was not Lucretius' main source for this passage. In the next section the character of Lucretius' passage and its possible relations to meteorological and other literature will be examined.

3.3 Terrestrial Phenomena Other Than Earthquakes

3.3.1 *Lucretius*

On p. 98 above it was suggested that Lucretius' section on the constant size of the sea was the first of a new class of problems, different in character from the preceding phenomena, both atmospherical and seismic. The clue as to what this difference might be is given right at the beginning of this new division

---

86   Galen, *De naturalibus facultatibus*, I 14 [2.45.4–52.2 Kühn] (= Epic. fr.293 Us.).
87   Diog. Laërtius X 28. In addition Herculaneum papyrus 1012, col. 22 p. 33 De Falco (Epicurus fr. 18 Arrighetti) ascribes to Epicurus a work titled Περὶ νόσων [καὶ θα]νάτου, *On diseases and death*, which may or may not refer to the same work.

(lines 608/9): 'Principio mare *mirantur* non reddere maius / naturam, ...'—'In the first place people *wonder* why nature doesn't make the sea bigger, ...' As Giussani and Bailey point out, this sense of *wonder* also characterises many of the problems that follow (explicitly so in 608 *mirantur*, 654–655 *mirari* & *miratur*, 850 *admirantur*, 910 *mirantur*, 1056 *mirari*). In this sense, then, these problems differ from the preceding ones, which may incite awe and fear, but are not said to cause wonder.[88]

Although Lucretius does not tell us explicitly in what way the second group of phenomena should inspire this sense of wonder, which the preceding do not, it is not hard to see that there *is* a difference in character between the phenomena in the first group and *most* of the second group. While the phenomena in the first group (including earthquakes) are all capable of occurring just about anywhere, the majority of the subjects discussed in the second part are concerned with *exceptional* and *local* phenomena, the kind of phenomena the ancients referred to as παράδοξα, θαυμάσια or θαύματα, and *mirabilia* or *miracula*,[89] i.e. 'paradoxes', 'marvels' or 'miracles'.[90] Such are the Etna, the river Nile (explicitly said to be 'unique'—713 *unicus*), the 'Avernian' places (one near Cumae, one in Athens and one in Syria), and the spring near the shrine of Hammon (in the Siwa-oasis in Egypt). Also local is the cold spring which kindles tow (lines 879–905), whose location Lucretius does not reveal, but which may be identified with either the spring of Jupiter in Dodona or the spring of the nymphs in Athamania,[91] about which similar stories were told. The magnet, too, may be perhaps counted among local phenomena, as it is found specifically—so Lucretius tells us—in the land of the Magnetes (in Lydia, Asia Minor). I have summarized all this in the following table (for the sake of completeness I have also included 703–711 which do not deal with a specific phenomenon, but with the method of multiple explanations in general):

---

88 This is not entirely true: in the introduction to book VI Lucretius speaks of people wondering (59 mirantur) about things that take place above our heads in the 'ethereal' regions (61 quae supera caput aetheriis cernuntur in oris), i.e. astronomical and atmospherical phenomena. Yet, in the body of the text the use of this verb is restricted to certain terrestrial phenomena only.

89 For ancient names for such phenomena see Ziegler (1949) cols. 1137–1138; Schepens & Delcroix (1996) 380–382; Wenskus (2000) col. 309.

90 Lists of such phenomena are found throughout Pliny's *Naturalis Historia*, where they are referred to as *miracula* and *mirabilia*.

91 See Ernout-Robin (1925–1928) ad loc. The spring of Jupiter in Dodona is described by Pliny *NH* 2.228.1–3 and Mela 2.43 and the spring of the nymphs in Athamania by Antigonus 148, the Paradoxographus Florentinus 11 and Ovid *Met.* 15.311–312.

TABLE 3.2 *Lucretius' account of terrestrial phenomena*

Lines	Subject		Marvellous / exceptional	Local
608–638	Why the sea does not grow bigger			
639–702	Etna		654 mirari	639 Aetna
703–711	Multiple explanations			
712–737	The summer flooding of the Nile		713 *unicus*	712 Nilus
738–839	Avernian places	Lacus Avernus (746–748)		747 Cumas apud
		Acropolis (749–755)		749 Athenaeis in moenibus
		Place in Syria (756–759)		756 in Syria ... locus
840–847	Water in wells colder in summer			
848–878	The spring at the shrine of Hammon		850 admirantur	848 apud Hammonis fanum
879–905	A cold spring which kindles tow			(Dodona / Athamania)
906–1089	The magnet		910 mirantur	909 Magnetum in finibus
			1056 mirari	
1090–1286	Diseases			1115 Aegypto
				1116 Achaeis finibus
				1117 Atthide

Three phenomena stand out in the above list: the constant size of the sea, wells being colder in summer, and diseases. The sea, which occupies such a large portion of the earth,[92] can hardly be called a *local* phenomenon. It is clear that the sense of wonder it is said to inspire is of a different kind from that inspired by, for instance, the Nile or the spring of Hammon. Thematically it seems to be more closely related to the atmospherical phenomena of the preceding section: one of the explanations offered (627–630)—viz. that a considerable portion of water is drawn up by the clouds—is the exact counterpart (as Lucretius himself points out in 627) of one of the causes of cloud formation (470–475) and of rain production (503–505).[93]

The account of why water in wells is colder in summer than in winter is also different. Whereas most phenomena in this section are somehow exceptional among their kind—the Etna among mountains, the Nile among rivers,

---

92 Cf. Lucr. *DRN* V 203.
93 For this reason Robin (Ernout-Robin ad loc.), ignoring Lucretius' own textual clues, prefers to include this passage with the atmospherical phenomena that precede it.

the spring of Hammon among springs and the magnet among stones—, this passage is about something generally attributed to all of its kind: all wells were believed to be colder in summer and warmer in winter.[94] Yet it is not hard to imagine why Lucretius included it in this section: the annual temperature fluctuation of wells is somewhat similar to the daily temperature fluctuation of the spring of Hammon.

This brings us to the last subject, not just of the 'terrestrial phenomena' but of the entire book: diseases. The language with which Lucretius introduces the subject does not suggest any major break with the preceding subjects. Soon, however, it appears that the present subject is somewhat different. Diseases, according to Lucretius, are produced in two different ways (1098–1102):

Atque ea vis omnis morborum pestilitasque aut extrinsecus, ut nubes nebulaeque, superne per caelum veniunt, aut ipsa saepe coorta de terra surgunt, ubi putorem umida nactast intempestivis pluviisque et solibus icta.	And all this might of diseases and this pestilence either comes from without, like clouds and mists, from above through the sky, or often, having gathered, they rise from the earth itself, when this, being moist, has come to rot, having been hit by out-of-season rains and suns.

Diseases either come from without through the sky,[95] like clouds and mist, or they arise locally from the earth itself. These two kinds seem to correspond to what ancient as well as modern medicine refers to as *epidemic* and *endemic* diseases.[96] In the following lines (1103–1118) Lucretius first deals with the sec-

---

94   Hipp. *Aer.* 7.43–44 (Littré); Cic. N.D. II 25.7–26.1; Sen. NQ VI 13.3–4; Plin. NH II 233.1–2.

95   This theory is anticipated in line 956 ('morbida visque simul, cum extrinsecus insinuatur') in the account of the magnet, and seems to look back to 483 ff. where the possibility of an extra-cosmic origin of clouds is suggested. Line 956, which explicitly refers to disease, seems to ascribe an extra-cosmic origin to diseases as well, but only if it is connected with the preceding line from which it is separated by a lacuna of unknown length (see the commentaries). In the present passage, however, apart from the reference to clouds, there is nothing to suggest that diseases might come from without the cosmos. The point rather seems to be that diseases are either innate to a certain region, or come from elsewhere. See Kany-Turpin (1997).

96   Bailey does observe the distinction but makes nothing of it. He calls all the diseases in this passage 'epidemic' and also 'marvel'. For the distinction between epidemic and endemic diseases, see Galen, *In Hippocratis de acutorum morborum victu*, 15.429–430 (Kühn); idem,

ond kind of diseases: those that are peculiar to certain regions and peoples, depending on the local climate, and which may also affect those who travel there. Lucretius ends this brief account with three examples that would not be out of place in our list of local marvels: elephantiasis which is unique to Egypt, a foot-disease peculiar to Attica, and an eye-disease typical of Achaea. Then (1119–1132) Lucretius goes on to discuss the other kind of diseases: those that travel with the air like clouds and mist and thus come upon us. Before coming to the 'finale', Lucretius recapitulates the major point of his account (1133–1137): it makes no difference whether we travel to an unwholesome place (*endemic* disease) or whether nature brings the unwholesomeness to us (*epidemic* disease). Lucretius ends the account, and the entire work, with a long description of an outbreak of the second type (*haec* ratio morborum), the famous plague of Athens.[97] In sum, Lucretius' account of diseases does not entirely blend in with the preceding accounts of mostly local marvels. While *endemic* diseases bear a certain resemblance to local marvels such as the Nile, the spring of Hammon and especially Avernian places,[98] which are accounted for in a similar manner,[99] *epidemic* diseases have more in common with the non-local atmospherical phenomena of the first half of book VI; they are repeatedly likened to clouds and mist.

Beside these three problems—the constant size of the sea, the paradoxical annual temperature fluctuation in wells, and diseases—all the other subjects discussed in the second half of book VI are concerned with exceptional local phenomena.

3.3.2   *Parallels in Meteorology and Paradoxography*
Although some of the local marvels discussed by Lucretius, and similar ones, are also mentioned in some of the other meteorologies, phenomena of this

---

  In Hippocratis epidemiarum libri, 17a.1–2 & 12–13 (Kühn); Ps.-Galen, *Definitiones medicae*, 19.391 (Kühn). See also the admirably concise accounts in Karl-Heinz Leven (ed.) *Antike Medizin. Ein Lexikon*, Munich 2005, lemmata 'Endemie' and 'Epidemie'.

[97]   In ancient medical literature (e.g. Galen, *In Hippocratis de acutorum morborum victu*, 15.429 (Kühn) & idem, *In Hippocratis epidemiarum libri*, 17a.13 (Kühn)) the *plague* (Greek λοιμός) is defined as a *deadly epidemic* (ἐπιδημία ὀλέθριος). See also Karl-Heinz Leven, op. cit., lemma 'Pest'.

[98]   Galen too notes the similarities between endemic diseases and the effects of Charonian places: *In Hippocratis epidemiarum libri*, 17a.10 (Kühn) & *In Hippocratis de natura hominis*, 15.117 (Kühn).

[99]   See esp. lines 769–780 on the earth containing both beneficial and harmful elements; and lines 781–817 listing a number of harmful substances and places.

class are more typically found in a genre of writings known as paradoxography.[100]

Paradoxography is the activity and the written result of collecting accounts of natural marvels. Such 'marvels' comprise 'unexpected features of the natural world (animals, plants, rivers and springs), but also the world of man, human physiology, unusual social customs, and even curious historical facts (...)',[101] which are drawn from all kinds of earlier writings, often with explicit acknowledgement of the source,[102] but without qualification as to their veracity, and without physical explanation.[103] In this respect paradoxography differs from the scientific writings from which many of the marvellous stories were culled. If we discard the *On marvellous things heard*, which is generally believed to be the work, not of Aristotle (to whom it is ascribed), but of a number of subsequent authors working between the 3rd cent. BC and 2nd cent. AD,[104] the oldest reported writer of paradoxography is Callimachus of Cyrene (ca. 310–240 BC),[105] part of whose work is reproduced in Antigonus' *Historiarum mirabilium collectio*.[106]

The place of natural marvels in ancient scientific literature—and especially meteorology—and the relationship between science and paradoxography are complex subjects. Although paradoxography as a separate genre seems to have originated only in the Hellenistic age, scientific works like Aristotle's not only provided it with many individual marvel stories, but actually set the example of producing *lists* of particular local phenomena. Aristotle's *Meteorology*, for instance, contains two lists of particular waters: at 350b36–351a18 of partially underground rivers, and at 359a18–b22 of salty and other-tasting waters, many particular instances of which later reappeared in paradoxographical literature. In its turn scientific literature of later ages reappropriated many marvellous accounts from paradoxography. In book 11 of Pliny's *Naturalis Historia*, for instance, (as in the rest of his work) several long lists of 'miracula' (as Pliny calls them) are produced, which are almost indistinguishable in

---

100 On paradoxography in general see e.g. Ziegler (1949); Schepens & Delcroix (1996); Pajón Leyra (2011).
101 Schepens & Delcroix (1996) 381.
102 Schepens & Delcroix (1996) 383–386; Pajón Leyra (2011), 29–32.
103 Pajón Leyra (2011), 30–32, citing Jacob (1983) 131–135.
104 Ziegler (1949) cols. 1150–1151; Schepens & Delcroix (1996) 427; Vanotti (2007) 46–53.
105 On Callimachus as the founding father of paradoxography see Ziegler (1949) col. 1140; Schepens & Delcroix (1996) 383; Wenskus (2000) col. 311 with refs.; Pajón Leyra (2011) 103–104.
106 Antigonus *Hist. mir.* 129 ff.

character from the accounts found in purely paradoxographical works, from which they seem to have been borrowed.[107]

Both the attitude towards, and the space devoted to, particular local phenomena in ancient meteorologies changes in the course of time. Aristotle's attitude towards such phenomena is ambiguous. Although, as we have just seen, the *Meteorology* includes several lists of such particular problems, these are never the actual *objects* of inquiry. The list of underground rivers at 350b36–351a18 only serves to illustrate why some people might—incorrectly—think that the sea itself is replenished from underground reservoirs, and the list of salty waters at 359a18–b22 serves to lend plausibility to the belief that the salty taste of sea-water too is due to admixture. In later meteorologies the amount of paradoxographical passages increases dramatically. In the *De mundo* the entire section on terrestrial phenomena, save for a long account of earthquakes, is a mere enumeration of local marvels. In Seneca's *Naturales Quaestiones* a large portion of book III 'On waters' is filled with lists of peculiar waters. Pliny's *Naturalis Historia* caps them all: book II contains extensive lists of earth miracles (206b–211), water miracles (224b–234) and fire miracles (235–238), often, as in paradoxography proper, with explicit reference to the source of each story. As a rule, in paradoxography as well as meteorology, such peculiar local phenomena are not explained, or only in the most general terms.

A curious exception to this rule is the summer flooding of the Nile. True to his own precept, in *Metaph.* VI 2, 1027a20–26, that science should not be concerned with particular, but only general problems,[108] Aristotle does not even mention the Nile flood in his *Meteorology*, where we only find the general and unqualified observation (I 13, 349b8) that *all* rivers flow higher in winter than in summer.[109] Interestingly, among the works ascribed to Aristotle there is a treatise, only preserved in a 13th century Latin translation, titled *Liber de inundacione Nili* ('Book on the flooding of the Nile'), which is entirely devoted to the question of why the Nile, in contrast to all other rivers, overflows in summer.[110] Although the work's precise authorship is still a matter of

---

107 On Pliny's paradoxographical passages see e.g. Ziegler (1949) cols. 1165–1166; Schepens & Delcroix (1996) 433–439; Naas (2002), ch. 5.
108 Cf. *An. post.* I 8, 75b21–36. See also Taub (2003) 83.
109 Arist. *Mete.* I 13, 349b8: Διὸ καὶ μείζους {sc. τοὺς ποταμοὺς} ἀεὶ τοῦ χειμῶνος ῥεῖν ἢ τοῦ θέρους ...—'Therefore also (they suppose that) rivers always flow higher in winter than in summer ...'.
110 Arist. (?) *Liber de Nilo* 2–4: 'Propter quid aliis fluminibus in hyeme quidem augmentatis, in estate autem multo factis minoribus, {sc. Nilus} solus eorum, qui in mare fluunt, multum estate excedit [...]?'—'For what reason, while other rivers rise in winter, and become much

TABLE 3.3 DRN VI 608ff. with parallels in meteorology and paradoxography

Subject	Work	Lucretius DRN VI	Arist. Meteor. I–III	Aët. Plac. III–IV 1
Constant size of the sea		608–638	355b20–32	
Volcanoes			(367a1–11)	
– Etna		639–702		
The Nile flood		712–737		IV 1
Poisonous exhalations		738–839		
a. Lacus Avernus		746–748		
b. Acropolis		749–755		
c. Syria / Phrygia (*)		756–759		
Temperature in wells		840–847		
Marvellous waters			(350b36ff.) (359a18ff.)	
– Spring of Hammon		848–878		
– Spring which kindles tow		879–905		
– Sweet spring off Aradus		(890–894)	(~351a14–16)	
Magnets		906–1089		
Diseases		1090–1286		

(* On the probable identity of Lucretius' 'place in *Syria*' with the Plutonium in *Phrygia* see p. 120 below)

dispute, most scholars agree that it should at least be assigned to Aristotle's school.[111] Apparently then, even in Aristotle's school such a particular problem could be worthy of scientific investigation after all. It need not surprise us, therefore, that in some subsequent meteorological accounts the problem of the Nile flood came to be included within the scope of their subject matter. This is

smaller in summer, does the Nile, alone of those that flow into sea, rise strongly in summer […]?'

111 The work is variously attributed to Aristotle and Theophrastus. An influential case for Aristotle's authorship has been made by J. Partsch (1909); and for Theophrastus' by P. Steinmetz (1964) 278–296. Both views have had their adherents until quite recently; for an overview see Sharples (1998) 197 with notes. Partsch's view is also accepted by R. Jakobi & W. Luppe (2000), while R.L. Fowler (2000) again favours the ascription to Theophrastus.

[Arist.] De mundo 4	Seneca NQ II	Pliny NH	Antigon. Hist. mir.	[Arist.] Mir. Ausc.
	III 4–8	166.9–11		
5b19–23	(V 14.4 etc.)	236–238	166–167	34–40
5b 21		236.1–3		38b & 40
	IV a			
5b26–30	(III 21) & (VI 28)	207.9–208.10	121–123, 152a–b	
			152b	102
			12, ~122	
5b 30		208.4–5	123	
	(VI 13.3–4) &	233.1–2		
	(IV a 26–27)			
	III 20 & 25–26	224b–232	129–165	53–57 etc.
		228.6–10	144	
		228.1–6	148	
		227.4–5	~129.2	
	(VI 27–28)			

the case with Aëtius and Seneca, who devote to this subject a whole chapter (IV 1) and a whole book (IV a), respectively. It is interesting to note that precisely this problem, perhaps the most famous of all local marvels, does not feature in any of the surviving paradoxographies. Perhaps the very fact that so many had studied it, and provided explanations for it, disqualified it as a 'mirabile'.

In table 3.3 Lucretius' account of—mostly local and particular—terrestrial phenomena is compared with the treatment of such phenomena in other meteorological as well as paradoxographical works. As some of the 'meteorologies' we have hitherto referred to do not discuss terrestrial phenomena (except for earthquakes) at all, I will here limit myself to those that do, viz. Aristotle's *Meteorology*, Pseudo-Aristotle's *De mundo* 4, Seneca's *Naturales Quaestiones* and Pliny's *Naturalis Historia* II. From the many paradoxographical works dealing with terrestrial phenomena, I have chosen to include in the table only Antigonus' *Historiarum mirabilium collectio* and Pseudo-Aristotle's *De mirabilibus auscultationibus*, which are the most extensive ones and present most parallels. In the accompanying text occasional references shall be made to other paradoxographies as well, such as Apollonius' *Historiae Mirabiles*, Claudius Aelianus' *De natura animalium* and the so-called 'Paradoxographus Vaticanus' and 'Paradoxographus Florentinus'. In the table I have included all terrestrial phenomena discussed by Lucretius in *DRN* VI 608 ff., to which I have added the sweet spring in the sea off Aradus, which Lucretius mentions (890–894) as a partial parallel to the spring that kindles tow (879–905), and which may well derive from the same stock. I have also added two more general categories, volcanoes and marvellous waters, of which Lucretius offers only some specimens. In the table a reference in bold characters indicates that the phenomenon in question is not merely mentioned but also physically accounted for. Brackets indicate that the phenomenon is not mentioned in its own right but only to serve as a parallel to, an example of, or a symptom of some other phenomenon. A wavy line (~) indicates an almost identical phenomenon at a different geographical location. References to the *De mundo* 4 are indicated by the last digit of the Bekker page only, the first two digits of the relevant pages being always 39.

Below I will briefly discuss each subject (or group of subjects), point out the parallels in meteorological and paradoxographical literature, and indicate to what extent Lucretius' treatment corresponds to, and differs from, these.

3.3.2.1   Constant Size of the Sea
This problem is not, as we have seen, a 'mirabile' in the technical sense, and accordingly not found in any ancient paradoxography. Thematically the problem is closely related to atmospherical phenomena and was probably discussed in this context long before Aristotle's *Meteorology*: in Aristophanes' *Clouds* (produced ca. 420 BC), in a parody of contemporary physical theory,[112] the constant size of the sea is one of several problems to be discussed (1278–1295), the others being the origin of rain (369–371), the causes of thunder (375–394), and the causes of the thunderbolt (395–407). Later the same problem is discussed in a meteorological context in Aristotle's *Meteorology* and Seneca's *Naturales Quaestiones*, and briefly touched upon in Pliny's *Naturalis Historia* II. Lucretius' discussion of this subject is entirely in line with this.

---

112   For the date see Dover (1968), pp. lxxx–xcviii. For the parody of contemporary physics see Dover's and other commentaries ad locc.

3.3.2.2 Volcanoes

Although earthquakes appear to have been a standard ingredient in ancient meteorologies, volcanoes were not.[113] Aristotle only briefly refers to them, as possible side-effects of earthquakes, not phenomena to be studied in their own right, and he does not even mention the most formidable of them all, the Etna. Seneca's attitude is similar. After some brief references to volcanic phenomena in book II (at 10.4; 26.4–6 and 30.1), in book V (14.4) he promises to discuss the subject more fully in connection with earthquakes. Book VI on earthquakes, however, contains only one, disappointingly brief, reference to volcanoes, in a catalogue of possible side-effects of earthquakes (VI 4.1). Aëtius omits the subject altogether. Things are different in the *De mundo* and in Pliny's *Naturalis Historia*. In both works volcanoes are clearly set apart from earthquakes, and both works mention a number of volcanoes by name, including the Etna. Neither work, however, attempts to explain volcanism, except in the most general terms. In this respect they resemble the paradoxographical accounts in Antigonus (166–167) and the Pseudo-Aristotelian *De mirabilibus auscultationibus* (34–40), only the latter of which includes the Etna itself. Lucretius' approach is entirely different: he not only mentions the Etna as a phenomenon to be studied in its own right, but also provides an extensive explanation.

3.3.2.3 The Nile Flood

From very early on the summer flooding of the Nile had aroused the interest of the Greeks. Whereas, in their experience, all other rivers rose in winter, the Nile alone did so in summer. This called for an explanation and many different theories were devised to account for this curious behaviour.[114] Aristotle does not deal with the subject in his *Meteorology*, but it is discussed at length in Seneca's *Naturales Quaestiones* and in the meteorological section of Aëtius' *Placita*. It would seem therefore that Lucretius, by including this local phenomenon, is perfectly in line with the meteorological tradition.

3.3.2.4 Poisonous Exhalations

Lucretius next discusses what he calls 'Avernian places' (738 *loca Averna*), places where poisonous exhalations rise from the ground.[115] Many instances

---

113 Hine (2002) 58–60.
114 Different theories concerning the summer flooding of the Nile are listed by Herodotus *Hist.* II 19–27, [Arist.] *Liber de Nilo*, Seneca *NQ* IVa, Aëtius IV 1, and several others. For a conspectus see Diels (1879), 228, or more recently Gambetti (2015).
115 Others referred to such places as 'Charonian' (Plin. *NH* 2.208; Strabo 12.8.17 & 14.1.44; Antigonus 123; Stoics *SVF* III 642 apud D.L. VII 123; Galen, *In Hippocratis epidemiarum*

of such phenomena are mentioned in paradoxography, and the subject also turns up in some of our 'meteorologies', such as the *De mundo* 4, Pliny's *Naturalis Historia* II and Seneca's *Naturales Quaestiones*. Seneca discusses the subject twice (at III 21 and VI 28) but gives no specific examples, and makes no real effort at explaining the phenomenon. The *De mundo* (395b26–30) offers a few examples, but no explanations. Pliny's account (207.9–208.10) is by far the longest, and yet the least scientific. It consists of a long list of instances, which is itself part of an even longer list of phenomena which Pliny describes as 'earth's wonders' (206.5 *terrae miracula*). Far from physically explaining these phenomena Pliny ends his account by simply attributing them to the 'divine power of nature' (208.9–10 *numen naturae*). Lucretius' attitude is entirely different from Pliny's: for Lucretius these phenomena, like any other, must and can be explained physically.

Lucretius offers three examples. The first example is the Lacus Avernus, where overflying birds are said to drop dead from the sky. Although the case was well known in antiquity, and is reported in several paradoxographies (Antigonus 152b, *Mir. ausc.* 102, Paradox. Vat. 13), it is not mentioned in any of our 'meteorologies'.[116] Lucretius' second example is the Athenian Acropolis which crows are said to avoid. This story, too, is not mentioned in any of our 'meteorologies',[117] but it is found in a number of paradoxographies: Antigonus 12, Apollonius 9, and Claudius Aelianus' *Natura animalium* V 8. None of these, however, relates this phenomenon to 'Avernian places'. The only ancient text, beside Lucretius', that makes the connection is Philostratus' *Life of Apollonius of Tyana* (II 10), where the absence of crows on the Acropolis is compared to the absence of birds on mount Aornus on the fringes of India, and to similar phenomena in Lydia and Phrygia.[118]

    libri, 17a.10.5 (Kühn); etc.) or 'Plutonian' (Cic. *Div.* 1.79; Strabo 5.4.5; 13.4.14 & 14.1.44.): cf. Lucr. VI 762 'ianua Orci'. There are no ancient parallels for Lucretius' generic use of the word 'Avernus', unless perhaps, knowing that lake Avernus was called Aornos (bird-less) in Greek, Lucretius felt he could render other instances of the word 'Aornos' as 'Avernus' as well. Other places of a similar nature bearing the name 'Aornos' or 'Aornon' were reported to exist near Thymbria in Caria (Strabo, 14.1.11) and in Thesprotis in northern Greece (Pausanias, 9.11.6). Philostratus (*Life of Apollonius of Tyana* 2.10) attributes a similar character to Mount Aornos in India.

116 Pliny knows the story (*NH* XXXI xviii 21), but does not mention it in his list of places with poisonous exhalation in *NH* II.

117 Pliny knows the story (*NH* X xiv 30), but does not mention it in his list of places with poisonous exhalation in *NH* II.

118 Antigonus' story of the crow-less Acropolis (12) is not part of his list of Charonian places

Lucretius' third example, a place in *Syria* where four-footed animals collapse and die, is not known from any other ancient text. Lucretius' description is, however, as Lambinus already observed, very similar to the stories told about a place near Hierapolis in *Phrygia*,[119] and Robin (Ernout-Robin ad VI 749 ff.) even quotes Lucretius with 'Phrygia' instead of 'Syria', without commenting on the change. It is possible that Lucretius' story is actually a garbled version of the reports about the site in Phrygia, transferred somehow to Syria, perhaps through confusion between the Phrygian Hierapolis and its Syrian namesake.[120]

3.3.2.5 Temperature in Wells

Lucretius next discusses the temperature variation in wells, which were commonly believed to be warm in winter, and cold in summer.[121] This problem is not a 'mirabile' in the technical sense of the word, as it is not confined to one or a few specific places, but common to all of its kind. Accordingly it does not normally feature in paradoxographical works, although in Pliny's *NH* II it is mentioned (233.1–2) as part of a long section on 'water miracles' (224b–234), consisting for the most part of true 'mirabilia'. The problem is not generally discussed in meteorology either. Beside Lucretius and Pliny, the only meteorologist to mention the phenomenon is Seneca, who refers to it twice: in book IVa in the context of Oenopides' account of the Nile flood,[122] and in book VI in the context of Strato's explanation of earthquakes. Curiously Seneca's attitude varies: in book VI he accepts the observation as well as Strato's explanation of it, but in book IVa he rejects not just Oenopides' theory, but the observation itself: wells and other underground recesses only *seem* warm in winter and cold in summer, because they are protected from external temperature fluctuations. It is possible, as Robin observes,[123] that the present subject's connection with

---

(121–123), which does, however, include the approximately parallel account of the Black-Sea island of Leuce, which was purportedly avoided by all birds (122).

119   Strabo 13.4.14; Apuleius *De mundo* 17.17 ff.; Pliny *NH* II 208.4–5; Paradoxogr. Vaticanus 36, etc. Especially this last account is very close to Lucretius': Ἐν Ἱεραπόλει τόπος ἐστὶ Χαρώνιος λεγόμενος, ἐν ᾧ οὐδὲν ζῷον δῆτα βαίνει· πίπτει γὰρ παραυτίκα. 'In Hierapolis there is a so-called 'Charonian' place, in which no animal goes: for it falls immediately.' Note that in this account 'Hierapolis' is not further specified, so that someone unfamiliar with the story might easily connect it to the wrong city.

120   The Syrian Hierapolis, a.k.a. Bambyce, is mentioned by Plut. *Ant.* 37 & *Crass.* 17; Strabo 16.1.27; Pliny *NH* V 81; Ael. *De nat. an.* 12.2; and Pseudo-Lucian *De dea Syria* (*passim*).

121   See n. 94 above.

122   See also Diodorus Siculus I 41.

123   Ernout-Robin ad *DRN* VI 840–847.

the Nile flood (as testified by Seneca and others) is what persuaded Lucretius to include it in his account of terrestrial phenomena.

3.3.2.6 Marvellous Waters

Lists of marvellous or peculiar waters were a standard ingredient of ancient paradoxography. Such lists are found for instance in Antigonus (129–165: 'borrowed', as Antigonus himself claims, from Callimachus), in the *De mirabilibus auscultationibus* (53–57), and in the Paradoxographus Florentinus. Similar lists, of various lengths and serving various purposes, are also found in some of our meteorologies. In Aristotle's *Meteorology* two such lists occur: in I 13, 350b36–351a18, Aristotle mentions a number of rivers that flow underground before emerging again, and in II 3, 359a18–b22 a number of salty and other-tasting waters; yet neither group is presented as something to be studied in its own right: the first group serves merely as a concession to those who believe that all rivers are fed from one or several underground reservoirs, a point of view rejected by Aristotle, and the second group serves to illustrate Aristotle's theory that the salty taste of sea water is due to the admixture of something. Longer lists are found in book III of Seneca's *Naturales Quaestiones* 20 & 25–26. Formally the purpose of Seneca's lists is to illustrate that tastes and other peculiar properties of local waters are acquired through contact with some other substance, but the second list ends in a mere enumeration of marvel stories. Finally, Pliny, who offers the longest list (CVI 224b–234), does not even pretend that these are anything other than marvels: in his own table of contents this section is described as 'Mirabilia fontium et fluminum' (Marvels of sources and rivers), and similar terms recur throughout the passage.[124]

Lucretius' approach is, again, very different. Instead of producing a long list of marvellous waters, he singles out just two instances for further discussion: the spring of Hammon, and a spring (not otherwise identified) which kindles tow. A third spring, the spring of sweet water in the sea off Aradus, is not discussed in its own right but only described to serve as a partial analogy for the spring which kindles tow. Yet, as we shall see, it is likely that all three water marvels stem from the same source.

The first marvellous water described by Lucretius is the spring of Hammon whose waters are cold in the day-time and hot at night. This story, first reported by Herodotus and repeated by many historians and other authors since, is also found in Pliny's section on marvellous waters in *NH* II (228.6–10), but

---

124 Pliny *NH* II 224 'mirabilius', 'miraculis', 227 'natura mira', 232 'mira oracula', 233 'permira naturae opera'.

not in any of the other meteorologies. A variant account, according to which the spring's waters are hot at midnight *and midday*, but cold at daybreak and sunset, is only found in the paradoxographical works of Antigonus (144) and the Paradoxographus Florentinus (19).

Lucretius next gives an account of a cold spring that kindles tow which is held above. He does not identify the spring but his account is very similar to the stories told of the spring of Jupiter at Dodona, reported in Pliny's section on marvellous waters (228.1–6) and in Pomponius Mela 2.37, and the spring at the shrine of the nymphs in Athamania, reported by Antigonus (148), the Paradoxographus Florentinus (11) and Ovid (*Met*. 15.311–312).

In the course of this account Lucretius also mentions the spring of sweet water in the sea near the island of Aradus off the coast of Phoenicia, a spring which—among the ancients—is only mentioned by Strabo (16.2.13) and Pliny (*NH* 2.227.4–5). In Pliny's account it is connected to a similar spring near the Chelidonian islands off the coast of Lycia in Asia Minor. Of these two springs Antigonus only mentions the latter (129.2). Other springs of this nature are reported to exist in the eastern part of the Black Sea in Aristotle's list of underground waters (*Mete*. I, 13.351a14–16).

3.3.2.7 Magnets

This is the first of Lucretius' 'terrestrial phenomena' that we know to have been discussed by Epicurus too.[125] Before him the subject had been dealt with by Thales, Empedocles and Democritus.[126] Unfortunately we do not know the contexts of these discussions. The subject of magnets is not found in any of our 'meteorologies'. The only link that I can see is with the subject of stones in general, which is briefly touched upon in Aristotle's *Meteorology* III 6, and even more briefly by Pliny in *NH* 207 (see table 3.1 on p. 100 above). Surprisingly, the subject does not occur in any of the surviving paradoxographies either, although it would have fit in nicely. The closest parallel in paradoxography is the strange claim in Apollonius *Hist. Mir*. 23 that magnets attract iron *only during the day-time*. Perhaps, as with the Nile-flood, the fact that so many had already investigated and explained the phenomenon barred it from inclusion in paradoxographies.

---

125 See nn. 86 and 87 on p. 109 above.

126 Thales A22 D–K (Arist. *De an*. A2, 405a19ff.), Empedocles A89 D–K (Alex. Aphr. *Quaest*. II 23), Democritus A165 D–K (Alex. Aphr. *Quaest*. II 23).

3.3.2.8   Diseases

This is another subject that we know to have been discussed by Epicurus himself.[127] There is no evidence, however, that Epicurus had somehow linked the subject to meteorology, as Lucretius does.

Above (on p. 112) we have seen that Lucretius distinguishes two kinds of diseases: *epidemic* diseases, which come upon a place and after a while abate again, and *endemic* diseases which are peculiar to a certain place and never wholly disappear. Lucretius gives us three examples of *endemic* diseases (1115–1117): elephantiasis in Egypt, a particular foot-disease in Attica and an eye-disease in Achaea. Although these and similar cases would not seem out of place in paradoxography, in fact they do not feature in any surviving work of that genre.

Nor are diseases generally discussed in meteorological works. The only meteorologist, beside Lucretius, who has something to say about diseases is Seneca. In *NQ* VI 27–28, remarking upon certain peculiar phenomena that accompanied the Campanian earthquake of 62 AD, Seneca also comments on the ensuing death of hundreds of sheep, which he attributes to a plague. It is said, he claims, that plagues often occur after great earthquakes. This he accounts for on the assumption that the earth contains many harmful and lethal seeds, which may be released by the force of the earthquake. That the earth should contain such pestilential seeds can be inferred also, writes Seneca, from the existence of certain places that emit poisonous exhalations all the time (VI 28.1). So, according to Seneca, plagues are related in their origin to poisonous places, and may occur as a symptom of earthquakes. One might be tempted to see the account in Seneca's *NQ* as an example of the kind of treatment that might have induced Lucretius to include diseases in a discussion of mostly meteorological phenomena. Yet, it seems more likely that the relationship went the other way round: Seneca's description of diseases and poisonous exhalations appears to owe a lot to Lucretius, who may well have been Seneca's source of inspiration.[128]

But if Lucretius' inclusion of diseases with meteorological phenomena was inspired by neither meteorology nor paradoxography, perhaps the link must be looked for outside these genres. An interesting parallel for the inclusion of diseases among a number of mostly meteorological occurrences is provided by Cicero. In *ND* II 13–15 (= *SVF* I 528), Cicero relates the four causes which, according to Cleanthes, are responsible for forming the notion of gods in the

---

127   See p. 109 above.
128   Piergiorgio Parroni (2002), for instance, refers to Lucr. VI 1093–1096 and VI 740–746 as possible sources of inspiration for Sen. *NQ* VI 27–28.

minds of men. The third cause (*ND* II 14) consists in the occurrence of irregular and unexpected phenomena. It is described as follows:

Tertiam quae terreret animos fulminibus tempestatibus nimbis nivibus grandinibus vastitate *pestilentia* terrae motibus et saepe fremitibus lapideisque imbribus et guttis imbrium quasi cruentis, tum labibus aut repentinis terrarum hiatibus tum praeter naturam hominum pecudumque portentis, tum facibus visis caelestibus tum stellis is quas Graeci κομήτας nostri cincinnatas vocant ..., tum sole geminato, quod ut e patre audivi Tuditano et Aquilio consulibus evenerat ..., quibus exterriti homines vim quandam esse caelestem et divinam suspicati sunt.	As a third cause (Cleanthes posited that) which might frighten the minds on account of thunderbolts, storms, clouds, snow, hail, devastation, *pestilence*, earthquakes and frequent rumblings (of the earth), showers of stones, blood-like drops of rain, then landslides or sudden chasms in the earth, then preternatural portents of man and beast, then the sight of celestial torches, then those stars which the Greeks call κομῆται (comets) and we long-haired stars ..., then the doubling of the sun ..., the fear of which has brought people to suspect that some celestial and divine power exists.

This cause together with the fourth—the orderly motion of the heavens and heavenly bodies (*N.D.* II 15)—is very similar to what Lucretius himself cites as the cause of mankind's *mistaken* belief in intervening gods (V 1183–1193). It is this *misconception* Lucretius sets out to dispel in book VI (as he promises in VI 48–90), pointing out that all these phenomena can be explained physically. Although diseases do not properly belong to meteorology, yet, just like meteorological phenomena, they frighten us, because of our inability to understand their causes, into believing that they are brought about by the gods, and to eradicate this fear diseases need to be physically accounted for no less than meteorological phenomena and local marvels.[129]

3.3.3 Conclusion

The inclusion of local marvels in works of meteorology, especially in sections concerned with terrestrial phenomena, is not exceptional. Save for the summer flooding of the Nile, which is treated on a par with other meteorological phe-

---

[129] In his commentary on Cic. *ND* II 14 'pestilentia' Pease (1958) seems to view the inclusion of diseases in Lucr. *DRN* VI in the same light.

nomena, local marvels are dealt with in meteorology in much the same way as in paradoxographical literature, to which they properly belong and from which they are probably drawn: in most meteorological works such local marvels as are dealt with are simply enumerated in long lists, with no, or hardly any, effort to explain them individually. Lucretius' treatment of local marvels differs from that of other meteorologies in two important ways. In the first place, instead of the long lists found in some of the other meteorologies, Lucretius offers just a small selection. In the second place, each of the marvels that has been selected is provided with an extensive explanation.

In addition to the virtual absence of such explanations in the extant meteorologies, there are several considerations to suggest that explaining local marvels was a relatively new affair. Firstly, excepting the problem of the Nile flood, local marvels are not dealt with in Aëtius' doxography, as they probably would have been if there had been a tradition of explaining them, resulting in divergent opinions. Secondly, in the case of local marvels Lucretius himself most often provides just one explanation, whereas problems of a more general nature, like thunderbolts and earthquakes, are accounted for with a number of alternative explanations,[130] which are most often drawn from earlier authors (see § 2.4 above); this suggests that in the case of local marvels not many explanations had been devised before, upon which Lucretius could have drawn.

It would be interesting to know whether this new approach to marvellous phenomena is due to Lucretius himself or perhaps to Epicurus or an intermediary writer. Again there are two considerations to suggest that the account of marvellous phenomena does not derive from Epicurus. In the first place, local marvels are not discussed in Epicurus' *Letter to Pythocles*, nor is there any fragment or testimony to suggest that he discussed such phenomena elsewhere, except for magnets and diseases.[131] In the second place, if—as I have suggested above—the passages on local marvels in Lucretius book VI are ultimately derived from paradoxographical works, and if—as is usually assumed—the paradoxographical genre was inaugurated by Callimachus (ca. 310–240),[132] then it is chronologically improbable (though not impossible) for Epicurus (341–270) to have written on these subjects. If this is true, the passage must have been conceived either by Lucretius himself or by another Epicurean from whom Lucretius subsequently borrowed it.[133] Lucretius or his source may have

---

130   For an overview of the number of alternative explanations per problem in *DRN* V and VI see APPENDIX 2 on p. 272 below.
131   See p. 109 above.
132   See n. 105 on p. 114 above.
133   Robert D. Brown (1982), 349, thinks that Lucretius may have been inspired by Callimachus.

been inspired to do so by the fact that other meteorologies too incorporated exceptional local phenomena, yet without explanations. He may have felt himself entitled to do so by Epicurus' own occasional exhortations to the reader to find out certain things for himself (*Hdt.* 45, 68, 83; cf. Lucr. *DRN* I 402–409; 1114–1117 and VI 527–534): a good Epicurean is expected to apply the principles of Epicurean physics to other, as yet unsolved problems. In this respect Lucretius' discussion of local marvels is the obverse of *DRN* VI 527–534. There Lucretius chooses *not* to discuss a number of phenomena which Epicurus *had* discussed (snow, wind, hail, hoar-frost and ice), instead inviting the reader to find out by himself; here he includes a number of phenomena which Epicurus had *not* discussed, accepting Epicurus' explicit invitation to do so.

3.4 Order of Subjects

3.4.1 *Introduction*

Above I have compared *DNR* VI with a number of other meteorologies as regards the *range* and *subdivision* of their subject matter. It is now time to have a look at another aspect as well. It has often been observed that some of the aforementioned works exhibit an especially close correspondence in the *order* of their subjects. This is the case with Lucretius VI, Aëtius III, and the *Syriac meteorology*, and to a lesser degree Epicurus' *Letter to Pythocles* (from chapter 17 [99] onwards). Several scholars have produced useful synopses to bring out these similarities.[134] Yet, most of these synopses suffer from lack of perspicacity and detail, and from a certain bias in their presentation, exaggerating the similarities by omitting some of the evidence, rather than letting the evidence speak for itself. Most of them also fail to include Epicurus' *Letter to Pythocles*, which one would expect to be very close to Lucretius VI as well. I will therefore repeat the exercise in some more detail, paying due attention to all the resemblances as well as differences in the order of presentation of each of the four works.

Incidentally, these same works also resemble each other in another respect. In all four of them, meteorological problems are generally accounted for by a number of different explanations. In Aëtius' *Placita* every single explanation is attributed by name to one or several earlier thinkers, while Lucretius, Epicurus and the *Syriac meteorology* instead present their explanations, without reference to their original authors, as equally possible alternatives. The subject of

---

134 Reitzenstein (1924) 34–35; Runia (1997a) 97; Sedley (1998a) 158. See also the lists in Kidd (1992) 305–306.

TABLE 3.4   *Order of subjects in Aëtius, the* Syriac meteorology, *Lucretius and Epicurus*

Aëtius, *Placita* III + IV 1		Syriac meteorology	
1.	Milky Way		
2.	comets and shooting stars		
3.	thunder, lightning, thunderbolts, *prēstēres* & *typhōnes*	1.	thunder
		2.	lightning
		3.	thunder without lightning
		4.	lightning without thunder
		5.	why lightning precedes thunder
		6.	thunderbolts
4.	clouds, rain, snow and hail	7.	clouds
		8.	rain
		9.	snow
		10.	hail
		11.	dew
		12.	hoar-frost
5.	rainbow		
6.	rods and mock suns		
7.	winds	13.	wind
		(*13.43–54: prēstēres*)	
8.	**winter and summer**	14.1–13: halo round the moon	
9–14.	THE EARTH	14.14–29: theological excursus	
15.	earthquakes	15.	earthquakes
16.	origin and bitterness of the sea		
17.	ebb and flood		
18.	halo		
IV 1.	flooding of the Nile		

[Stob. 39 'On waters']*

(* On the possibility that Stobaeus 39 goes back to a lost Aëtian chapter see p. 86 above)

Lucretius, *DRN* VI	Epicurus, *Letter to Pythocles* 17–35	
	17A.	clouds
	17B.	rain
96–159: thunder	18.	thunder
160–218: lightning	19.	lightning
(*164–172: why lightning precedes thunder*)	20.	why lightning precedes thunder
219–422: thunderbolts	21.	thunderbolts
(*379–422: theological excursus*)		
423–450: *prēstēres*	22.	*prēstēres*
451–494: clouds	23.	**earthquakes**
495–523: rain	24.	**subterranean winds**
	25.	hail
	26.	snow
	27A.	dew
	27B.	hoar-frost
	28.	ice
524–526: rainbow	29.	rainbow
	30	halo round the moon
527–534: snow, winds, hail, hoar-frost, ice	31.	comets
	32–34.	**STARS & PLANETS**
	35.	shooting stars
535–607: earthquakes		
608–638: constant size of the sea		
639–702: the Etna		
712–737: flooding of the Nile		
738–839: poisonous exhalations		
840–905: wells and springs		
906–1089: the magnet		
1090–1286: diseases		

multiple explanations has already been explored in Chapter Two of the present work. Yet, this further similarity enforces the impression that these four texts are somehow more closely related to each other than to other writings of this genre. In this section, however, the focus will be on the similarities in the order of subjects.

3.4.2 *The Table*
In table 3.4 the subjects of each of the four works are presented, in the order in which they occur, in four parallel columns. Items whose inclusion or placement in one of the works seems deviant with respect to the others are printed in bold, while subjects that are subordinated to, or included in, another, more general subject are bracketed and italicized. The subjects of the *Letter to Pythocles* are indicated by the chapter numbers of the edition of Bollack and Laks, which I have occasionally subdivided into an A and a B part.

3.4.3 *Some Observations*
The order of subjects of each of the four works (including Epicurus' *Letter to Pythocles*) is so similar that one can hardly escape the impression that they must have derived this order, directly or indirectly, from one and the same work. None of the four works seems to have completely preserved this original order, but each one deviates from it as a result of conscious decisions made by their respective authors as well as unconscious mistakes made in the course of each text's transmission. Below I will explore the extent to which, and the possible reasons why, each of the four works deviates from the others as regards their order of subjects, and what the underlying original order may have been.

3.4.3.1 Milky Way, Comets and Shooting Stars
Working his way down through the phenomena of the Aristotelian sublunary world, Aëtius starts his account of τὰ μετάρσια with the phenomena that Aristotle had assigned to the topmost part of the sublunary sphere just below the realm of the stars: the Milky Way, comets and shooting stars. All three subjects are absent from the *Syriac meteorology* and *DRN* VI. Two of them, shooting stars and comets, are discussed in Epicurus' *Letter to Pythocles*, but only at the end, where they straddle a number of indisputably astronomical subjects. This suggests that Epicurus may have associated *comets* and *shooting stars* with astronomical rather than atmospherical phenomena (see p. 94 above), and the same consideration may explain why Lucretius in *DRN* VI and the *Syriac meteorology* omit them altogether.

3.4.3.2   Problems Pertaining to the Earth as a Whole

Whereas in the other three works earthquakes appear to be closely linked to, and even numbered among atmospheric phenomena, in Aëtius' account they are delegated to a different section—τὰ πρόσγεια—and separated from τὰ μετάρσια by the intrusion of several chapters dealing with the earth as a whole. By including these subjects in his meteorology Aëtius deviates, not just from the three works mentioned above, but from almost every other ancient work on meteorology and most importantly from Aristotle's *Meteorology* (see p. 103 above). The inclusion of these chapters by Aëtius therefore most likely reflects a deliberate departure from his source.

3.4.3.3   Theological Excursus

The final part of Lucretius' account of thunderbolts is an argument against the view that thunderbolts come from the gods. The argument consists of a series of rhetorical questions like: why do thunderbolts strike high mountains, why do they fall in uninhabited regions, why do they sometimes strike good, god-fearing people, and leave the evil-doers alone, etc.,[135] all leading up to the inevitable conclusion that the falling of thunderbolts cannot be attributed to the gods. A very similar argument is also found in the *Syriac meteorology*, in a passage commonly referred to as 'the theological excursus'.[136] Here too the core of the argument is a series of rhetorical questions all concerned with *thunderbolts*, and in most cases closely matching the rhetorical questions and their development in Lucretius. At the end of the excursus, however, the author rather unexpectedly concludes that (14.25–26) "it is thus not right to say ⟨about⟩ hurricanes[137] [sic!] that they come from God", even though the entire argument was about *thunderbolts*, not *hurricanes*. To all likelihood, therefore, 'hurricanes' was just a slip of the pen for 'thunderbolts'. Equally strange is the fact that the excursus does not, as in Lucretius, follow the exposition of *thunderbolts* (ch. 6), but is appended to the chapter on *haloes* (ch. 14), to which it has no bearing at all. Daiber therefore suggests that the passage "actually [belongs] to the chapters on thunderbolts (ch. 6) and on εὖρος and *prēstēr* (13.33–54)".[138] That it should belong to the chapter on thunderbolts seems

---

135   Mansfeld (1992a), 320, points out that the argument as such is an old one, mentioned already in Aristophanes' *Clouds* 398–402. See also n. 112 on p. 118 above, and text thereto.

136   See Daiber (1992) 280–281, Mansfeld (1992a) and Van Raalte (2003), who rejects the passage as a later interpolation.

137   Arabic الزوابع [*az-zawābiʿ*], plural of الزوبعة [*az-zaubaʿa*], 'storm', 'hurricane', the same word that was used in 13.45, 47 and 51 as a synonym for the *prēstēr*.

138   Daiber (1992) 280.

obvious, but Daiber's reference to εὖρος and *prēstēr* was probably elicited by the mention of *hurricanes* in the conclusion of the excursus, which—I argued—may be just a mistake for 'thunderbolts'. Mansfeld, ignoring Daiber's mention of εὖρος and *prēstēr*, interprets his words as implying that the excursus "should probably be reallocated to [...] the chapter on thunderbolts".

Mansfeld himself opts for a different solution. In the introduction to the excursus its subject is stated as (14.14) "the thunderbolt" and "anything that has been mentioned". This means that the excursus, given its present location after the discussion of the halo but before earthquakes, should apply to haloes (among other things), but not to earthquakes, which is strange because haloes are quite harmless and earthquakes are not. Mansfeld therefore conjectures that the excursus would originally have been the concluding chapter of the whole treatise, which would make the backward reference apply to earthquakes as well.[139]

Of these two proposals for reallocation of the excursus I prefer the first one. Mansfeld's proposal assumes that the excursus was meant to apply to earthquakes as well. Yet, the excursus does not say anything to this effect. Instead its entire argument is about thunderbolts, and therefore the excursus would have been most naturally placed directly following the physical account of thunderbolts, just as the corresponding passage in Lucretius. Another parallel for this position is provided by Seneca, who incorporates a similar passage (*NQ* II 42) in his overall account of thunder and lightning. The words "[and] anything that has been mentioned" in the statement of the excursus' subject matter (in 14.14) may simply have been added to make sense of the curious position the passage later came to occupy.

3.4.3.4   *Prēstēres* (Whirlwinds)

The *Syriac meteorology* differs from the other three works in its placement of the discussion of the *prēstēr*. Whereas in the accounts of Aëtius, Lucretius and Epicurus the *prēstēr* is attached to *thunder, lightning* and *thunderbolts*, in the *Syriac meteorology* it is appended to the discussion of *wind*.[140] In order to fully appreciate what is happening here it will be necessary to cast the net a bit wider, and include some other meteorological writings.

In Aristotle's *Meteorology*, and in most subsequent meteorologies, two types of whirlwind are distinguished: the *typhōn* and the *prēstēr*.[141] Both types are

---

139   Mansfeld (1992a) 316 & 318.
140   As observed by Kidd (1992) 303, and Sedley (1998a) 159 &182.
141   Arist. *Mete.* III 1, 371a8–18; [Arist.] *De mundo* 4, 395a21–24; Arrian (fr.3. p. 187.4–11 Roos-

closely linked to the *thunderbolt*, so much so that all three are considered manifestations of the same phenomenon, differing only by degree, the thunderbolt being wholly fiery, the *prēstēr* half, and the *typhōn* not at all. Aëtius conforms entirely to this tradition, discussing under one single heading not just *thunder, lightning* and *thunderbolts*, but also *prēstēres* and *typhōnes*. Theophrastus too can be placed in this tradition: in the *De igne* (1.8–9) *prēstēres* and *thunderbolts* are mentioned together, as examples of *fire* being produced through violent motion.[142] In other words, for Theophrastus, as for Aristotle, *prēstēres* are *fiery* and closely related to *thunderbolts*.[143]

At first sight Epicurus and Lucretius may seem to belong to the same tradition, for they too discuss the *prēstēr* immediately following the *thunderbolt*. However, their account of the *prēstēr* differs from more traditional accounts in three significant ways: (1) there is no mention of its fiery nature, (2) there is no reference to the *typhōn* as its fireless counterpart,[144] and (3) there is nothing in the text to suggest that the *prēstēr* and the *thunderbolt* (as well as the *typhōn*) are varieties of the same phenomenon. The same three observations also apply to the discussion in the *Syriac meteorology*, which closely matches the accounts of Epicurus and Lucretius. However, whereas Epicurus and Lucretius, while severing the traditional ties between the *prēstēr* and the *thunderbolt*, still maintain the traditional order of subjects, the *Syriac meteorology* goes one step further and reassigns the *prēstēr* to the chapter on *wind*.[145] It seems reasonable to assume that in this case the *Syriac meteorology* has departed from the original order of subjects, which has been preserved by Aëtius, Epicurus and Lucretius.[146]

Wirth) apud Stob. *Ecl*. 29.2.4–7; Stoics apud Diog. Laërt. VII 154.4–6; Chrysippus *SVF* II 703 apud Aët. III 3.13; Seneca *NQ* V 13.1–3 (who calls the *typhōn* 'turbo', and defines the *prēstēr* as an 'igneus turbo', i.e. a fiery *typhōn*); Pliny *NH* II 133–134 (who mentions beside the *turbo/typhōn* and the *prēstēr* a number of other similar phenomena: *vertex, procella, columna, aulon*).

142 Text quoted on p. 68 above. See however Theophrastus *De ventis* 53 where the *prēstēr*, without reference to its fiery nature, is said to be produced from the conflict of two opposed winds.
143 As observed by Kidd (1992) 303.
144 Lucr. VI 438 uses 'turbo' simply as a synonym for '*prēstēr*'.
145 Sen. *NQ* V 13.1–3 and Plin. *NH* II 133–134, too, deal with whirlwinds in the context of winds, yet, like Aristotle and several others, they use the word '*prēstēr*' in the limited sense of 'fiery whirlwind' (see n. 141 above).
146 My conclusion is basically that of Sedley (1998a) 159 and 182.

3.4.3.5 Rainbow, Halo, Rods and Mock Suns

As we observed above (see p. 86), Aët. III 18 on the halo is out of place. In almost every other meteorology the halo is related to the rainbow and to rods & mock suns, which in Aët. III make up chapters 5 and 6 respectively. It is very likely, therefore, that Aëtius, too, discussed the halo contiguously to these two subjects. Jaap Mansfeld (2005) argues for the order rainbow—halo—rods & mock suns.¹⁴⁷ This sequence is partly preserved in Epicurus' *Letter to Pythocles* which discusses the rainbow and the halo in two subsequent chapters. In *DRN* VI and the *Syriac meteorology* only one of these phenomena is discussed: the rainbow (very briefly) in *DRN* VI, and the halo in the *Syriac meteorology*.

3.4.3.6 Winds

The subject of winds is differently placed in each of the four works. In Lucretius VI (527–534) wind is mentioned amidst a number of subjects (snow, winds, hail, hoar-frost and ice) which the reader is invited to investigate for himself. This does not necessarily mean that Lucretius found all these subjects consecutively in whatever source he used; he may have simply collected at the end of his exposition of atmospherical phenomena all those subjects which he chose not to elaborate. We should not, therefore, attach too much weight to the location of the subject of *wind* in Lucretius.

Another problem presents itself in Epicurus' *Letter to Pythocles*. Immediately following the chapter on *earthquakes*, Epicurus goes on to discuss *winds* (πνεύματα). There has been much debate about this chapter. Usener thinks it is a dislocated fragment of the account of *prēstēres*,¹⁴⁸ and Bailey that it is a relic of a chapter on *volcanoes* (matching Lucr. VI 639–702), most of which would have been lost in a lacuna.¹⁴⁹ Others prefer to see it as an explanation of the origin of *atmospherical winds*,¹⁵⁰ thus providing the counterpart of Aëtius III 7 and *Syriac meteorology* ch. 13. However, there is no textual evidence to suggest a lacuna, and if the chapter had been about *atmospherical winds*, one would have expected Epicurus to use the normal Greek word for *wind*, ἄνεμος, instead of πνεῦμα, which may refer to any gust of air.¹⁵¹ The use of a definite article

---

147 See n. 34 on p. 86 above.
148 Usener (1887) p. 48 in apparatu critico ad loc.: "nam haec adduntur superiori de turbinibus loco."
149 Bailey (1926) 310; id. (1947) 1655.
150 Arrighetti (1973) 533.
151 In Aristotle's *Meteorology* (I 13 & II 4–6) and the *De mundo* (4, 394b7–395a10) the word ἄνεμος is used specifically to refer to atmospherical wind (cf. *De mundo* 4, 394b13: τὰ δὲ ἐν ἀέρι πνέοντα πνεύματα καλοῦμεν ἀνέμους ...), while πνεῦμα may refer to any (supposed) gust

and a connecting particle at the beginning of the chapter (τὰ δὲ πνεύματα ...) in fact suggests that Epicurus, far from starting something new, is expanding on something he mentioned just before, viz. the *subterranean wind* (πνεύματος ἐν τῇ γῇ) of the previous chapter, which he holds responsible for the production of *earthquakes*.[152] Epicurus' chapter 24, therefore, is probably *not* about *atmospherical winds* at all, and does *not* form a counterpart to Aët. III 7 and *Syr.* 13, but constitutes an appendix to the preceding chapter on earthquakes. The only drawback of this conclusion is that it robs the *Letter to Pythocles* of its chapter on atmospherical winds, which otherwise seems to have been a standard subject in ancient meteorology (see table 3.1 on p. 100 above).

This leaves us with only two texts out of four, from which we might hope to learn something about the possible original position of the section on winds. Unfortunately the two texts diverge on this point: in the *Syriac meteorology* the halo is discussed *after* winds, but in Aëtius III the halo was probably discussed in connection with the rainbow and rods and mock suns (see previous heading) *before* the chapter on winds. On the basis of this evidence alone it is therefore impossible to say with certainty which of the two works may have preserved the more traditional order. On the whole, however, it seems more likely that winds were originally dealt with after the halo.

Above (on p. 104) we saw that Seneca and his fellow Stoics include earthquakes with atmospherical, rather than terrestrial, phenomena. The reason for this, Seneca states, is that earthquakes, being caused by *air*, should be dealt with in connection with other phenomena of the *air*. In this respect he closely follows Aristotle who attributes earthquakes to subterranean winds and discusses them immediately after atmospherical winds. It is probably because of this close connection that Diogenes Laërtius' overview of Stoic meteorology, which is otherwise limited to atmospherical phenomena, also includes earthquakes. I also argued that the inclusion of earthquakes among 'lofty' phenomena in the *Syriac meteorology* and Epicurus' *Letter to Pythocles* must be due to these works going back to this same tradition, even though they no longer endorse the underlying assumption that earthquakes are uniquely caused by winds. However, if both works include earthquakes with atmospherical phenomena because of their traditional connection with winds, it seems likely that originally the section on earthquakes would have *immediately followed*

---

      of air, including e.g. thunder, lightning, thunderbolts and whirlwinds (*Mete.* II 9 – III 1; *De mundo* 4, 395a11–24). Aëtius' chapter on atmospherical winds (III 7) is called Περὶ ἀνέμων, as is Theophrastus' treatise on winds.

152   Bollack & Laks (1978) 240–241.

the chapter on winds (as in Aristotle's *Meteorology*). If this is true the most likely original sequence of subjects would have been: *rainbow—halo—rods & mock suns—winds—earthquakes,* which is precisely the sequence we find in Aëtius III (accepting the reallocation of the chapter on the halo as proposed above). There is only one problem: in Aëtius III the chapters on winds and on earthquakes are separated from each other by no less than seven intervening chapters: one on winter and summer, and six dealing with the earth as a whole. However, as we saw above, Aëtius' inclusion of these six chapters most likely reflects a deliberate departure from his source, and it is quite possible that the chapter on winter and summer, which consist of just one lemma and has no counterpart in the other three works, was a later addition as well. Stripped of these possible later additions Aëtius III appears to have preserved the presumed original order of subjects, with earthquakes following winds. If this reconstruction is correct, the chapter on haloes in the *Syriac meteorology* must at some stage have been relocated from its original position before the chapter on winds to its present location after.[153]

3.4.3.7 Terrestrial Phenomena Other Than Earthquakes

While in the *Syriac meteorology* and Epicurus' *Letter to Pythocles* earthquakes are the only 'terrestrial' phenomena to be dealt with, Aëtius and Lucretius discuss several more:

TABLE 3.5   *Terrestrial subjects other than earthquakes in Aëtius and Lucretius*

Aëtius *Placita* III 16 – IV 1	Lucretius *DRN* VI 608–1286
16. origin and bitterness of the sea	
17. ebb and flood	608–638: constant size of the sea
–	639–702: the Etna
IV 1. the flooding of the Nile	712–737: the flooding of the Nile
–	738–839: poisonous exhalations
(Stobaeus 39 'On waters'?)	840–905: wells and springs
–	906–1089: the magnet
–	1090–1286: diseases

---

153   Sedley (1998a), 159, also believes the *Syriac meteorology*'s chapter on the halo to be misplaced, on the grounds that in its present location it spoils the top-down sequence of phenomena.

Although the correspondence is far from complete, the two sequences show some remarkable similarities: both works deal with the sea (though not with the exact same phenomena), with the flooding of the Nile, and—if we accept the Aëtian origin of Stobaeus *Ecl. Phys.* 39 (see p. 86 above)—with 'waters' (even though here too the precise subjects may not be the same). If we take into account the large degree of correspondence in the order of subjects in the atmospherical sections of both works, the parallelism in the subsequent terrestrial section can hardly be a matter of coincidence.

3.4.4 *Proposed Original Order of Subjects*

Table 3.6 below presents a reconstruction of the original order of subjects that may underlie both Aëtius III, the *Syriac meteorology*, Epicurus' *Letter to Pythocles* 17 ff. and Lucretius *DRN* VI. Subjects whose inclusion in this original order seems uncertain have been bracketed; three dots indicate that other subjects have or may have been dealt with in between; subordinate subjects are indented; and subjects whose placement deviates from the proposed original order are printed in bold.

3.4.5 *Deviations from the Proposed Original Order*

Despite a strong general correspondence, each of the four works also deviates from the proposed original order of subjects in its own way. Below I will briefly discuss the most important omissions, additions and transpositions each of the four works presents.

Aëtius III: The original order of subjects seems best preserved by Aëtius. The only discrepancy is the chapter on haloes, which is illogically placed in the section on terrestrial phenomena, probably due to a scribal error. Aëtius' account further differs from the other works by the transfer of its chapter on earthquakes from atmospherical to terrestrial phenomena and the insertion of a number of other chapters before it, viz. one on the alternation of the seasons, and several dealing with the earth *as a whole*, subjects that were usually discussed in the context of cosmology and astronomy. Both additions should probably be ascribed to Aëtius' wish to organize the subject matter strictly according to each phenomenon's location.

Syriac meteorology: The Milky Way, comets & shooting stars are not discussed in the *Syriac meteorology*. Their absence may indicate that the author did not consider these phenomena atmospherical, as Aristotle had done, but astronomical. Also absent is the rainbow, which was a standard ingredient in every other ancient meteorology. Its absence is even more conspicuous because the

TABLE 3.6   *Proposed original order of subjects*

	Aet. III	*Syr. met.*	Ep. *Pyth.*	Lucr. VI
Milky Way	1			
Comets	2		31	
Shooting stars	2		35	
Thunder	3	1	18	96–159
Lightning	3	2	19	160–218
(Thunder without lightning)		3		
(Lightning without thunder)		4		
(Why lightning precedes thunder)		5	20	164–172
Thunderbolts	3	6	21	219–422
(Theological excursus)		14.14–29		379–422
Whirlwinds (*prēstēres*)	3	13.43–54	22	423–450
Clouds	4	7	17A	451–494
Rain	4	8	17B	495–523
Snow	4	9	26	*527–534*
Hail	4	10	25	*527–534*
Dew		11	27A	
Hoar-frost		12	27B	*527–534*
(Ice)			28	*527–534*
Rainbow	5		29	524–526
Halo round the moon	18	14.1–13	30	
Rods and mock suns	6			
Winds	7	13		*527–534*
Earthquakes	15	15	23–24	535–607
(…)				
Subjects relating to the sea	16–17			608–638
(…)				…
The flooding of the Nile	IV 1			712–737
(…)				…
(Other terrestrial waters)		(Stob. 39)		840–905
(…)				…

less well known halo, which is traditionally related to the rainbow, *is* dealt with. I cannot imagine any reason why the rainbow should be omitted, except for an accident in the text's transmission. Three passages, moreover, seem to have been moved to a different position. The section on *prēstēres* has been transplanted from its traditional place next to thunderbolts to a new position at the end of the chapter on winds, perhaps in recognition of the changed view of the *prēstēr's* nature (see p. 132 above). Two other moves are harder to account for. The halo, which, as we have seen, was originally dealt with in connection with the rainbow, has probably been moved from a position before, to a position after winds. Moreover, the theological excursus, in which the divine provenance of thunderbolts is refuted, has been separated from the physical account of thunderbolts, to which it obviously belongs, and been appended to the chapter on the halo, with which it has nothing to do. Whatever the reasons for the reallocation of the three passages, it is remarkable that they ended up together, separating the chapters on winds and earthquakes.

Lucretius *DRN* VI: Like the *Syriac meteorology* and possibly for the same reason (see above) Lucretius omits the Milky Way, comets and shooting stars. He further skips snow, hail, dew, hoar-frost and ice—perhaps so as not to bore his readers with accounts that would for the most part be only variations of what was already said about clouds and rain. He goes on to discuss the rainbow very briefly (524–526), but omits the halo and winds. Before moving on from lofty to earthy phenomena, he first sums up most of the phenomena he earlier omitted (527–534)—snow, winds, hail, hoar-frost and ice—inviting the reader to examine the possible explanations for himself.

Epicurus *Letter to Pythocles*: The order of subjects in the *Letter to Pythocles* is so strange and irregular that the correspondence with the other three works is easily overlooked.[154] The strangest feature is the return, after a long intermezzo on atmospherical phenomena, to astronomy. This unexpected return also makes it hard to decide whether Epicurus meant comets and shooting stars to go with astronomy or meteorology, although the evidence seems slightly in favour of the first option (see p. 94 above). As for the truly meteorological part of the *Letter*, i.e. chs. 17–30, although the order of its subject-matter may seem very different from the order found in the other three works, on closer inspection the differences amount to just two major transpositions. First, for some unknown reason clouds and rain have been detached from the other types of precipitation and placed before the accounts of thunder, lightning, thunderbolts and the *prēstēr*. Secondly, earthquakes and subterranean winds have been moved from

---

154 Runia (1997a) 97, comparing Lucr. VI and Aët. III, and Sedley (1998a) 157–158, comparing Lucr. VI, Aët. III and the *Syriac meteorology*, both ignore the order of subjects of the *Letter to Pythocles*.

their presumably original place at the end of the passage, after the rainbow and the halo, to a position immediately following the account of the *prēstēr*, perhaps because of their similarly destructive effects.[155] Only after the chapters on earthquakes and subterranean winds does Epicurus return to the other kinds of precipitation. Atmospherical winds are omitted altogether. Although these transpositions have much disturbed the *Letter*'s order of subjects, two long sequences out of the original order remain virtually intact: *thunder—lightning—thunderbolts—prēstēr*; and (with the minor transposition of snow and hail): *hail—snow—dew—hoar-frost—ice—rainbow—halo*. Especially significant is the fact that in the first sequence the *prēstēr* has retained its traditional position after the thunderbolt, even though—as we have seen—in Epicurus' view the two phenomena are not related (see p. 132 above).

It appears to be possible to reconstruct an original order of subjects from which all four works derive, each of them deviating from it in its own special way. It would be interesting to know what work this order of subjects originally came from and in what way our four texts relate to this original and to each other. In order to come closer to answering these questions it will be useful to have a closer look at the structure of our four texts on the level of individual chapters and sections.

3.4.6 The Internal Structure of Chapters and Sections

Although the order in which the different subjects are presented in each of the four works is very similar, the internal structure of chapters and sections is not always the same. A first point of difference can be gleaned from table 3.4, on p. 128 above. While Epicurus, Lucretius and the *Syriac meteorology* all tend to deal with each phenomenon in isolation, Aëtius sometimes collects a number of related subjects into one chapter. Chapter 3, for instance, deals with thunder, lightning, thunderbolts, *prēstēres* and *typhōnes*, which are accounted for by a number of integrated theories, and chapter 4 deals with clouds, rain, snow and hail in the same way.

Of the three remaining works Epicurus' exhibits the simplest structure. The account of almost every phenomenon is reduced to the question of its causation, which is accounted for with a list of alternative theories, sometimes followed by a brief methodological remark. In a few cases other aspects of the phenomenon under investigation are dealt with separately: in this way

---

155 Note that the overview of Stoic meteorology in Diog. Laërt. VII 151–154, whose order of subjects is even more garbled than Epicurus', also jumps from *prēstēres* to *earthquakes*.

the account of lightning in general (ch. 19) is followed by an account of why lightning precedes thunder (ch. 20); the account of earthquakes (ch. 23) by an account of the subterranean winds responsible for earthquakes (ch. 24); the account of hail (ch. 25-a) by an account of the round shape of hailstones (ch. 25-b), the account of the rainbow (ch. 29-a) by an account of its round shape (ch. 29-b); and the account of the halo (ch. 30-a) by an account of the circumstances that may lead to a halo (ch. 30-b).

A more complex structure is found in the *Syriac meteorology* (see APPENDIX 3 on p. 274 below). Although some of its chapters are limited, as in Epicurus' *Letter to Pythocles*, to listing a number of alternative explanations for the phenomenon under investigation, most of them deal with several related questions and aspects as well. The most outspoken examples are chapter 6 on thunderbolts, and chapter 13 on winds. A very similar structure, though less explicit (since there are no chapters), underlies Lucretius' meteorology, which in its more complex sections deals with many of the same questions, and in a somewhat similar order, as the *Syriac meteorology*. In the table below I have printed in parallel columns the contents of the Syriac's and Lucretius' sections on thunderbolts.[156] Items that have no match in the other text are italicized.

TABLE 3.7   Syriac meteorology *and Lucretius VI on thunderbolts*

Syriac meteorology chapter 6		Lucretius *DRN* VI 219–422	
2–9:	The nature of thunderbolts	219–224:	The nature of thunderbolts
10–16:	Their subtlety and penetration *and speed*	225–238:	Their subtlety and penetration
	–	*239–245:*	*Introduction to the causes of thunderbolts*
16–21:	Causes of thunderbolts	246–268:	Necessary conditions
21–28:	Necessary conditions	269–280:	Causes of thunderbolts
28–36:	Their escape from the cloud	281–294:	Their escape from the cloud
	–	*295–322:*	*More causes of thunderbolts*
36–67:	*Reasons for their downward movement*	323–347:	*Reasons for their speed*
67–74:	Why they are more frequent in spring	348–356:	Their effects
74–85:	*Why they are more frequent in high places*		–
85–91:	Their effects	357–378:	Why they are more frequent in *autumn and* spring
	(topic discussed in 14.14–29)	379–422:	Thunderbolts not the work of the gods

---

156   The close correspondence between the Syriac meteorology's and Lucretius' sections on thunderbolts was already pointed out by Mansfeld (1992a) 326–327.

Although the corresponding sections of the *Syriac meteorology* and DRN VI are structured very similarly, neither text can be reduced to the other: amidst matching items each text also includes items that have no counterpart in the other one.

Similar observations can be made on the level of the individual items of each section. For instance, the last item of Lucretius' section on thunderbolts, which deals with the question whether thunderbolts are instruments of the gods, offers virtually the same arguments as the corresponding section of the *Syriac meteorology* (14.14–29), but in a different order.[157] Also the lists of alternative explanations for individual phenomena correspond to a large degree—sometimes even including the analogies used to illustrate each explanation—,[158] although the order in which they are presented often differs. As an example, in Table 3.8 below I have printed in two parallel columns the alternative explanations of thunder as offered in the *Syriac meteorology* and by Lucretius, with the corresponding illustrative analogies added in square brackets.

Four explanations appear to be common to the *Syriac meteorology* and Lucretius, two of which are even illustrated by the same analogy (the exploding bladder and red-hot iron being quenched in water). Yet, the order in which these four explanations occur is entirely different between the two works. Comparison of other lists of alternative explanations will yield similar results, revealing a significant correspondence in content, which, however, rarely extends to the order of presentation.

In sum: a comparison of the internal structure of the corresponding chapters and sections of our four meteorologies shows different degrees of similarity. While the Syriac and Lucretius' meteorologies appear to be very similar, and hence probably very closely related, Epicurus' *Letter to Pythocles* is structured in a much simpler way, confining itself to the investigation of the causes of each phenomenon, which possibly testifies to the *Letter*'s summary nature. The chapters of Aëtius' account, by contrast, are organized according to a different rationale, often combining a number of related phenomena, which suggests that Aëtius' book III is more distantly related to the other works.

---

157 Mansfeld (1992a) 326–327.

158 The degree of correspondence in the illustrative analogies is variously assessed: Kidd (1992) 301 sees 'close parallels including the illustrative analogies' between Lucretius and the Syriac meteorology, while Garani (2007) 97 observes a 'remarkable lack of correspondence between Theophrastean [i.e. the Syriac meteorologist's] and Lucretian analogies.'

TABLE 3.8 *Alternative explanations of thunder in the* Syriac meteorology *and* DRN VI

Syriac meteorology chapter 1		Lucretius DRN VI 96–159	
2–5:	(1) collision of concave clouds [clapping hands]	96–115:	(1) collision of clouds [flapping and tearing of a canvas awning]
6–8:	(2) wind whirling in hollow cloud [wind in caves and large jars]	116–120:	(2) friction of clouds [–]
9–11:	(3) thunderbolt quenched in a moist cloud [white-hot iron quenched in cold water]	121–131:	(3) exploding cloud [explosion of inflated bladder]
12–14:	(4) wind hitting and breaking an icy cloud [flapping paper]	132–136:	(4) wind blowing through ragged clouds [wind blowing through trees]
15–17:	(5) wind blowing through crooked cloud [butchers blowing into guts]	137–141:	(5) wind rending cloud [wind uprooting trees]
18–20:	(6) exploding cloud [explosion of inflated bladder]	142–144:	(6) waves breaking in the clouds [waves breaking in rivers and the sea]
21–23:	(7) friction of clouds [millstones being rubbed together]	145–149:	(7) thunderbolt quenched in a moist cloud [white-hot iron quenched in cold water]
		150–155:	(8) cloud burnt by thunderbolt [forest fire]
		156–159:	(9) frozen cloud breaking up [–]

3.5   Relations between the Four Texts

Four meteorological texts, viz. Aëtius' *Placita* III, Epicurus' *Letter to Pythocles* (from ch. 17 onwards), the *Syriac meteorology* and Lucretius' *DRN* VI, resemble each other closely in the order of their subjects, while the latter two show a large degree of correspondence in the structure of individual chapters and sections as well. Although it is likely that these four texts are somehow related, it appears to be impossible to simply reduce them to each other. It is unlikely, therefore, that they are all directly related, but the similarities in order and structure must have been transmitted by still other texts, which are no longer extant. In this section I will try to specify these missing links as best as the evidence allows. In order to do this it will also be necessary to deal with the identity of the *Syriac meteorology*, whose equation with (a part or a summary of) Theophrastus' *Metarsiology* has hitherto been accepted too readily.

3.5.1 *Epicurus'* Letter to Pythocles *and His "Other Meteorology"*

In the *Letter to Pythocles* Epicurus claims to be merely summarizing what he wrote "in other places" (ἐν ἄλλοις; see also p. 92 ff. above).[159] As far as the cosmology and astronomy are concerned, these "other places" have been identified as (parts of) books XI and XII of the *On nature*. As for the meteorological portion of the *Letter* these "other places" have not yet been identified, but we can be certain that there existed some other, more elaborate, account of meteorology, whether this was part of book XII or XIII of the *On nature*, as Sedley suggests,[160] or of some other of Epicurus' numerous works. This "other meteorology", as I shall call it, may well be the source of the reports on Epicurus' seismology in Seneca's *NQ* VI 20 and Aët. III 15.11, and on Epicurus' views on rain and hail in Aët. III 4.5, which provide details not present in Epicurus' *Letter to Pythocles*. In this "other meteorology" Epicurus will have discussed at least those subjects that are also dealt with in the meteorological section of the *Letter* (i.e. chs. 16–30), and perhaps more (e.g. such subjects as are dealt with in the terrestrial parts of Aëtius III and Lucretius' VI).

Since the *Letter to Pythocles* is only a summary of Epicurus' "other meteorology", it is likely that the traces of the original order (see p. 137 ff. above) it still preserves were transmitted to it from this more extensive work.

3.5.2 *Lucretius* DRN VI *and Epicurus' "Other Meteorology"*

There has been much debate about Lucretius' sources. Any investigation into these sources must begin with Lucretius' own statement on the matter: in the whole body of the *DRN* Lucretius acknowledges only one source: the writings of Epicurus.[161] I do not think this necessarily means that Epicurus' writings were Lucretius' only source, but this is where any investigation should start.[162] As for the sources of Lucretius' meteorology in book VI, the first work that comes to mind is, of course, Epicurus' *Letter to Pythocles*, which deals with many of the same subjects. However, even a superficial comparison will show that the succinct treatment in the *Letter to Pythocles* cannot have been the main source for Lucretius' much richer account. Yet, as I have argued above,

---

159   Jaap Mansfeld points out to me, as he once did to David Sedley, that the plural ἐν ἄλλοις was sometimes used to refer to a single passage: see Sedley (1998a) 120 n. 68.

160   See n. 64 on p. 95 above and text thereto.

161   *DRN* III 9–12: "... tuisque ex, inclute, chartis / floriferis ut apes in saltibus omnia libant / omnia nos itidem depascimur aurea dicta ..."—"... and from your pages, illustrious man, just as bees in the flowery woods sip everything, we likewise feed on all your golden words ...".

162   So already Reitzenstein (1924) 37.

the *Letter* itself only summarizes what Epicurus had written on these matters elsewhere. Although this other, more detailed, account of meteorology is no longer extant, it is reasonable to assume that this was Lucretius' main source, at least for the subjects up to and including earthquakes. Moreover, if both the meteorological part of Epicurus' *Letter to Pythocles* and book VI of Lucretius' *DRN* derive from Epicurus' "other meteorology", and if both works have each in their own way adapted the 'original order of subjects' established above, it seems likely that such traces of the original order as each of the two works preserves have come to them through Epicurus' "other meteorology". If so, Epicurus' "other meteorology" must have been closer to the original order of subjects than either of the two works derived from it.

3.5.3 Authorship and Identity of the Syriac Meteorology

It is now time to address a question I have been postponing for some time: is the *Syriac meteorology*, as it claims to be, a work by Theophrastus? The majority view nowadays seems to be that it is, that it is in fact a translation of (either the whole, or possibly a part of) Theophrastus' *Metarsiology*,[163] a work in two books of which otherwise only the title,[164] a 'table of contents' (possibly incomplete),[165] and a few paraphrasing fragments survive.[166] Another option, that the work goes back to Epicurus, was suggested and then swiftly rejected by Bergsträßer, the first editor of the first text found, and this procedure was repeated by several other early commentators.[167] The most recent editor of all three versions, Hans Daiber, makes no reference to this option but confidently claims that the work is an unabridged translation of Theophrastus' *Metarsiology*.[168]

---

163 See n. 186 on p. 70 above.
164 Diog. Laërt. 5.44: Μεταρσιολογικῶν α' β'.
165 Proclus *In Platonis Ti.* 35A (II 121.3 Diehl = Theophr. fr.159 FHS&G): ... ζητοῦντος, πόθεν μὲν αἱ βρονταί, πόθεν δὲ ἄνεμοι, ποῖαι δὲ αἰτίαι κεραυνῶν, ἀστραπῶν, πρηστήρων, ὑετῶν, χιόνος, χαλάζης, ἃ δὴ καλῶς ποιῶν ἐν τῇ τῶν μετεώρων αἰτιολογίᾳ (so Steinmetz (1964), 216–217, ms.: ἀπολογίᾳ) τῆς πρεπούσης εἰκοτολογίας καὶ αὐτὸς ἠξίωσεν—"... by investigating whence come thunders, whence winds, what are the causes of thunderbolts, lightnings, *prēstēres*, rains, snow, hail, which he too in his discussion of meteorological phenomena quite properly thought deserving of a fitting conjectural account ...".
166 Theophrastus' meteorological fragments have been collected in FHS&G (1992) 356–365 as frgs. 186A–194. Of these only 186B and 192 are explicitly attributed to Theophrastus' *Metarsiology*.
167 See n. 192 and text thereto on p. 71 above.
168 Daiber (1992) 285–286, 287.

I think there is still some reason for doubt. Above (§ 2.5.5 on p. 70) I have already indicated that the systematic application of multiple explanations, such as we find in the *Syriac meteorology*, is very different from Theophrastus' practice in other, undisputed, works. The best that recent commentators have been able to come up with is the observation that Theophrastus *occasionally* offers multiple explanations. Yet, as I have also shown, even in those cases Theophrastus usually offers far less explanations than the *Syriac meteorology* does, and generally does not support each alternative explanation with analogies from everyday experience, or derive his alternative explanations from the views of earlier thinkers. In all these respects the *Syriac meteorology* is closer to Epicurus' *Letter to Pythocles* and the astronomical and meteorological passages of Lucretius *DRN* (V 509–770 and VI respectively), than to any undisputed work by Theophrastus. Significantly, Theophrastus was never cited in ancient literature as a champion of multiple explanations, as Epicurus was. On the contrary, one ancient witness, Seneca, explicitly claims that regarding the causes of earthquakes Theophrastus held the same, single, view as Aristotle,[169] while citing Democritus and Epicurus for having offered multiple alternative explanations.[170]

Another trait of the *Syriac meteorology* that does not sit easily with Theophrastus' authorship, is its exclusion of the Milky Way, comets and shooting stars, which Theophrastus' teacher Aristotle had all assigned to meteorology. If Theophrastus had in fact transferred these phenomena from meteorology to astronomy and so redefined the boundary between the supra- and sublunary spheres, one would have expected this fact to have left at least some traces in later meteorologies or in the commentaries to Aristotle's work, but no such traces exist.[171]

The *Syriac meteorology* also differs from Theophrastus' undisputed works in another respect. As we have seen above (p. 132), the way *prēstēres* are viewed in the *Syriac meteorology* is very different from what we find in Theophrastus' *De igne*, yet very similar to the corresponding passages in the works of Epicurus and Lucretius. In *De igne* 1.8–9 Theophrastus views the *prēstēr*, like Aristotle before him, as a fiery whirlwind, closely related to the thunderbolt. In the *Syriac meteorology*, on the other hand, *prēstēr* seems to be the generic word for

---

169   Sen. *NQ* 6.13.1: "In hac sententia licet ponas Aristotelem et discipulum eius Theophrastum." (= Theophrastus fr.195 FHS&G).
170   Sen. *NQ* 6.20 (= Democritus fr.A98 D–K / Epicurus fr.351 Us.)
171   Steinmetz (1964) 167–168 argues that Theophrastus did in fact reject Aristotle's assignment of the Milky Way to meteorology; see, however, Sharples (1985) 584–585 and id. (1998) 108–111.

whirlwind, which is nowhere said to be fiery or to be somehow related to the thunderbolt. The *Syriac meteorology* even goes one step further than Epicurus and Lucretius, by relocating the discussion of the *prēstēr*, from its traditional position after the thunderbolt, to the end of the section on winds. It might of course be argued that Theophrastus changed his view of *prēstēres* after writing his *De igne* and before composing his *Metarsiology*. That Theophrastus did in fact change his position might seem to be confirmed by a passage in his *De ventis*, which is supposed to have been written after the *Metarsiology*.[172] Here, in ch. 53, the origin of *prēstēres* is attributed, without any reference to their fiery nature or their connection with *typhōnes* and thunderbolts, to the conflict of contrary winds. However, the brief mention of *prēstēres* and thunderbolts in the *De igne* explicitly refers the reader to an earlier exposition of these phenomena, which is usually interpreted as a reference to the *Metarsiology*.[173] Yet, if this is true, the *Metarsiology* must have presented the more traditional view of the *prēstēr* as a fiery whirlwind related to the thunderbolt, and not the innovative concept of a *prēstēr* as a special (and fireless) type of wind, as advanced by Epicurus, Lucretius and the *Syriac meteorology*, and perhaps also Theophrastus' *De ventis*.

These, to my view, are the most important objections against the identification of the *Syriac meteorology* with Theophrastus' *Metarsiology*. None of them seems absolutely fatal: Theophrastus could have dealt with meteorological matters differently from other subjects (it will not do, however, to state that using multiple explanations was characteristic of him), Seneca could have been misinformed about Theophrastus' account of earthquakes, Theophrastus could have rejected Aristotle's assignment of the Milky way, comets and shooting stars to astronomy, just as he could have changed his view of *prēstēres* between writing his *De igne* and his *Metarsiology* (in which case the *De igne* must have been written earlier, and the supposed backward reference to the *Metarsiology* be explained in some other way). Yet, the effect of all these objections is cumulative, and shows that the alternative hypothesis of an Epicurean origin should at least have been taken more seriously.

Yet, the assumption of an Epicurean origin is not without its difficulties either. Below I will discuss four passages that seem to testify to a Theophrastean (or at least Peripatetic) rather than an Epicurean origin.

---

172   Steinmetz (1964) 9 n1, 56; Daiber (1992) 286.
173   Steinmetz (1964) 9 n2, 114; Daiber (1992) 273, 286.

(1st) Chapter 6 of the *Syriac meteorology*, on thunderbolts, contains a section (36–67) concerning the downward motion of thunderbolts, which has no counterpart in Lucretius' book VI. According to the *Syriac meteorology* (36–41), a thunderbolt reaches us (i.e. *moves downward*) either because winds beat the cloud on top, or because the cloud is split at the bottom. In both cases the movement is the result of a force external to the thunderbolt.[174] The fact that the Syriac meteorologist finds it necessary to explain the thunderbolt's downward movement is significant. Although Lucretius does not deal with the subject in the corresponding portion of book VI, he refers to the question elsewhere. In *DRN* II 203–215 he warns his reader not to believe that things can move upward of their own accord. Everything material, including fire, has a natural tendency to move downwards, a tendency which can only be checked or reversed by an external force. When, under normal circumstances, fire is seen to rise, this must be attributed to the surrounding air, which, by being heavier, squeezes the fire upwards. Therefore, when a thunderbolt (which is fiery) moves downward, this is due primarily to its own nature, and would seem to need no further explanation. However, it can be argued that even from an Epicurean perspective a further explanation may still be asked for. If, under normal circumstances, fire is seen to rise, even though this is not to be attributed to its own nature but to upward pressure from the surrounding air, one may reasonably ask why in the case of thunderbolts these normal circumstances do not apply. Therefore, an explanation of the downward movement of thunderbolts need not be un-Epicurean *per se*, even though it is absent from both Epicurus' *Letter to Pythocles* and Lucretius' book VI.

However, the continuation of the explanation in the *Syriac meteorology* (41–48) is distinctly *un*-Epicurean:

> The reason that the cloud is split from the bottom and not from (42) the top is as follows: Those *two vapors* which ascend from the earth are joined, (43) namely the thick vapor and the fine vapor. If they ascend, (44) then the fine one of both (kinds) moves quickly *upwards*, because it approaches its *natural place*. (45) And that is because each one of the bodies, when it is distant (46) from its (natural) place, has a weak and slow movement; but if it is near to its (natural) place, (47) (its movement)

---

174 Cf. Arist. *Mete.* I 4, 342a13–27 on natural vs. forced motion of shooting stars (and thunderbolts), and II 9, 369a20–30 on the downward motion of thunderbolts, fallwinds (*eknephiai*) etc.

is quick and strong. Therefore, whenever the fine vapor (48) ascends, it has a *much quicker*[175] movement.

tr. DAIBER (1992), modified

Two elements in this passage bear witness to its Peripatetic origins: the theory of the two vapours and the notion of a natural place. The two vapours, which play a major part in Aristotle's *Meteorology*,[176] do not seem to be essentially incompatible with Epicurus' atomism, but in fact neither Epicurus, in the *Letter to Pythocles*, nor Lucretius, in *DRN* VI, ever refers to this theory.

The notion of a *natural place* is, however, typical of the Peripatos and entirely alien to Epicureanism. According to Aristotle, the sublunary sphere is subject to two opposing natural motions: heavy bodies move naturally downwards and light ones upwards; both natural motions have their end-point in a natural place: the downward motion ends in the absolute *down* which is the centre of the (finite) universe, and the upward motion ends in the absolute *up* which is the circumference of the sublunary sphere.[177] Both motions accelerate when the natural place is approached.[178] According to Epicurus, however, there is only one natural motion, viz. downward, to which there is no limit, as the universe itself is unlimited, and which is uniform (i.e. not accelerated).[179] The notion of a natural place is therefore entirely un-Epicurean. We may conclude therefore that the source of *this* passage, at least, cannot have been Epicurus.

Still, there is something strange about the passage. While here the formation of clouds is attributed to the interaction of the two vapours, the next chapter (ch. 7), which deals specifically with cloud formation, says nothing about these *two* vapours or exhalations. Could it be that the present passage and the chapter on clouds have different origins?

(2nd) The two vapours, which we first encountered in ch. 6 on thunderbolts, make a second appearance in ch. 13 on winds. In fact the whole chapter, except

---

175 Daiber (1992), following the Arabic version of (presumably) Ibn al-Khammār, has '*quick*', but Wagner, in Wagner & Steinmetz (1964), who follows the Syriac version, has '*much quicker*' ('viel schneller'). Unfortunately, the section in missing in the Arabic version of Bar Bahlūl.
176 Arist. *Mete.* I 4, 341b6–13 *et passim*. Cf. also *Phys.* IV 1, 208b9–10 and 210a3–6; *Cael.* IV 3, 310a30–35. See Taub (2003) 78, 86–91.
177 Arist. *Cael.* IV (esp. chs. 3–5 = 310a16–313a13) and *Phys.* IV 1, 208b9–22.
178 Arist. *Cael.* I 8, 277a28–29 & b5–9; ibid. II 6, 288a17–22.
179 Epic. *Hdt.* 60–61.

for its final section on *prēstēres* (43–54), has a decidedly Peripatetic flavour.[180] Unfortunately, we are not able to compare it with the Epicurean position, since both Epicurus and Lucretius fail to include a section on atmospherical winds in their meteorological overviews (see p. 134 above).

(3rd) Although, as we just observed, ch. 7 on clouds does not mention the Peripatetic theory of the *two vapours*, in another respect it is quite Theophrastean. It is worth quoting the relevant lines (2–9 & 27–29):

> (2) The clouds come into existence for two causes: because of the accumulation (3) and thickness of air and its transformation into *a watery substance*[181] or because of much vapor (4) which ascends and with which the ascending vapors of the seas as well as the remaining (5) fluids become mixed.
>
> Air comes together and becomes thick for two reasons: (6) because of *coldness* or because of contrary winds which *squeeze* it and bring it together. (7) We can observe something similar amongst us: When ascending vapor in (8) the bath encounters the roof and cannot penetrate this because of its thickness, (9) it accumulates and becomes water.
> {…}
> The clouds (28) turn into water, when they become very thick; their thickness is (29) caused by the *pressure* of hard winds or by *coldness*.[182]

In this case we are in a position to compare this with the views elsewhere attributed to Theophrastus. In his commentary on Aristotle's *Meteorology* Olympiodorus writes:[183]

> One should know, that while Aristotle says that cooling is the only cause of clouds turning into water, Theophrastus says that *cooling* is not the only cause of the production of water, but also *compression*. Note, for instance, that in Aethiopia, where there is no cooling, rain still pours down because

---

180 See Daiber (1992) 278–280.
181 The Arabic version of (presumably) Ibn Al-Khammār has '*the nature of water*'. According to Daiber (1992), 219, this is probably a misinterpretation of the Syriac (352b29), '*a watery substance*'. Unfortunately, the section is missing in the Arabic version of Bar Bahlūl.
182 Translation by Daiber (1992), modified.
183 Olympiodorus *In Arist. Mete.* I 9, 346b30 (p. 80.30–81.1 Stüve) = fr.211B FHS&G (for the Greek text see p. 69 above). See also Proclus *In Plat. Tim.* 22E (= Theophr. fr.211A FHS&G); Galen *In Hippocr. Aer.* 8.6 (= fr.211C ibid.), and Theophrastus *De ventis* 5.1–5.

of compression: for he says that there are very high mountains over there, against which the clouds collide, and that subsequently rain pours down, because of the resulting compression. But in cauldrons too, says he, moisture runs down again, and in the vaults of baths, where there is no cooling; this clearly coming about through compression.

The text of the *Syriac meteorology* agrees with this report in two ways: firstly they both name the same two causes of water-formation in clouds, viz. cooling and compression, and secondly, they have one example in common, viz. the example of water-formation against the roof of a bath-house.[184]

Again, neither explanation—cooling or compression—seems to be essentially incompatible with Epicurus' atomism, but the fact is that for some reason both Epicurus and Lucretius, while including *compression* among a number of alternative causes,[185] entirely ignore *cooling*.[186] In this respect, then, the *Syriac meteorology* is closer to Theophrastus than to Epicurus and Lucretius.

(4th) The second half of ch. 14 is devoted to a refutation of divine interference in the case of thunderbolts. As was first observed by Jaap Mansfeld, the passage is very close in content and structure to Lucretius VI 379–422.[187] Both passages offer a list of rhetorical questions (why do thunderbolts strike high mountains, why do they fall in uninhabited regions, why do they sometimes strike good, god-fearing people, and leave alone the evil-doers, etc.), all leading up to the conclusion that the falling of thunderbolts cannot be attributed to god or the gods. Lucretius leaves this conclusion unspoken, but in the *Syriac meteorology* the passage is introduced as follows:

---

184 Based on these correspondences Drossaart Lulofs (1955), 442, and Steinmetz (1964), 55, identify the two passages and then use Olympiodorus' testimony as a proof of the incompleteness of the *Syriac meteorology*. Daiber (1992), 276 & 283–284, on the other hand, suggests that Olympiodorus may have added material from Theophrastus' lost Περὶ ὑδάτων.

185 Epicurus on cloud formation (*Pyth*. 17A): Νέφη δύναται γίνεσθαι καὶ συνίστασθαι καὶ παρὰ πιλήσεις ἀέρος πνευμάτων συνώσει ..., and on the production of rain (*Pyth*. 17B): "Ἤδη δ' ἀπ' αὐτῶν ᾗ μὲν θλιβομένων ..., and Lucretius on cloud formation (VI 463–464) 'venti / portantes *cogunt* ad summa cacumina montis', and on the production of rain (VI 510–512): 'nam vis venti *contrudit* et ipsa / copia nimborum turba maiore coacta / *urget* et e supero *premit* ac facit effluere imbris'.

186 Also observed by Montserrat & Navarro (1991) 301 with n. 78.

187 Mansfeld (1992a), 326–327.

(14) Neither the thunderbolt (pl.) nor anything that has been mentioned has its origin in God. For it is (15) not correct (to say) that God should be the cause of disorder in the world; nay, (He is) the cause (16) of its arrangement and order. And that is why we ascribe its arrangement and order to God (17) [mighty and exalted is He!] and the disorder of the world to the nature of the world.

tr. DAIBER (1992)

The first part of this remark is not unlike Epicurus. The sentence: "For it is not correct to say that God should be the cause of disorder in the world" is, as Mansfeld observes,[188] very close to what Epicurus says in his *Letter to Menoeceus*, 134: "for nothing is done by a god in a disorderly way" (οὐθὲν γὰρ ἀτάκτως θεῷ πράττεται). However, what the Syriac meteorologist goes on to say is very un-Epicurean. Whereas he leaves God in charge of everything orderly in the world, the Epicureans explicitly deny the gods' involvement in orderly and disorderly matters alike.[189] The *Syriac meteorology* repeats its un-Epicurean position in the concluding part of the excursus (25–29).

In sum, the following aspects of the *Syriac meteorology* seem to exclude an Epicurean origin of the text:

(i) Section 6.36–67 accounts for the downward motion of thunderbolts, which in Epicurean physics does not need to be accounted for.
(ii) Section 6.41–48 appeals to the Peripatetic theory of natural place which is incompatible with Epicurean physics.
(iii) Section 6.41–48 and chapter 13 appeal to the Peripatetic theory of the two vapours, which is never found in Epicurus and Lucretius. In the first of these two passages the two vapours are invoked to account for the formation of clouds. Surprisingly, in ch. 7 which is entirely devoted to the subject of clouds the interaction of the two vapours plays no role at all.
(iv) In chapter 7 (lines 6 and 29) the *Syriac meteorology* ascribes cloud formation and rain production to cooling and compression, just as Theophrastus does in fr.211B FHS&G. Yet, cooling does not feature among the several explanations offered by Epicurus and Lucretius.

---

188 Mansfeld (1992a), 325–326.
189 Mansfeld (1992a), 325. Cf. Epic. *Hdt.* 76–77; Lucr. *DRN* II 1090–1104; ibid. V 1183–1193; 1204–1210; Cic. *ND* I 52.

(v) In the theological excursus (14.14–29) the Syriac meteorologist, like Epicurus and Lucretius, denies God's responsibility for disorder in the world, but, unlike Epicurus and Lucretius, leaves God in charge of everything orderly (lines 15–17 and 27–29).

Two of these objections (ii and v) are positively fatal to the assumption of an Epicurean origin of the *Syriac meteorology* (or at least the pertinent portions of it), a conclusion strengthened by the three remaining objections. A purely Epicurean origin of the entire document must therefore be rejected. It is possible, however, that the *Syriac meteorology* is a compendium of some sort, derived for the most part from Epicurus' meteorology, but supplemented and 'corrected' on the basis of other, possibly Peripatetic or even specifically Theophrastean theories. This hypothesis would account for the overall Epicurean, rather than Theophrastean, character of the work, and also for the curious fact that those features which object most strongly to an Epicurean origin are concentrated in just a few passages.

It would seem, then, that we are left with two possible scenario's: either the *Syriac meteorology* is, as the *communis opinio* would have it, a version of Theophrastus' lost *Metarsiology* (allowance made for a certain amount of shortening, omissions, transpositions and perhaps also additions), or it is a compendium largely, but not exclusively, dependent on Epicurean meteorology, into which certain peripatetic elements have been incorporated. How such a compendium came to be transmitted under the name of Theophrastus I could not say, but there is certainly nothing exceptional about such false ascriptions.[190]

3.5.4 *Lucretius, Epicurus and the* Syriac Meteorology

As we have seen above, Lucretius *DRN* VI 96–607 and the *Syriac meteorology* resemble each other to such a degree, that it is hard to avoid the conclusion that they must be very closely related. Yet, we have also seen that the relevant portion of *DRN* VI probably derives directly from a more extensive account of meteorology by Epicurus, which Epicurus himself later summarized in his *Letter to Pythocles*. If this is true, the close kinship between Lucretius VI and the *Syriac meteorology* (whichever view we take of this work's identity) must run via Epicurus' more extensive meteorology.

---

[190] In the Arabic tradition think e.g. of "Aristotle's" *Theology* and *Liber de causis*, which go back to works by Plotinus and Proclus respectively. See D'Ancona (2015), chapters 2.1 'The pseudo-Theology of Aristotle and the Liber de causis' & 2.2 'Other pseudepigrapha'.

In the previous section two hypotheses about the status of the *Syriac meteorology* have been proposed. If the *Syriac meteorology* is, as is commonly believed, a—possibly shortened and garbled—version of Theophrastus' *Metarsiology*, then this work of Theophrastus' is likely to have been the immediate source for Epicurus' more extensive meteorology,[191] and thus an indirect source for Lucretius *DRN* VI 96–607 and Epicurus' *Letter to Pythocles* 17–30.

If, on the other hand, the *Syriac meteorology* is a compendium largely based on Epicurean meteorology, its most likely source is Epicurus' more extensive account of meteorology, on which *DRN* VI 96–607 and Epicurus' *Letter to Pythocles* 17–30 depend as well.

The two possible scenarios for the relations between Epicurus, Lucretius and the *Syriac meteorology* are illustrated below. Texts that are no longer extant are bracketed; arrows indicate the influence these works may have exercised on each other *with respect to their structure and order of subjects* (they are not meant to exclude the possibility of other, external, sources for particular problems and theories).

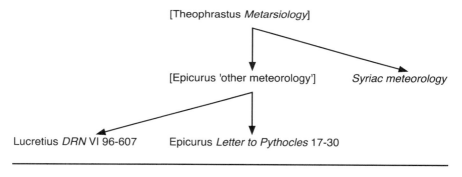

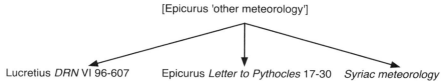

FIGURE 3.1 *Possible relations between Epicurus, Lucretius and the* Syriac meteorology: *scenario 1 (above) and scenario 2 (below)*

---

191 So Podolak (2010) 56, 64. On the other hand, Sedley (1998a), 182, while acknowledging the close similarity between Lucretius VI and the *Syriac meteorology*—which he believes to be Theophrastus' *Metarsiology*—nevertheless maintains that not this work but the corresponding section of Theophrastus' *Physical Opinions* was Epicurus' and therefore Lucretius' ultimate principal source.

3.5.5 Aëtius' Placita *and Theophrastus'* Physical Opinions

This leaves us with only one more work whose relation to the rest needs to be established. Above we have seen that the original sequence of subjects which seems to underlie both Aëtius III and the three works mentioned above, is best preserved by Aëtius. He alone, like Aristotle before him, includes the Milky Way, comets and shooting stars among atmospherical phenomena, and he alone preserves the original, Aristotelian, sequence *wind—earthquakes* (even though several chapters on the earth as a whole have been interposed). Yet, it is also clear, and not just for chronological reasons, that Aëtius III could not have been the original source from which the other three works derive their order and structure. It is more probable that it derives its order from the same source as the other three works, but independently.

An attractive candidate for this common origin would be Theophrastus' *Physical opinions*,[192] a doxographical work from which Aëtius' *Placita* is commonly believed to be ultimately derived,[193] and from which Epicurus is often assumed to have culled most of his alternative explanations in astronomy and meteorology (see p. 58 ff. above).[194]

It should be noted, however, that the evidence for Theophrastus' *Physical opinions* is slight. Only a handful of fragments remain,[195] which do not permit any strong conclusion about the work's structure and order of subjects. If the *Syriac meteorology* really is, as is commonly believed, a version of Theophrastus' *Metarsiology*, then we can safely assume that the corresponding portion of the *Physical opinions*, by the same author, had roughly the same order of subjects, which, therefore, it could have transmitted to Aëtius III. Yet, as I have shown, the attribution of the *Syriac meteorology* to Theophrastus is by no means certain, and neither, therefore, is the structure and order of subjects of Theophrastus' *Physical opinions*. Its relation to Aëtius' *Placita* and Epicurus' meteorological works is, therefore, at best conjectural.[196]

---

192   On Theophrastus' work being called Φυσικαὶ δόξαι (*Physical opinions*) rather than Φυσικῶν δόξαι (*Opinions of the physicists*) see Mansfeld (1990) 3057–3058 n. 1 and id. (1992b) 64–65.

193   Diels (1879) 102 ff. For an overview of Diels' views on the matter see e.g. Burnet (1892) 33–36; Regenbogen (1940) cols. 1535–1539; and, more critically, Mansfeld & Runia (1997) 78–79.

194   For Epicurus' dependence on Theophrastus' *Physical opinions* see Usener (1887) xl–xli. Cf. also Sedley (1998a) 166–185.

195   The collections of fragments made by Usener (1858); and Diels (1879) 473–495 are too inclusive: see e.g. Regenbogen (1940) 1536.68–1537.14; Steinmetz (1964) 334–351; Runia (1992) 116–117.

196   For a more cautious view concerning the relation between Aëtius' *Placita* and Theophrastus' *Physical opinions* see e.g. Mansfeld (1989) esp. 338–342; id. (1992b); id. (2005).

3.5.6   *Summary*

Above we have explored the possible relationships between Epicurus' *Letter to Pythocles* (17–30), Lucretius' *DRN* VI (96–607) and the *Syriac meteorology* as regards their scope and order of subjects. We have also, tentatively, indicated how Aëtius III may relate to the other three works. It is now time to bring together the main threads of the argument and summarize our findings:

1. Both Lucretius' *DRN* VI 96–607 and Epicurus' *Letter to Pythocles* 17–30 derive their scope and order of subjects from a more extensive account of meteorological phenomena by Epicurus, now lost, that may have been part of book XII or XIII of his *On nature*.
2. The *Syriac meteorology* is either [scenario 1] a—possibly shortened and garbled—version of Theophrastus' *Metarsiology*, which Epicurus all but reproduced as his own in his more extensive meteorological account, and from which he adopted the use of multiple explanations, which he subsequently extended to astronomical phenomena as well; or [scenario 2] it is a compendium largely based on Epicurus' more extensive meteorological account, in which case the use of multiple explanations may well have been Epicurus' own invention, and another source for the order of subjects and the individual explanations in Epicurus' meteorological work must be assumed.
3. In scenario 1, Theophrastus' *Metarsiology* is the most likely direct source of Epicurus' more extensive meteorology, with Theophrastus' *Physical opinions* as a possible secondary source.[197] In scenario 2, there is no reason to assign any role to Theophrastus' *Metarsiology* (which in this case we know hardly anything about) and Theophrastus' *Physical opinions* may well have been Epicurus' primary source.
4. In both scenarios, Theophrastus' *Physical opinions*, from which Aëtius' work is believed to be derived, is the most likely link between Aëtius on the one hand and Epicurus, Lucretius and the *Syriac meteorology* on the other.

The two scenarios are illustrated below. Texts that are no longer extant are bracketed; arrows indicate the influence these works may have exercised on each other *with respect to their structure and order of subjects*, and dashed arrows indicate a possible alternative or additional influence.

With respect to the two scenarios presented in figure 3.2 a number of remarks and reservations need to be made:

---

197   See n. 191 above.

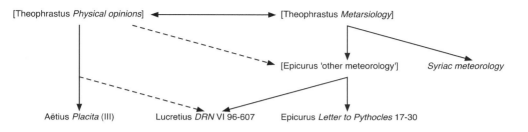
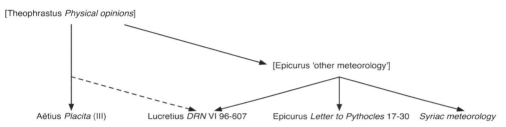

FIGURE 3.2 *Possible relations between Aëtius, Epicurus, Lucretius and the* Syriac meteorology: *scenario 1 (above) and scenario 2 (below)*

1. It must be borne in mind that the two schemas only indicate the major influences the works involved may have exercised on each other *with respect to the scope and order of subjects*, and must not be read so as to exclude other, possibly external, sources for particular problems and theories.
2. There is one aspect which both of the above scenarios fail to explain. Whereas the *Syriac meteorology* and Epicurus' *Letter to Pythocles* 17–30 restrict themselves to the explanation of atmospherical phenomena (and earthquakes), Aëtius III and Lucretius' *DRN* VI both continue their accounts with a number of terrestrial problems, one of which is common to both works, viz. the summer flooding of the Nile, while two other accounts, viz. those on the sea and on waters are at least thematically related (see table 3.5 on p. 136 above). This is sometimes taken as an indication that Lucretius drew directly upon the doxographical tradition as well (hence the dashed arrows in the illustrations above).[198] This assumption, however, still leaves the majority of the—mostly exceptional and local—phenomena in the latter part of Lucretius' book VI unaccounted for. Their inclusion may have been, as I suggested above (§ 3.3 on p. 109 ff.), a personal innovation by Lucretius in answer to the increasing popularity of such marvel stories in the paradoxographical as well as meteorological literature of his time.

---

198 Cf. Runia (1997a), esp. pp. 98–99.

3. It may be argued that the two scenarios are overly simplistic, and in a way they are: for as soon as we admit other sources for particular problems and theories beside the ones proposed above (see points 1 and 2), we must also admit the possibility that these other sources influenced the scope and order of subjects as well. On the other hand, although this possibility cannot be absolutely ruled out, it must also be observed that no other sources, beside the ones suggested above, are needed to account for the similarities in the scope and order of subjects of the four works involved.

3.6 Conclusions

In this chapter Epicurus' *Letter to Pythocles* and the sixth book of Lucretius' *DRN* have been compared to a number of other ancient meteorological writings as to the range and order of their subject matter. One of these writings was the Syriac meteorology commonly ascribed to Theophrastus, which in many ways resembles the strictly meteorological portions of the *Letter to Pythocles* and *DRN* VI. The inclusion of a large number of local marvels in the later part of *DRN* VI also led me to compare Lucretius with a number of paradoxographical works, while a comparison of the order of subjects of the *Letter to Pythocles*, the *DRN*, Aëtius III and the *Syriac meteorology* provided an occasion to examine the degree and nature of their mutual relations, and to re-examine the claims about the authorship of the *Syriac meteorology*. The most important findings of this chapter are the following.

The range of subjects covered by the *Syriac meteorology*, seems to be complete (except perhaps for an account of the rainbow). The absence of a passage dealing with comets and shooting stars, though perhaps surprising in a meteorological work ascribed to a pupil of Aristotle's, has a parallel in Lucretius' account of meteorological phenomena in book VI of the *DRN*. Epicurus too, in his *Letter to Pythocles*, seems to have associated comets and shooting stars with the stars rather than with atmospheric phenomena. In this respect, then, the Syriac meteorology is closer to the two Epicurean accounts than to Aristotle's. Another interesting feature Epicurus, Lucretius and the Syriac meteorology have in common, is the inclusion of earthquakes with atmospherical phenomena. Although in this respect they do not seem to differ from Aristotle's and several other meteorological accounts, the crucial difference is that in the latter accounts the inclusion of earthquakes is warranted by their close association with atmospherical winds, whereas in the three former works causation by wind is only one of several alternative explanations. This suggests that the inclusion of earthquakes among atmospherical phenomena in these three works was inherited from a tradition, initiated by Aristotle, whose principles they no longer endorsed.

The later part of *DRN* VI differs from the earlier part as well as from corresponding sections of other meteorological works in predominantly discussing particular local phenomena (marvels). Although other meteorologies too sometimes refer to such phenomena, as a rule these are not themselves the objects of inquiry but only serve to convey some general point. The only notable exception is the summer flooding of the Nile which had since long been the object of physical speculation, and was sometimes included in discussions of meteorological phenomena. However, particular local phenomena are more typically found in works of the literary genre known as paradoxography, which is not about physical inquiry at all, but simply contents itself with recounting marvellous stories. In *DRN* VI Lucretius deals with such phenomena in a way that is different from the approach in other meteorological as well as paradoxographical works, by making them the principal objects of inquiry rather than simply listing them. For chronological reasons it is unlikely that Lucretius' treatment of these phenomena derives from Epicurus. It seems more likely that the passage on marvellous phenomena was added by Lucretius himself, either of his own initiative or prompted by an intermediate Epicurean source.

Four meteorological texts, viz. Lucretius *DRN* VI, Epicurus' *Letter to Pythocles* (17–30), the *Syriac meteorology*, and book III of Aëtius' *Placita*, exhibit a remarkable similarity in the order of their subjects. On the basis of a detailed comparison I have made an attempt at reconstructing an original order, from which each of these four texts could have somehow, directly or indirectly, derived its own order. In addition I have tried to chart the mutual relations of these works. It is very likely that the meteorological portions of Epicurus' *Letter to Pythocles* and Lucretius' *DRN* VI both go back to a more extensive account of meteorology by Epicurus, probably part of his magnum opus *On nature*. It seems probable that this more extensive meteorological account is in turn related to the *Syriac meteorology*. However, our view of the nature of this relation depends on our position with respect to the identity of the latter work. For this reason I submitted the *Syriac meteorology* to some closer scrutiny. In addition to its use of multiple explanations (see previous chapter), certain details, such as its treatment of the *prēstēr*, sit more easily with the assumption of an Epicurean, rather than Theophrastean, origin. A wholly Epicurean origin of the *Syriac meteorology* is, however, precluded by the presence in this work of certain unmistakably Aristotelian and Theophrastean theories, such as natural place, the two exhalations, the view that condensation may be due to both pressure and cooling, and that order is attributable to the gods. It is remarkable, however, that these

Peripatetic theories are concentrated in just a few passages and not consistently implemented throughout the work. If, therefore, we take the view that the *Syriac meteorology* is Epicurean in origin, we must at least admit that it also contains some Peripatetic additions or modifications. This leaves us with two possible views of the nature of the *Syriac meteorology*, and hence of its relation with Epicurean meteorological writings. If we accept its commonly accepted identification with Theophrastus' *Metarsiology*, it seems likely that this work was the main source for Epicurus' more extensive account of meteorology. If, on the other hand, we take the view that the *Syriac meteorology* is not Theophrastus' *Metarsiology*, but a compendium of mainly Epicurean meteorology, there is no reason to assign any special role to Theophrastus' meteorological treatise in the formation of Epicurean meteorology. A final problem that needed to be addressed is the relation of Aëtius' doxographical work to the *Syriac meteorology*, Epicurus' *Letter to Pythocles* and book VI of Lucretius' *DRN*. It is clear that, as far as the order of subjects is concerned, all four works must go back, directly or indirectly, to a single archetype. In view of the presumed doxographical origin of most the alternative theories presented in the latter three works (see also pp. 58 ff. above) it seems likely that this common origin was itself a doxographical work, which may perhaps be identified with Theophrastus' *Physical opinions*, a work from which Aëtius' doxography is commonly supposed to be ultimately derived.

3.7 Epilogue: Epicurean Cosmology and Astronomy

In Epicurus' *Letter to Pythocles* the study of τὰ μετέωρα—"lofty phenomena"—comprises not just atmospherical, but also cosmological and astronomical problems. Both with respect to the completeness of our coverage of τὰ μετέωρα and in view of next chapter's subject—the shape of the earth—it would, therefore, seem logical to extend the present chapter's exercise to cosmological and astronomical problems as well. However, whereas there was ample material for our investigation into the range and order of subjects in works of ancient meteorology, a similar exploration of ancient astronomical and cosmological writings has much less to go on. Not only is there is no astronomical counterpart to the *Syriac meteorology*, to which the Epicurean accounts might be fruitfully compared, but even Aëtius' *Placita*, whose third book, on atmospherical and terrestrial problems, proved to be such a close parallel to the corresponding portions of Lucretius' and Epicurus' works, shows but little agreement when it comes to the range and order of cosmological and astronomical subjects. Also other ancient astro-cosmological accounts such as Aristotle's *De caelo*, or

Cleomedes' *Caelestia* show but a small degree of correspondence. For this reason the present study has been confined to meteorology proper.

CHAPTER 4

The Shape of the Earth

4.1   Introduction

In Plato's *Phaedo* (97d), which claims to report the conversations held by Socrates on the last day of his life in 399 BC, Socrates first tells us how, as a young man, about to start reading Anaxagoras, he expected to be told, among other things, whether the earth is flat or round. Later in the same dialogue (108e, 110b) Socrates states his present conviction that the earth is shaped like a sphere. While we cannot be certain that Socrates really thought any of these things, it is clear that Plato himself at least was convinced of the earth's sphericity, as were most of his Greek contemporaries and those who came after; the last Greek philosopher reported to have advocated a flat earth is Democritus, who died around 370 BC. From this date onwards the earth's sphericity, supported by an ever increasing amount of evidence, was accepted by all.

Well, perhaps not all. It is often claimed that, in spite of all the evidence, Epicurus (341–270 BC) and his Roman follower Lucretius (99–55 BC) rejected the earth's sphericity and stubbornly clung to the antiquated concept of a flat earth,[1] a view that has earned them such qualifications as "singularly behind the time",[2] "alarmingly retrograde",[3] "queerly indifferent to scientific truth",[4] "scarcely [deserving] to be mentioned in the history of science",[5] "ludicrous"

---

1   The Epicureans' rejection of the earth's sphericity or, conversely, their advocacy of a flat earth is stated for a fact by, for instance, Dreyer (1906) 171–172; Thomson (1948) 167–168; Rist (1972) 47; Sedley (1976) 49; Schmidt (1990) 33, 215; Jürss (1994) 246; Furley (1996) 119; Sedley (1998b) 346; Furley (1999) 420–421, 429; Milton (2002) 184; Chalmers (2009) 52; Sedley (2013) 'Lucretius' in *The Stanford Encyclopedia of Philosophy*; Konstan (2014) 'Epicurus' in *The Stanford Encyclopedia of Philosophy*; and the article 'Flat Earth' in *Wikipedia, The Free Encyclopedia* (30 October 2015, 03:04 UTC), accessed 30 October 2015 https://en.wikipedia.org/w/index.php?title=Flat_Earth&oldid=688172248.
2   Dreyer (1906), loc. cit.
3   *Thriceholy*, 'Flat Earth and the Ptolemaic System', accessed 30 October 2015 http://thriceholy.net/desert.html#Flat%20Earth.
4   Thomson (1948), loc. cit.
5   Dreyer (1906), loc. cit.

(a pun on the name 'Lucretius'),[6] and "sadly un-Greek".[7] By contrast, I know of only one serious attempt to defend the Epicureans. In a brief paper, published in 1996,[8] David Furley sets out to answer three questions:

1. Why did the Epicureans hold on to the claim that the earth is flat?
2. Were they familiar with contemporary astronomy?
3. Did they know of its arguments and put up a reasoned defence of their own position?

In answer to the first question Furley points out, firstly, that a flat earth follows inevitably from the Epicurean theory of atomic motion, and, secondly, that this shape also naturally commended itself to the Epicureans, who set such a high value on perception: the earth looks flat, and therefore we must start from the assumption that it is flat. With respect to the second question Furley refers to the fragments of Epicurus' *On nature*, where Epicurus engages critically with the methods and pretentions of contemporary astronomy. In answer to the third question Furley argues that another controversial Epicurean theory, about the sun and the other heavenly bodies being as small as they appear to us, may have been devised partly in order to reconcile the Epicureans' flat-earth cosmology with Aristotle's observation that the aspect of the sky changes with latitude.

Although this attempt to defend or at least to better understand Epicurus' views must be applauded, Furley, like so many other scholars, omits to mention one important point: nowhere in Epicurus' remaining works and fragments, and nowhere in Lucretius' *De rerum natura* do we find any explicit statement about the shape of the earth.[9] What evidence there is, is at best circumstantial and not without some ambiguity. Even Epicurus' ancient critics, who were always ready to make fun of him—e.g. for his claim that the sun is as big as it seems, or for seriously considering the possibility that the sun might be extinguished at night—, remain silent on this point. A more thorough investigation of Epicurus' views concerning the shape of the earth should therefore start with

---

6 Ethical Atheist, *The Flat Earth*, 'Chapter 5: Analysis of 7000 Years of Thinking Regarding Earth's Shape', posted 26 June 2007, accessed 30 October 2015 http://atlantisonline.smfforfree2.com/index.php?topic=1792.0.
7 Thomson (1948), loc. cit.
8 Furley (1996). See also Furley (1999) 420–421, 429.
9 As is rightly observed by Munro (1864) vol. 2, p. 341 ad *DRN* V 534; Woltjer (1877) p. 123 ad *DRN* V 534ff.; Ernout-Robin (1925–1928) vol. 3, p. 72 ad *DRN* V 534–536; Bailey (1947), vol. 3, p. 1403 ad *DRN* V 534ff. Taub (2009) 114–115 (referring to Conroy (1976) 110).

a survey of the evidence for attributing a flat-earth cosmology to the Epicureans. In the commentaries this attribution is usually inferred from such passages as Epicurus *Hdt.* 60, Lucretius' *DRN* I 1052 ff. and V 534–563, in which a parallel natural motion is argued for, or implied. Although it is hard to see how a parallel motion could be combined with anything other than a flat earth, it is methodologically wiser to distinguish such passages from those where the shape of the earth can be directly inferred from astronomical observations or theories. Another limitation of Furley's paper is that it focuses exclusively on one piece of astronomical evidence, viz. the changing aspect of the sky when one moves to the north or to the south, while in fact several such proofs were known in antiquity. One wonders how Epicurus and Lucretius would or could have dealt with those.

In this chapter I propose to conduct a thorough investigation into the views of the Epicureans concerning the shape of the earth, their motivation for these views and their attitude towards the relevant astronomical theories and proofs. Based on a number of passages from Epicurus, Lucretius and other Epicureans, I will try to answer the following questions:

1. What natural motion did the Epicureans assign to the atoms and bodies in general?
2. Does this natural motion imply a flat earth?
3. Do Epicurean astronomical views presuppose a flat earth?
4. Why (if they did) did the Epicureans hold on to the claim that the earth is flat?
5. Were they familiar with contemporary astronomy?
6. Did they know of its arguments for the earth's sphericity and put up a reasoned defence of their own position?

I will go about this investigation as follows: first, in §4.2, I will deal with some preliminary issues that may serve as a background to the investigation. Then, in §4.3, I will discuss a number of passages from Epicurus, Lucretius and other Epicureans that are (or may seem to be) relevant to one or more of our questions. I will take my lead from Lucretius, for two reasons. In the first place Lucretius provides most of the evidence, while such evidence as is furnished by other sources is mostly parallel to what we find in Lucretius (and will be so presented). Secondly, following Lucretius allows us to read most of the relevant passages as part of one continuous account, which may provide additional clues as to the underlying argument. Finally, in §4.4, I will summarize and combine my findings and try to present a balanced answer to the questions formulated above.

4.2 Historical and Conceptual Context

4.2.1 *The Shape of the Earth in Antiquity: A Historical Overview*

The world picture that arises from the earliest works of Greek literature, the epics of Homer and Hesiod, is fairly simple.[10] The earth is a flat disk, encircled by the waters of Ocean, with the heavens as an inverted bowl above, and the Tartarus in a corresponding position below.[11] The views of Thales of Miletus (624–547), the first philosopher, were still relatively close to the archaic view. According to Thales, the earth is a disk floating like a log or a vessel on a primordial sea.[12] If our reports are true, the first important step towards a fundamentally different worldview was made by Thales' younger compatriot Anaximander (610–546).[13] While sticking to the image of a flat earth (a column drum or a cylinder),[14] he dispensed with the need for an underprop, such as Thales' water. Instead he claimed that the earth, being equably related to every portion of the surrounding heavenly sphere, has no reason to move in any direction, and therefore does not fall down.[15] If this account of Anaximander's views is correct,[16] he did not convince his successor Anaximenes (584–526), who returned to the old view of an earth in need of support: according to Anaximenes, the earth is flat like a table, and rides upon the air on account of its flatness.[17]

It was at the other end of the Greek world, in southern Italy, that the idea first arose that the earth might be a sphere, an idea variously attributed to Pythago-

---

10 On the history of ancient astronomy in general, see e.g. Heath (1913) 7–129; id. (1932) xi–lv; Dicks (1970), Evans (1998). On the shape of the earth in particular, see e.g. Thomson (1948) 94–122; Evans (1998) 47–53.

11 Dicks (1970) 29–30; Heath (1913) 7; id. (1932) xi; Furley (1989a) 14. Cf. also Geminus *Isagoge* 16, 28.

12 Arist. *Cael.* II 13, 294a28–32 [fr.A14 D–K] and Sen. *NQ* III 14 [fr.A15 D–K]. The ascription to Thales of a spherical earth in Aët. III 10, 1 cannot be correct, because (1) it is incompatible with its floating, (2) it is anachronistic with respect to his successors' views, and (3) it is contradicted by the explicit claims that either Pythagoras or Parmenides was the first to make the earth spherical (see n. 18 below).

13 For a detailed account of Anaximander's theory see Kahn (1960) 76–81.

14 Hippol. *Ref.* I 6.3 [fr.A11 D–K]; Aët. III 10, 2 [fr.B5 D–K]; Ps.-Plut. *Strom.* 2 [fr.A10 D–K].

15 Arist. *Cael.* II 13, 295b10–16 [fr.A26 D–K]; Hippol. *Ref.* I 6, 3 [fr.A11 D–K].

16 Furley (1989a), 17–22, is inclined to think that the ascription of this theory to Anaximander is anachronistic.

17 Arist. *Cael.* II 13, 294b13–30 [fr.A20 D–K]; Hippol. *Ref.* I 7, 4 [fr.A7 D–K]; Ps.-Plut. *Strom.* 3 [fr.A6 D–K]; Aët. III 10.3 [fr.A20 D–K]; Aët. III 15.8 [fr.A20 D–K].

ras (ca. 580–500) and Parmenides (ca. 500).[18] We are not told what brought them to this idea, but they may have been inspired by Anaximander's equilibrium theory, which, according to our reports, was also held by Parmenides.[19] Perhaps they thought that if the earth was equably related to every portion of the surrounding heavenly sphere, the earth's shape too should be so related, i.e. spherical. The idea of a spherical earth remained current in subsequent centuries among Pythagoras' followers,[20] and it is quite likely that the sphericity of the earth was also accepted by Empedocles (ca. 490–430), who is said to have been a follower of both Pythagoras and Parmenides,[21] and who in one of his fragments testifies to a *centrifocal* conception of up and down (see § 4.2.4 below).[22]

In the meantime the eastern Greeks still adhered to the traditional view. The historian Herodotus (484–425) probably believed the earth to be flat,[23] and

---

18 Diog. Laërt. VIII 48 [Parmenides A44 D–K]. Cf. Diog. Laërt. VIII 25 [Pythagoristae A1a D–K], and IX 21 [Parmenides A1 D–K].
19 Aët. III 15.7 [Parmenides A44 D–K]. See also Kahn (1960) 79–80.
20 Arist. *Cael.* II 2, 285b22–27 ascribes to the Pythagoreans the view that the north pole and the inhabitants of the northern hemisphere are above, whereas the south pole and the inhabitants of the southern hemisphere are below, and Arist. *Cael.* II 13, 293b25–30 describes how the Pythagoreans defend their claim that the earth is not at the centre by pointing out that even on the supposition of a centrally placed earth, its inhabitants are not at the centre but half the diameter away from it, which nevertheless doesn't seem to affect our observations of the heavens. Both these theories presuppose a spherical earth. See also Dicks (1970) 72–73.
21 Diog. Laërt. VIII 54–56.
22 Fragment B35 D–K, lines 3–4: [...] ἐπεὶ Νεῖκος μὲν ἐνέρτατον ἵκετο βένθος / δίνης, ἐν δὲ μέσῃ Φιλότης στροφάλιγγι γένηται, [...] = 'when Strife reached the *lowest depth* / of the vortex, and Love came to be in the *middle* of the whirl ...' ('the lowest depth' and 'the middle' seem to denote *the same place*), and lines 9–10: ὅσσ' ἔτι Νεῖκος ἔρυκε μετάρσιον· οὐ γὰρ ἀμεμφέως / τῶν πᾶν ἐξέστηκεν ἐπ' ἔσχατα τέρματα κύκλου = 'all the things which Strife retained *up high*: for it had not (yet) altogether retreated perfectly from them to the *outermost boundaries of the circle.*' ('up high' and 'the outermost boundaries of the circle' seem to indicate the same thing). On the other hand, in Aëtius II 8 concerning the inclination of the cosmic axis with respect to the flat earth, Empedocles is mentioned alongside the flat-earthers Anaxagoras and Diogenes of Apollonia (see p. 249 with n. 206 below).
23 Furley (1989a) points to Herodotus' report in III 104 that in India the hottest time of the day is in the morning, and that from then on the temperature steadily drops until at sunset it is extremely cold. This, according to Furley, indicates a flat earth, where in the east the sun arrives vertically overhead very soon after sunrise. Another proof may be found in IV 42 where Herodotus, reporting on the circumnavigation of Africa, expresses disbelief at the sailors' claim that during their westward passage around the southern portion of the

Anaxagoras of Clazomenae (500–428) and Diogenes of Apollonia (ca. 460) both adopted Anaximenes' view of a flat earth floating on air.[24] They were followed in this respect by the atomists from Abdera, Leucippus (early 5th century) and Democritus (ca. 460–370).[25] By this time, however, word of the new theory had reached Athens, where for a time both theories existed side by side. In Plato's *Phaedo* (97d) we hear that young Socrates (470–399), upon taking in hand a volume of Anaxagoras', expected to be told, among other things, whether the earth is flat or round.[26]

In the end the second view won the day. In the same dialogue (108e–109a) Plato (427–347) has Socrates, much older now, explain that he has been convinced by someone that the earth is a sphere and remains where it is, because, being placed equably in the centre of heaven, it has no reason to move in this direction rather than that—the same set of theories that was also attributed to Parmenides. From then on the idea of a spherical earth spread rapidly. In fact, there is no explicit information of anyone later than Democritus propounding a flat-earth cosmology.[27] Even within the circle of his followers, there seems to have been some dissent. If we may believe the scanty information provided by Diogenes Laërtius,[28] a certain Bion (early fourth century BC?), who 'was a follower of Democritus and a mathematician, from Abdera, {...} was the first to say that there are regions where the night lasts six months and the day six months.' The only regions for which this statement applies are the north and south poles, and Bion could only have arrived at such a claim on the basis of a firm understanding of all the implications of a spherical earth.[29] Also Eudoxus (408–355), the leading mathematical astronomer of his day, probably accepted the earth's sphericity. This much at least can be inferred from his theory about the Nile flood (which he generously attributed to the priests of Heliopolis), viz.

---

continent they had the sun on their right, i.e. to the north. With the image of a spherical earth before him, Herodotus would have had no reason for doubting the sailors' statement.

24 On Anaxagoras see Arist. *Cael.* II 13, 294b13–30; Hippol. *Ref.* I 8, 3 [fr.A42 D–K]; Diog. Laërt. II 8 [fr.A1 D–K]; Exc. astron. cod. Vatic. 381 (ed. Maass *Aratea* p. 143) [fr.A87 D–K]. On Diogenes see *Schol. in Basil. Marc.* 58 [fr.A16c D–K] and Aët. II 8.1.

25 On Leucippus see Aët. III 10.4 [fr.A26 D–K]; Diog. Laërt. IX 30 [fr.A1 D–K]. On Democritus see Arist. *Cael.* II 13, 294b13–30; Aët. III 10.5 [fr.A94 D–K].

26 See also p. 162 above.

27 That is, until around 300 AD, when the Christian writer Lactantius (*Div. Inst.* III 24) rejected the spherical earth on account of its incompatibility with the Holy Scripture.

28 Diog. Laërt. IV 58.4–6. See Hultsch (1897) 485–487; Abel (1974) 1014; Jürss (1994) 246–247.

29 For the theoretical background of Bion's claim see e.g. Achilles 35.23–38; Cleom. I 4, 219–231; Ptol. *Alm.* II 6, 116.21–117.9; Plin. *NH* II, 186–187.

that the Nile had its sources in the southern hemisphere where it is winter when we have summer,[30] a theory that presupposes a spherical earth. Moreover, his famous theory of concentric *spheres* on which the fixed stars and planets move also seems to presuppose a *spherical* earth at the centre of the cosmos.

To Aristotle (384–322) we owe the first thorough argument concerning the earth's shape.[31] Aristotle accounts for the earth's sphericity and its stable position in the centre of the cosmos on the assumption of a general centripetal tendency of all heavy matter. He also offers two astronomical proofs for the earth's sphericity—the circular shape of the earth's shadow during a lunar eclipse, and the changes in the sky's aspect when one moves to the north or the south—, and he informs us that mathematicians had calculated the earth's circumference, arriving at a number of 400,000 stades (≈ 72,000 km). Not much later the voyages of Pytheas of Marseille (ca. 325 BC) in the northern Atlantic, and Alexander's conquests of Egypt, Persia and parts of India (334–323) provided geographers and astronomers with a large body of new observations on which to base their theories and calculations. Especially Pytheas is important in this respect: he is the first person on record to have used the length of the shadow of the gnomon to determine latitude,[32] and also the first to record the maximum day length for a number of different latitudes (up to 24 hours in Thule on the arctic circle).[33] This, then, was the state of affairs when, around 300 BC, Epicurus (341–270) devised his (presumably) flat-earth cosmology.

---

30 Aët. IV 1.7 (Eudoxus fr.288 Lasserre).
31 Arist. *Cael.* II 14, 297a8–298b20; cf. also *Mete.* I 3, 340b35–36.
32 Strabo I 4, 4: '… for as to the ratio of the gnomon to its shadow, which Pytheas has given for Massilia, this same ratio Hipparchus says he observed at Byzantium, at the same time of the year as that mentioned by Pytheas.' [transl. Horace Leonard Jones (1917), modified] & Strabo II 5, 8: 'But if the parallel through Byzantium passes approximately through Massilia, as Hipparchus says on the testimony of Pytheas (Hipparchus says, namely, that in Byzantium the ratio of the gnomon to its shadow is the same as that which Pytheas gave for Massilia), …' [transl. Horace Leonard Jones, modified].
33 Gem. *Is.* VI 9 states that Pytheas reported on places where the night was extremely short, only 2 or 3 hours; Plin. *NH* II 186–187 ascribes to Pytheas the incorrect view that in Thule (six days sailing north of Britain) the day and the night each last 6 months (this is in fact only true of the geographical north and south poles), but at *NH* IV 104, apparently referring back to the previous passage but not mentioning Pytheas by name, he corrects his report, saying that in Thule *at the summer solstice* there is no night at all. Strabo II 5, 8 may be referring to the same observation when he states that according to Pytheas in Thule the arctic circle (i.e. the circle that comprises the ever visible stars) coincides with the summer tropic, a view that is astronomically equivalent to what Plin. *NH* IV 104 says.

THE SHAPE OF THE EARTH 169

But the story does not end here. At around the same time Zeno of Citium (333–264), whose followers—the Stoics—were to become the Epicureans' most fervent adversaries, opted for a spherical earth.[34] Not much later Eratosthenes of Cyrene (276–194) made his famous and remarkably accurate calculation of the earth's circumference.[35] Then, around 150 BC the Stoic scholar Crates of Mallos constructed the first terrestrial globe,[36] while Hipparchus (ca. 190–129) prescribed using simultaneously observed eclipses to establish differences in longitude.[37] Finally, during Lucretius' life-time (ca. 95–55), Posidonius (135–151), another Stoic scholar, made a new, very influential, estimate of the earth's circumference.[38] Nor was familiarity with the sphericity of the earth restricted to the Greeks: among the Romans too the theory had found currency. Among Lucretius' contemporaries both Varro (116–127) and Cicero (106–143) accepted the earth's sphericity for a fact.[39]

4.2.2 Ancient Proofs of the Earth's Sphericity

In his *De caelo* II 14, 297b24–298a10, as we just saw, Aristotle provides two astronomical proofs of the earth's sphericity. In subsequent centuries many more proofs were devised. Lists of proofs are provided by the Roman poet Manilius (before 14 AD), the Roman encyclopedist Pliny the elder (23–79 AD), the Greek philosopher Theon of Smyrna (ca. 70–135 AD), the Greek astronomer Ptolemy (ca. 85–165 AD) and the Greek philosopher Cleomedes (ca. 200 AD). One proof is also reported by the Greek geographer Strabo (64 BC – 24 AD).[40] The longest lists are those of Cleomedes (five proofs) and Pliny (six proofs).

---

34 Aët. III 10.1 [*SVF* II 648]; Diog. Laërt. VII 145.6–7 [*SVF* II 650]; Achilles *Isagoge* 4 [*SVF* II 555]; Cic. *N.D.* II 98.5–6 [not in *SVF*]. Cf. Diog. Laërt. VIII 48.10–12 [*SVF* I, 276] where Zeno claims Hesiod's authority for his own theory.

35 Cleomedes I 7, 49–110. For other ancient reports as well as modern literature on this measurement see Bowen-Todd (2004) ad loc.

36 Strabo II 5, 10.3–5.

37 Strabo I 1, 12: 'In like manner, we cannot accurately fix points that lie at varying distances from us, whether to the east or the west, except by a comparison of the eclipses of the sun and the moon. That, then, is what Hipparchus says on the subject.' [transl. Horace Leonard Jones]

38 Cleomedes I 7, 7–48. For other ancient reports as well as modern literature on this measurement see Bowen-Todd (2004) ad loc.

39 Varro *Men.* fr.516 Bücheler, apud Non. 333, 25 'in terrae pila'; Cic. *Rep.* VI 15 = *Somn. Scip.* 3, 7. Cf. Cic. *Tusc.* I 68.12 and *ND* II 98.5–6.

40 The relevant texts and passages are given in table 4.2 below.

TABLE 4.1   *Time-line of ancient theories on the shape of the earth*

Flat earth		Spherical earth
Homer Hesiod	800 BC	
	700 BC	
Thales Anaximander Anaximenes	600 BC	Pythagoras? Parmenides
Anaxagoras Herodotus Leucippus Democritus	500 BC	Empedocles? Pythagoreans
	400 BC	Bion of Abdera? Plato Eudoxus Aristotle
Epicurus?	300 BC	Pytheas Zeno of Citium Eratosthenes
	200 BC	Crates of Mallos Hipparchus
Lucretius?	100 BC	Posidonius Cicero Varro

Below I will discuss each of the ancient proofs, briefly explaining how it works, who reported it, when it may have been devised, and how—if at all—it might have been refuted by someone wishing to uphold the assumption of a flat earth:

1. According to Aristotle, the convex shape of the earth's shadow as it passes over the moon during a lunar eclipse proves that the earth must be spherical. In fact, however, such a shadow could be produced by many different forms, even by a flat, disk-shaped earth, as Furley observes.[41] For this reason, perhaps, this 'proof' was not repeated by subsequent authors. The argument depends, of course, on the assumption that the moon receives its light from the sun, and is sometimes robbed of this light by the interposition of the earth. Another possible way to escape the consequences of this argument would therefore be to deny that lunar eclipses are produced in this way.
2. Another proof, mentioned by Strabo, Pliny, Theon, Ptolemy and Cleomedes, is based on observations of, and from, approaching and departing ships. If a ship approaches land, then from the ship the mountain-peaks are seen first, and then gradually the lower-lying portions of the land seem to rise from the water. In the same way, from the land the top of the mast is seen first, and then gradually the rest of the ship appears to rise from the water. And when the ship departs then conversely the land and the ship are seen to sink under water. These effects can only be explained on the assumption that the surface of the water is curved. The oldest known report of this proof is provided by Strabo, who postdates both Epicurus and Lucretius. It is possible, therefore, that Epicurus and Lucretius were not familiar with this proof. If they had been familiar with it, and had wished to refute it, they might have dismissed such observations as mirages of the kind that is often observed above water.
3. As Aristotle observed,[42] and many others repeated afterwards, the position of the stars changes according to the observer's *latitude*. Some stars that are seen in the south are invisible in more northerly countries, and some stars that are continuously visible in the north, are seen to set and rise in the south. Such observations are commonly explained on the assumption

---

41 Furley (1999) 421.
42 Furley (1996), 121 (referring to Dicks (1970) pp. 87–88, who does not say so, and Vlastos (1975) pp. 38–40), claims that there is some evidence that this proof may have been known to Euctemon in the last half of the fifth century. Dicks (1970) n. 380 (referring to Eudoxus frs.75a+b Lasserre) suggests that Eudoxus may have been Aristotle's source for this proof. For our purposes it is enough to know that the formulation of this proof predates Epicurus.

that the earth's surface is curved from north to south. David Furley (1996) suggests that observations of this kind can be reconciled with a flat earth, if the heavenly bodies are assumed to be relatively close by. He compares this to the effect of someone walking in a big room under a painted ceiling: as he walks he will observe the same paintings from different angles. In fact, however, this model only explains part of Aristotle's observation: while it may account for the fact that observers at different latitudes see the same stars at different heights, it fails to explain how certain stars can disappear from sight altogether. Yet, even so the conclusion that the earth is spherical is not inevitable. The observation in question could be accounted for, at least in a qualitative sense, with reference to the unevenness of the earth's surface.

4. A special case of the previous proof, briefly mentioned by Pliny and discussed more fully by Cleomedes, concerns the height of the celestial north pole. Cleomedes observes that in northern countries, such as Britain, the celestial north pole is seen high above the northern horizon, but in southern regions, such as southern Egypt and Aethiopia,[43] appears low in the sky. These observations, he claims, can only be accounted for on the assumption that the earth's surface is curved from north to south. Although the use of such observations as proof of the earth's sphericity cannot be dated to anyone before the time of Pliny the elder, the essential ingredients were known to Hipparchus (ca. 190–120 BC),[44] and may well go back even further. Chronologically, therefore, there is no reason why Lucretius could not have known of the observations relating to this proof. The consequences of this proof are, however, easily dissipated on the assumption that the heavenly bodies, including the celestial north pole, are relatively close by, as Furley suggests.

5. Not just the position of the celestial north pole, but also the position of the sun at noon on a given day of the year varies with latitude. In our part of the world, the midday sun stands generally higher for observers in southern countries, and lower for observers in northern countries, and accordingly shadows are longer in the north and shorter in the south. This observation, which is another special case of proof 3, is presented as a further proof of the earth's sphericity by Pliny the elder. The relevant observations are much

---

43 Text and translation are (partially) quoted on p. 250 and p. 253 below.
44 In his *In Arati et Eudoxi Phaen.* I 3, 6–7, Hipparchus provides the average values for Greece of: (a) the ratio of the gnomon to its shadow at noon during the equinox (cf. proof 5), (b) the maximum day-length (cf. proof 6), and (c) the polar height (τὸ ἔξαρμα τοῦ πόλου) (cf. proof 4), thereby implying that these three values are different at other latitudes.

older, however. Pytheas of Marseille (ca. 325 BC) already used the length of the shadow of the gnomon to determine latitude.[45] Consequently, Epicurus and Lucretius could have known of these observations. However, just as with the previous proof, these observations *could* be reconciled with a flat earth on the assumption that the heavenly bodies, more specifically the sun, are relatively close by.

6. Another proof of the earth's sphericity is based on the observation that the maximum day length increases with increasing latitude. Although Pliny is the only author to present this observation as a formal proof of the earth's sphericity, the facts were widely known.[46] The first person on record to measure latitude by the maximum day length was, again, Pytheas of Marseille.[47] Chronologically, therefore, Epicurus and Lucretius could have known the facts which underlie this proof. In this case, however, there is no easy way to escape its consequences. As Cleomedes already observed, on a flat earth sunrise and sunset would be the same for everyone, and accordingly there would be no latitudinal variation of day length.[48] The only escape route I can think of is to contest the validity of the observations themselves: clocks in antiquity were notoriously inaccurate, and to compare times at different places was well nigh impossible.

7. The last proof I wish to address concerns the time difference between places lying at different longitudes. Pliny reports that on several occasions when warning-fires were alighted successively from west to east, the last one was observed to be alighted at a much later (local) hour than the first. (Pliny's argument would have been more convincing if the fires had been alighted from east to west, and the last one been observed to be alighted at an earlier local time than the first.) However, a much more secure way to establish time-differences between places consisted in simultaneous observations of lunar eclipses. Hipparchus had already prescribed this procedure as a means to accurately measure differences in longitude,[49] and the same procedure was cited as a proof of the earth's sphericity by Manilius, Pliny, Theon, Ptolemy and Cleomedes. Concrete examples are offered by Pliny and Ptolemy, who both refer to the lunar eclipse of September 20, 331 BC, which was observed at a certain hour near Arbela in Mesopotamia, sev-

---

45 See n. 32 above.
46 See e.g. Strabo II 5.38–42; Geminus *Isag.* VI 7–8, Cleom. II 1, 438–451; Ptol. *Alm.* II 6, Mart. Cap. VIII 877.
47 See n. 33 above.
48 Concerning this proof see also subsection 4.2.3 below.
49 See n. 37 above.

TABLE 4.2 *Ancient proofs of the earth's sphericity*

	Aristotle 384–322 BC *Cael.* II, 14	Manilius before 14 AD *Astron.* I	Strabo 64 BC–24 AD *Geogr.* I, 1
1. Convexity of the earth's shadow during a lunar eclipse	297b24–31		
2. Observations of, and from, departing and approaching ships			20.18–27
3. Aspect of the sky varying with latitude	297b31–298a10	215–220	
4. Elevation of the pole varying with latitude			
5. Length of shadows varying with latitude			
6. Max. day-length varying with latitude			
7. Longitudinal time difference established by eclipses		221–235	

eral days before Alexander's famous battle, and at a much earlier hour in Carthage (so Ptolemy) and Sicily (so Pliny).[50] Although the argument cannot be dated with certainty till before the time of Hipparchus, Pliny's and Ptolemy's reference to the eclipse of 331 BC suggests that the argument may have been devised not much afterwards. Lucretius and Epicurus could have known it. As with the previous proof it is hard if not impossible to avoid its consequences. Here too the only way out, it would seem, is to contest the observations themselves. One might point to the fact that lunar eclipses are extremely rare, that simultaneous observations from different locations are even rarer, and that the historical records (such as Pliny's and Ptolemy's) are unreliable.

Table 4.2 provides a summary of the ancient proofs of the earth's sphericity, with reference to the authors, texts and passages reporting them, and the earliest datable reference to the principles behind each proof.

---

50   Pliny, *N.H.* II 180; Ptol., *Geogr.* I 4, 2.

| Pliny | Theon Smyrn. | Ptolemy | Cleomedes | Earliest datable |
| 23–79 AD | ca. 70–135 | ca. 85–165 | ca. 200? | reference |
N.H. II	*Expos.* III	*Alm.* I, 4	*Cael.* I, 5	
				Aristotle
				384–322 BC
164.1–5	3.1–14	16.13–18	114–125	Strabo
				64 BC–24 AD
177.8–179.4	2.10–26	15.23–16.13	49–54	Aristotle
				384–322 BC
179.4–11			44–49	Hipparchus
				ca. 190–120 BC
182–185				Pytheas
				ca. 325 BC
186–187			54–56	Pytheas
				ca. 325 BC
180–181	2.1–10	14.19–15.23	30–44	Hipparchus
				ca. 190–120 BC

Of the proofs mentioned above, the 1st, 3rd, 5th and 6th could chronologically have been known to Epicurus, while Lucretius could have known the 4th and 7th as well. If, as Furley suggests, the Epicureans did try to put up a reasoned defence of their own flat-earth cosmology against contemporary astronomy, one would expect them to betray a knowledge of, and engage with, the proofs the astronomers offered for the sphericity of the earth. In our survey of Epicurean passages in section 4.3 below, we will therefore also look for clues that might tell us whether Lucretius, Epicurus or other Epicureans were aware of these observations, and whether or how they managed to reconcile them with their own views concerning the shape of the earth.

4.2.3 *Epicurus' Ancient Critics*

In antiquity, Epicurus was criticized and even ridiculed for many of his doctrines. He was attacked, for instance, for assigning upward and downward directions to the infinite universe,[51] for assigning a random and uncaused motion,

---

51 Chrysippus (*SVF* II 539) apud Plut. *De Stoic. repugn.* 44, 1054b. See also on p. 216 below.

the παρέγκλισις or 'swerve', to the atom,[52] for holding that the sun is the size it appears to be,[53] and for contemplating the possibility that the sun might be extinguished at sunset and rekindled at dawn.[54] If Epicurus had also claimed, in contrast to everybody else, that the earth is flat, would his critics not have seized the opportunity to attack him for yet another 'stupidity'? Yet, no such criticism has come down to us.

One of Epicurus' most fervent critics was Cleomedes, who in the second book of his astronomical treatise attacks Epicurus by name for two of his theories.[55] Yet, in the first book, in a section where the theory that the earth might be flat is explicitly refuted, Cleomedes fails to mention Epicurus (or anyone else, for that matter). If he had known Epicurus to be a flat-earther, would he not have named him?

And yet Cleomedes comes very close to actually identifying Epicurus here: in I 5.11–13 those who believe the earth to be flat are said to be 'following only the sense presentation based on sight', which is almost exactly how the Epicureans are characterized in II 1.2–5.[56] What is more, one of Cleomedes' actual arguments against the earth being flat might as well have been aimed directly at the Epicureans. In I 5.30–37 Cleomedes argues that those who believe the earth to be flat must also believe that the sun rises and sets for everyone at the same time, which is observably wrong. However, in II 1.426–451 he ridicules Epicurus for believing that the sun may be extinguished at sunset and rekindled at sunrise, pointing out that on a spherical earth sunset and sunrise occur at different times in different regions. Cleomedes might at this point have concluded that Epicurus must be thinking of a flat earth, but he does not. Instead he opts for a *reductio ad absurdum*, suggesting that according to Epicurus' theory the sun must be at the same time extinguished for some observers and rekindled for others—and this incalculably many times. Again one wonders why Cleomedes does not simply accuse Epicurus of believing the earth to be flat.

---

52 Cicero *De fin.* I 18–20 (Epic. 281a Us.); id. *De fato* 46 (Epic. 281c Us.). See also on p. 215 below.
53 Cleomedes II 1.1–413; Cicero, *De fin.* I 20; *Acad.* II 82. See also on p. 236 below.
54 Cleomedes II 1.426–466; Servius *in Verg. Georg.* I 247 (Epic. 346a Us.), id. *in Verg. Aen.* IV 584 (Epic. 346b Us.). See also on p. 241 below.
55 Cleomedes II 1.1–413 on the size of the sun; II 1.426–466 on the sun's daily extinction and rekindling.
56 Cp. Cleom. I 5.11: ... αὐτῇ τῇ κατὰ τὴν ὄψιν φαντασίᾳ κατακολουθήσαντες ... and Cleom. II 1.2–5: ... αὐτῇ τῇ διὰ τῆς ὄψεως φαντασίᾳ κατακολουθήσαντες ...

The answer must be that Cleomedes (or his source) could not find any explicit statement to this effect in Epicurus' works. All the other criticisms, those of Cleomedes as well as other critics, were aimed at theories that could be explicitly ascribed to Epicurus. The view that the earth is flat, on the other hand, could not. Thus the critics' failure to address this subject is another indication, in addition to Lucretius' silence on the matter, that Epicurus never explicitly opted for a flat earth.

What we will be looking for, then, in our survey of relevant Epicurean passages, is circumstantial evidence, consisting in passages which either entail or presuppose a flat earth.

4.2.4   *The Direction of Natural Motion and the Shape of the Earth*

In his article 'The Dynamics of the Earth', David Furley distinguishes two fundamentally different ancient theories of natural motion.[57]

In one of these, which he calls *linear* or *parallel*, natural motion takes place along parallel lines. *Up* and *down* are defined in accordance with this *parallel* motion of fall, and the cosmos can be divided into an *upper* and a *lower* hemisphere. The tendency of heavy objects to fall down also applies to the earth as a whole, which, in order not to fall, must be supported by something underneath, be it water (Thales), or air (Anaximenes, Anaxagoras and Democritus), or the earth itself, extending downwards to infinity (Xenophanes).[58]

In the other theory, which Furley calls *centrifocal*, all natural motion is defined by the *centre*: *down* is motion towards the centre, *up* motion away from the centre, and the circular motions of the heavenly bodies are motions around the centre. The earth itself, being heavy, is also prone to move towards the centre, but being already agglomerated around the centre, it will not move; therefore there is no need for external support. This downward tendency applies to animals and human beings as well, allowing them to stand on every side of the earth, with their feet pointing towards the centre. People living diametrically opposite us are called *antipodes*, i.e. 'having their *feet* pressed *against* ours'.[59] Furley's prototype of a centrifocal cosmology is the one propounded by Aristotle, who assigned three different natural motions to the elements: earth

---

57   Furley (1989a), esp. p. 15. See also Furley (1986) 234–235.
58   Arist. *Cael.* II 13, 294b13–30. See also Arist. *Mete.* II 7, 365a20–37, where the parallel cosmology of Anaxagoras is explicitly opposed to Aristotle's own centrifocal system.
59   Cf. Cic. *Luc.* 123, 7: "Vos etiam dicitis esse e regione nobis, e contraria parte terrae qui adversis vestigiis stent contra nostra vestigia, quos *antipodas* vocatis." See also Kaufmann (1894).

and water moving *towards the centre*, air and fire moving *away from the centre*, and aether or the first element (or the fifth as it is commonly called) moving in circles *around the centre*. Other centrifocal cosmologies are those of Plato, Strato and the Stoics. Strato, for instance, although he seems to have dispensed with Aristotle's first element and assigned a natural downward motion to all four sublunary elements (attributing the apparent upward tendency of air and fire to extrusion by the heavier elements), still defined this natural downward motion as motion towards the centre.[60]

Plato's cosmology differs from Aristotle's in one important respect. In the *Timaeus*, 62c–63a, Plato observes that everyone speaks of 'up' and 'down' with respect to themselves, and that accordingly for someone standing on the other side of the earth 'up' and 'down' are exactly reversed. For this reason he rejects the use of these terms in a cosmological context altogether. In the *De caelo*, IV 1, 308a14–24, Aristotle criticizes this view and attempts to save the terms 'up' and 'down' in an absolute sense by redefining them as 'away from the centre' and 'towards the centre' respectively. After Aristotle these definitions of 'up' and 'down' were accepted by most *centrifocalists*.[61] This does not mean, however, that all *parallel-linear* terminology was banished from these centrifocal systems altogether. When speaking of a star's *setting* (δύσις / δυσμαί) and *rising* (ἀνατολή), or its *height* (ὕψος) or *elevation* (ἔξαρμα), they were in fact, as we still are, using a parallel-linear spatial reference system, where 'height' is used to denote the (angular) distance from the plane of the horizon, and not the distance from the centre of the earth.

Each of the two cosmological systems is associated with one particular shape of the earth: in *parallel* cosmologies the earth is usually considered *flat*, while *centrifocal* theories are thought to imply a *spherical* earth. In the *De caelo*, Aristotle offers two arguments in support of this implication. In the first place the all-sided centripetal pressure exerted by many individual chunks of earth, all seeking the centre of the universe, would naturally result in a spherical shape. Aristotle illustrates this by demonstrating that this shape would also have resulted if the earth had been generated at some time in the past (which he does not believe) (*Cael.* II 14, 297a12–b18). A similar argument was later put forward by the Stoics (Cic. *ND* II 116). In the second place a spherical shape

---

60 For the evidence on Strato see n. 133 on p. 208 below.
61 For the Stoics, see e.g. Ar. Did. fr.31, apud Stob. *Ecl.* I, 21.5 p. 184, 8 w. (= *SVF* II 527): τὸ μέσον σημεῖον τοῦ κόσμου […], ὃ δὴ τοῦ παντός ἐστι κάτω, ἄνω δὲ τὸ ἀπ' αὐτοῦ εἰς τὸ κύκλῳ πάντῃ, Cic. *ND* II 84: 'in medium locum mundi, qui est infimus', ibid. 116: 'id autem medium infimum in sphaera est', and esp. Cleomedes, I 1,158–192 (partly quoted as *SVF* II 557).

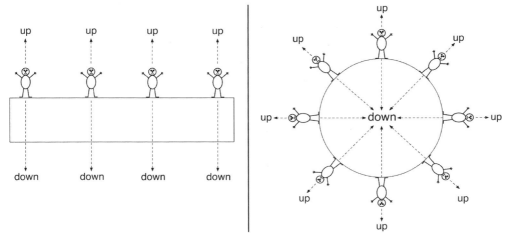

FIGURE 4.1 Linear *or* parallel *versus* centrifocal *cosmology*

logically follows from the observation that all heavy bodies fall at right angles to the surface of the earth, which in a centrifocal universe is only possible if the earth is spherical (*Cael.* II 14, 297b17–21). As regards the link between parallel natural motion and a flat earth no such arguments have come down to us, unless we count the view of Anaximenes, Anaxagoras and Democritus, who claimed that in order to be able to float on air the earth had to be flat and broad. However, a very plausible argument is provided by Furley, who points out that, as long as we accept the observed fact that heavy objects everywhere fall at right angles to the earth's surface (cf. Aristotle's second argument above), a parallel motion of fall requires a flat earth.[62]

The two cosmological systems with their implied shapes of the earth are illustrated in figure 4.1.

I have dealt with these two cosmological systems at some length because, as we shall see, the case for imputing a flat earth on the Epicureans largely depends on the assumed logical connection between the direction of natural motion and the shape of the earth. A clear understanding of the two theories of natural motion and their assumed implications for the shape of the earth will therefore be a useful tool in our survey of Epicurean passages.

---

62   Furley (1976) 90; (1981) 10, 12; (1983) 91–92, 99; (1999) 42.

4.3   Discussion of Relevant Passages

After these preliminaries I will now turn to the promised discussion of relevant Epicurean passages. In my search for such passages I have scanned through the whole of Lucretius' *DRN*, Epicurus' *Letters to Herodotus* and *Pythocles*, the fragments of book XI of Epicurus' *On nature*, and the cosmological and astronomical fragments of Diogenes of Oenoanda's inscription, as well as the commentaries thereto, looking for passages that might be, or have been, interpreted as somehow implying or presupposing a specific shape of the earth.

Lucretius appears to be the richest source of such passages, and for this reason I have chosen to take my lead from him. Such passages as can be found in the works and fragments of Epicurus and Diogenes are mostly parallel or supplementary to these, and will therefore be discussed in connection with them. In my discussion of these passages I will basically follow Lucretius' order, which may help us to understand how the different passages hang together. I have also come across an interesting fragment of Philodemus which seems relevant to my purpose and has no exact parallel with Lucretius or other Epicureans. A discussion of this fragment will be appended to the list. The following passages will be dealt with.

TABLE 4.3   *Passages to be discussed*

Lucretius	Parallel passages	Subject	Discussed in:	
I 1052–1093	–	*Rejection of centrifocal natural motion*	4.3.1	(p. 181)
II 62–250	Epic. *Hdt.* 60 + fr.281 Us.	*Parallel downward motion*	4.3.2	(p. 210)
IV 404–413	–	*Apparent proximity of the sun*	4.3.3	(p. 220)
V 204–205	Cic. *ND* I, 24	*Climatic zones?*	4.3.4	(p. 221)
V 449–508	Aëtius I 4	*Centrifocal cosmogony*	4.3.5	(p. 223)
V 534–563	Epic. *Nat.* XI fr.42 Arr.	*Stability of the earth*	4.3.6	(p. 235)
V 564–591	Epic. *Pyth.* 6 [91]	*Size of the sun*	4.3.7	(p. 236)
V 621–636	Diog. Oen. fr.13 I.11–13	*Centrifocal terminology*	4.3.8	(p. 239)
V 650–679	Epic. *Pyth.* 7 [92]	*Sunrise and sunset*	4.3.9	(p. 241)
V 762–770	Epic. *Pyth.* 13 [96] + scholion	*The conical shadow of the earth*	4.3.10	(p. 242)
VI 1107	–	*The limp of the cosmic axis*	4.3.11	(p. 245)
–	Philod. *Sign.* 47.3–8	*The varying length of shadows*	4.3.12	(p. 254)

4.3.1 The Rejection of Centrifocal Natural Motion (DRN I 1052 ff.)

4.3.1.1 Introduction

The first passage that seems to be relevant to our subject, and one that is often cited in support of the claim that the Epicureans held the earth to be flat, is found at the end of the first book of *DRN* (1052 ff.). Here Lucretius rejects a theory of *centrifocal* natural motion, which, as we have seen, is generally assumed to imply a spherical earth (see p. 177 ff. above). On the following pages I will analyse this passage, paying due attention to its context, to the intended target, to Lucretius' refutation of the theory, and to the positive conclusions regarding Lucretius' own views that may be drawn from the passage.

4.3.1.2 Context of the Passage

The passage itself begins in line 1052, but in order to fully understand it we must first deal with the preceding passage. In line 951 Lucretius had broached a new subject: the infinity of the universe and of its two constituent parts: matter and empty space. As such Lucretius' argument is an extended version of the more condensed argument in Epicurus' *Letter to Herodotus*, 41–42:

(i) Ἀλλὰ μὴν καὶ τὸ πᾶν ἄπειρόν ἐστι· τὸ γὰρ πεπερασμένον ἄκρον ἔχει· τὸ δὲ ἄκρον παρ' ἕτερόν τι θεωρεῖται· ⟨ἀλλὰ μὴν τὸ πᾶν οὐ παρ' ἕτερόν τι θεωρεῖται·⟩[63] ὥστε οὐκ ἔχον ἄκρον πέρας οὐκ ἔχει· πέρας δὲ οὐκ ἔχον ἄπειρον ἂν εἴη καὶ οὐ πεπερασμένον. Καὶ μὴν καὶ τῷ πλήθει τῶν σωμάτων ἄπειρόν ἐστι τὸ πᾶν καὶ τῷ μεγέθει τοῦ κενοῦ·

(ii) εἴ τε γὰρ ἦν τὸ κενὸν ἄπειρον, τὰ δὲ σώματα ὡρισμένα, οὐθαμοῦ ἂν ἔμενε τὰ σώματα, ἀλλ' ἐφέρετο κατὰ τὸ ἄπειρον κενὸν διεσπαρμένα, οὐκ ἔχοντα τὰ

(i) Moreover, the universe is infinite. For that which is bounded has an extremity, and the extremity is seen against something else: ⟨but the universe is not seen against something else: ⟩[63] therefore, since it has no extremity, it has no boundary; and, since it has no boundary, it must be infinite and not bounded. Furthermore, the sum of things is infinite both in the multitude of the bodies and the magnitude of the void.

(ii) For if the void were infinite and the bodies limited in number, the bodies would have no place to stay, but would be carried down and

---

63   Addition proposed by Usener (1887), xviii, based on Cicero's rendering of the same argument in *Div.* II 103 (Epic. fr.297 Us.): 'Quod finitum est, habet extremum; quod autem habet extremum, id cernitur ex alio extrinsecus; *at quod omne est, id non cernitur ex alio extrinsecus*; nihil igitur cum habeat extremum, infinitum sit necesse est.' See commentaries ad loc.

ὑπερείδοντα καὶ στέλλοντα κατὰ τὰς ἀνακοπάς·	scattered through the infinite void, having no other bodies to support and check them by means of collisions.
(iii) εἴ τε τὸ κενὸν ἦν ὡρισμένον, οὐκ ἂν εἶχε τὰ ἄπειρα σώματα ὅπου ἐνέστη.	(iii) And if the void were limited, the infinite bodies would have no space to be.

Instead of Epicurus' single argument for the infinity of the universe (i), Lucretius offers no less than four arguments (I follow Bailey's analysis of the passage[64]):

1. (958–967): What is finite, must have a boundary, but for it to have a boundary there must be something outside to bound it. Yet, (by definition) there is nothing outside the universe. Therefore, the universe cannot be finite.
2. (968–983): If there were a boundary, what would happen if someone standing close to the boundary were to throw a spear towards it? If it stops, there must be matter outside to stop it, but if it proceeds there must be space outside for it to move in. In either case there turns out to be something outside, to the effect that the assumed boundary cannot be the ultimate boundary of the universe. And this procedure can be repeated wherever you assume the boundary to be. Therefore the universe can have no boundary and cannot be finite.
3. (984–997): If *space* were finite, all matter would since long by force of its weight have been heaped up at the bottom, and nothing would ever happen below the vault of heaven, nor would there even be a heaven or a sun. Yet, in fact the atoms are forever in constant motion. Therefore, *space* cannot be finite.
4. (998–1001): In our everyday experience everything is always bounded by something else, but with the universe (by definition) there can be nothing else to bound it. Therefore, the universe cannot be finite.

Of these four arguments for the infinity of the universe, the first is identical to the one offered by Epicurus. The earliest report of this argument is found in Aristotle, who rejects it.[65]

---

64 Bailey (1947) pp. 763–764.
65 Arist. *Phys.* III 4, 203b20–22: ἔτι τῷ τὸ πεπερασμένον ἀεὶ πρός τι περαίνειν, ὥστε ἀνάγκη μηδὲν εἶναι πέρας, εἰ ἀεὶ περαίνειν ἀνάγκη ἕτερον πρὸς ἕτερον.

The second argument is a famous thought-experiment, ultimately deriving from the Pythagorean Archytas.[66] An essentially identical argument to Lucretius' was also used by the Stoics.[67]

The third argument is exceptional from several points of view. In the first place it seems to be concerned with the infinity of *space* rather than the *universe*. This has led several editors to apply all kinds of transposition to the text in order to 'restore' the logical sequence of the argumentation.[68] However, Bailey has quite rightly demonstrated that the argument is precisely where it should be. Since the universe is composed of matter and space, the infinity of either part—in this case *space*—automatically implies the infinity of the whole, i.e. the *universe*. The argument may therefore serve as an argument for the infinity of the universe. In the second place, however, by using this argument

---

66   Archytas, fr.A24 D–K (Simpl. *In Arist. Phys.* 467.26–35 Diels): Ἀρχύτας δέ, ὥς φησιν Εὔδημος, οὕτως ἠρώτα τὸν λόγον· 'ἐν τῷ ἐσχάτῳ οἷον τῷ ἀπλανεῖ οὐρανῷ γενόμενος πότερον ἐκτείναιμι ἂν τὴν χεῖρα ἢ τὴν ῥάβδον εἰς τὸ ἔξω ἢ οὔ;' καὶ τὸ μὲν οὖν μὴ ἐκτείνειν ἄτοπον· εἰ δὲ ἐκτείνω, ἤτοι σῶμα ἢ τόπος τὸ ἐκτὸς ἔσται (διοίσει δὲ οὐδέν, ὡς μαθησόμεθα). ἀεὶ οὖν βαδιεῖται τὸν αὐτὸν τρόπον ἐπὶ τὸ ἀεὶ λαμβανόμενον πέρας καὶ ταὐτὸν ἐρωτήσει, καὶ εἰ ἀεὶ ἕτερον ἔσται ἐφ' ὃ ἡ ῥάβδος, δῆλον ὅτι καὶ ἄπειρον.—'Archytas, as Eudemus reports [fr.65 Wehrli], approached the argument as follows: "having reached, for instance, the edge of the fixed heaven, could I stretch my hand or staff outward, or not? Not being able to stretch it out would be absurd, but if I stretch it out, the outside will be either body or place. It will not make a difference, as we shall learn." He {i.e. Archytas} will then always in the same manner proceed towards the assumed boundary and ask the same question, and if there will always turn out be another thing towards which the staff is stretched out, it will be clear that the universe is infinite.' On this argument and its Stoic and Epicurean counterparts, see Sorabji (1988) 125–126.

67   SVF II 535a (Simpl. *In Arist. De caelo* 284.28–285.2 Heiberg): Οἱ δὲ ἀπὸ τῆς Στοᾶς ἔξω τοῦ οὐρανοῦ κενὸν εἶναι βουλόμενοι διὰ τοιαύτης αὐτὸ κατασκευάζουσιν ὑποθέσεως. ἔστω, φασίν, ἐν τῷ ἐσχάτῳ τῆς ἀπλανοῦς ἑστῶτά τινα ἐκτείνειν πρὸς τὸ ἄνω τὴν χεῖρα· καὶ εἰ μὲν ἐκτείνει, λαμβάνουσιν ὅτι ἔστι τι ἐκτὸς τοῦ οὐρανοῦ εἰς ὃ ἐξέτεινεν, εἰ δὲ μὴ δύναιτο ἐκτεῖναι, ἔσται τι καὶ οὕτως ἐκτὸς τὸ κωλῦσαν τὴν τῆς χειρὸς ἔκτασιν. κἂν πρὸς τῷ πέρατι πάλιν ἐκείνου στὰς ἐκτείνῃ, ὁμοία ἡ ἐρώτησις· εἶναι γὰρ δειχθήσεται κἀκείνου τι ἐκτὸς ὄν.—'Those of the Stoa want there to be a void outside the heavens and prove it through the following assumption: let someone stand, they say, at the edge of the fixed heavenly sphere and stretch out his hand upwards {i.e. away from the centre = outwards}. If he does stretch it out, they take it that there is something outside the heavens into which he has stretched it, and if he cannot stretch it out, there will still be something outside which prevents his hand from being stretched out. And if he should next stand at the limit of this and stretch out his hand, the question is similar: for it will be demonstrated that there is something outside of that point too.'

68   See Bailey's commentary for an overview.

at this point Lucretius is wilfully anticipating something he has not yet proved, the proof of which partly depends on the outcome of the present passage. The argument assumes the laws of atomic motion, which Lucretius does not discuss until II 62 ff.—and especially the atoms' natural *downward* motion, which is not discussed until II 184 ff. However, (and this is something the commentators, including Bailey, do not seem to have appreciated) the acceptance of these laws of atomic motion—and again especially the atoms' natural *downward* motion—depends implicitly upon the rejection of the alternative theory of natural motion put forward in I 1052 ff. (to be discussed presently), a passage which in turn depends on the *infinity of space*, which is precisely the subject of the present argument. In sum, Lucretius' third argument is *circular* to a certain extent. Fortunately, it is only one of several arguments, and we do not therefore depend on it for the conclusion that space and the universe are infinite.

The fourth argument is basically a restatement of the first. A very similar argument is also brought forward by the Stoic Cleomedes.[69]

After these four arguments, in lines 1002–1007 Lucretius sums up his conclusion. One would have expected this conclusion to be that the *universe* is infinite, but instead Lucretius seems to say that *space* is infinite. However, the continuation suggests that this is something Lucretius does not yet consider to be quite established. We may perhaps solve the problem if we interpret 'space' here as the 'spatial extent of the universe', i.e. space, both empty and filled with matter.[70]

After this conclusion the argument moves on (1008–1013): given the infinity of the universe two theoretical options present themselves: (a) its two constituents, matter and space, are *both* infinite, or else (b) *either one* of these two is infinite.

At this point, to all likelihood, the text presents a lacuna. Option b seems to imply two further options: (b1) either matter is infinite and empty space is not, or (b2) empty space is infinite, while matter is not. Both these options are in fact explicitly discussed and refuted by Epicurus, in the passage quoted above (sections iii and ii respectively), but in Lucretius' text, as we have it, both seem to have been lost in the lacuna. Right after the lacuna we find ourselves in the middle of a discussion of what would happen if space were infinite, but matter not, i.e. the elaboration of option b2, corresponding to part ii of Epicurus' account. This means that the lacuna must have contained both the statement and refutation of b1, as well as the statement of b2. The text of the

---

69   Cleomedes I 1, 112–122.
70   See Bailey ad loc.

lacuna must have run something like this (using Bailey's words): "⟨But (b1) if space were limited, it could not contain the infinite bodies of matter; and (b2) if matter were limited⟩ ...". There follows the account of lines 1014 ff.

Before we go on to discuss this passage, it may be a good idea for us to pause briefly and take our bearings. When we arrive at line 1014, just after the lacuna, the infinity of the *universe* has been firmly established, but also, and more importantly, the infinity of *space*. It may be interesting to note that up to this point the position of the Epicureans is in no way different from that of their most ardent rivals, the Stoics: see e.g. Cleomedes I 1.39–149.[71] As we have seen, even some of Lucretius' arguments for the infinity of the universe are identical to Stoic ones: his second argument corresponds to the Stoic argument reported by Simplicius (SVF II 535a),[72] and his fourth to Cleomedes I 1.112–122.

Lucretius has only to fulfil one more promise, and this is where the Epicureans and Stoics part company: *to prove the infinity of matter*, given the infinity of space. To this task he devotes the rest of book I. The argument in lines 1014–1051 consists of two stages. First, in lines 1014–1020, Lucretius argues that without an infinite amount of matter the world would not be able to survive and would not even have been created, because all the atoms would be scattered through infinite space. This stage corresponds to Epicurus' argument in section ii of the text quoted above.[73] Then, in lines 1021–1051, Lucretius goes on to explain how the world *is* actually created and preserved. According to Lucretius, everything we see around us is created through chance meetings of atoms, which being (infinitely) many and having tried every kind of motion and combination during infinite time, finally happened to combine into such a structure as is our world. Yet, even now our world is still being preserved through the agency of atoms from outside, which either replace the atoms that are constantly being

---

71 On the Stoic conception of extra-cosmic void see Algra (1993) and id. (1995) 261–336.
72 See n. 67 above.
73 Cf. also Diog. Oen. fr.67: πεπερασμέ[ναι] τοιγ[α]ροῦν ἡμεῖν [ὑπ]οκείμεναι κατὰ τὸ [π]λῆθος αἱ ἄτομοι φύ[σεις κ]αὶ δι' ἃς εἰρήκαμεν [α]ἱ[τ]ίας ἀσυνέλευστοι τυγχάνουσαι (μετὰ γὰρ αὐτὰς ἄλλαι φύσεις οὐκέτ' εἰσίν, αἵ, περιλαβοῦσαι τὸ πλῆθος αὐτῶν, κάτωθέν τε ὑπερείσουσι καὶ ἐκ τῶν πλαγίων συνάξουσιν αὐτάς), πῶς ἀπογεννήσωσι τὰ πράγματα χωρὶς ἀλλήλων; ὥστ' οὐκ ἂν ἦν οὐδ' ὅδε ὁ κόσμος. εἰ γὰρ ἦσαν πεπερασμέναι, συνελθεῖν οὐκ ἠ[δύναντ' ἄν].—Therefore if the atoms are assumed by us to be finite in number and for the [reasons] we have stated are incapable of coming together (for there are no longer other atoms behind them to surround their number and support them from below and bring them together from the sides), how are they to engender things, when they are isolated from one another? The consequence is that not even this cosmos would exist. For if the number of atoms were finite, they [would] not [be able] to come together. (tr. Smith, slightly modified)

lost, or prevent them from escaping by battering them back into line. In order to accomplish this an infinite amount of matter is required.

May we take it then that the infinity of matter has now been proved? Not quite. In the following section Lucretius is forced to retrace his steps and suspend his desired conclusion about the infinity of matter, because an alternative theory must be removed first.

4.3.1.3 The Rival Theory

In lines 1052 ff. Lucretius describes and refutes a rival theory which would account for the cosmos staying together without the need for extra-cosmic matter. Because the passage involves the rejection of centrifocal cosmology, it is important to study it from every possible angle. I will therefore print the text in full, with a translation and a select commentary of my own, followed by a schematic presentation of the structure of the passage. Except for some minor changes, which—if relevant—will be indicated in the commentary, the text is as printed by Bailey (1947), while the additions to lines 1068–1075 are Munro's (1864).

Illud in his rebus longe fuge credere,
    Memmi,
in medium summae quod dicunt omnia niti,
atque ideo mundi naturam stare sine ullis
1055 ictibus externis neque quoquam posse
    resolvi,
summa atque ima quod in medium sint
    omnia nixa
(ipsum si quicquam posse in se sistere
    credis),
et quae pondera sunt sub terris omnia
    sursum
nitier in terraque retro requiescere posta,
1060 ut per aquas quae nunc rerum simulacra
    videmus.
Et simili ratione animalia suppa vagari
contendunt neque posse e terris in loca
    caeli
reccidere inferiora magis quam corpora
    nostra

In these matters, Memmius, flee far from believing this, what they say: that all things tend to the centre of the universe, and that therefore the cosmos stays together without any external blows, and cannot be dissolved in any direction, because the upper and the lower parts all tend to the centre (if you believe that anything can stand on itself), and that, what weights there are below the earth, these all tend upwards, and rest inversely placed against the earth, like images of things we now see in the water.
And they claim that in the same way animals wander upside down, and cannot fall back from the earth into the lower regions of the heavens, any more than these our bodies can of their own

THE SHAPE OF THE EARTH

sponte sua possint in caeli templa volare;
1065 illi cum videant solem, nos sidera noctis
cernere, et alternis nobiscum tempora caeli
dividere et noctes parilis agitare diebus.

Sed vanus stolidis haec ⟨error falsa probavit⟩
amplexi quod habent perv⟨ersa rem
  ratione⟩
1070 nam medium nihil esse potest ⟨quando
  omnia constant⟩
infinita; neque omnino, si iam ⟨medium sit⟩,
possit ibi quicquam consistere ⟨eam magis
  ob rem⟩
quam quavis alia longe ratione ⟨repelli⟩
omnis enim locus ac spatium, quod in⟨ane
  vocamus⟩,
1075 per medium, per non medium, concedere
  ⟨debet⟩
aeque ponderibus, motus quacumque
  feruntur.
nec quisquam locus est, quo corpora cum
  venere,
ponderis amissa vi possint stare in inani;
nec quod inane autem est ulli subsistere
  debet,
1080 quin, sua quod natura petit, concedere
  pergat.
haud igitur possunt tali ratione teneri
res in concilium medii cuppedine victae.

Praeterea quoniam non omnia corpora
  fingunt
in medium niti, sed terrarum atque liquoris
1085 —umorem ponti magnasque e montibus
  undas,
et quasi terreno quae corpore
  contineantur—,
at contra tenuis exponunt aëris auras
et calidos simul a medio differrier ignis,

accord fly into the temples of heaven; and that when they see the sun, we see the stars of the night and that they divide their heavenly seasons alternate with ours and have their nights equal to our days.

But idle ⟨belief suggested⟩ these things to the idiots, because they have embraced them with perv⟨erse reasoning⟩, for there can be no centre ⟨since the universe is⟩ infinite, nor—even if ⟨there were a centre⟩—could anything at all stand still there ⟨for that reason, rather⟩ than in a different manner ⟨be driven⟩ far ⟨away⟩; for every place and space, which we ⟨call void⟩—whether or not the centre—, ⟨must⟩ yield equally to weights, wherever their movements tend. nor is there any place, where bodies upon arriving lose the force of weight and stand still in the void; nor, on the other hand, must that which is void, resist to anything, but, as its nature demands, continue to yield.

Therefore, not in such a way can things be held in union, overcome by longing for the centre.

Besides, since they suppose that not all bodies tend towards the centre, but ⟨only⟩ those of earth and water—the moisture of the sea and mighty waves from the mountains, and such things as are contained as it were in earthly body—, while on the other hand they expound that the subtle breezes of air and hot fires are at the same time

atque ideo totum circum tremere aethera signis	borne away from the centre, and that therefore the whole aether twinkles all
1090 et solis flammam per caeli caerula pasci,	around with stars and the flame of the
quod calor a medio fugiens se ibi conligat omnis,	sun grazes through the blue of the sky, because all heat fleeing from the
nec prorsum arboribus summos frondescere ramos	centre gathers itself there, and that the topmost branches of the trees would
posse, nisi a terris paulatim cuique cibatum	not at all be able to sprout leaves, if not slowly from the earth to each food
* * * * * * * * * * *	⟨were distributed⟩
	* * * * * * * * * * *

Commentary

1053 quod dicunt: There is some disagreement about the right construction of 'quod dicunt'. According to some commentators,[74] it depends on 'in medium summae': 'the centre of the universe, *as they call it*', which would be an anticipation of Lucretius' criticism in lines 1070–1071 that the universe, being infinite, cannot have a centre. This would make the whole train of *AcI* ('Accusativus cum Infinitivo') constructions up to line 1060 (except 1057) appositional to 'illud' and therefore dependent on 'fuge credere'. According to others,[75] 'quod dicunt' goes directly with 'illud': 'this, *what they say*, viz. that …', making all the *AcI*'s effectively dependent on 'dicunt'. Although both options are syntactically viable, there is good reason to prefer the second one, since all the *AcI*'s in the rest of the passage depend on *verba declarandi* as well: 1061–1067 on 1062 'contendunt', 1083–1086 on 1083 'fingunt', and 1087–1093 on 1087 'exponunt'. This has serious consequences for the interpretation of the passage: if all the *AcI*'s in lines 1053–1060 depend on 'dicunt', they must all be interpreted as part of what the unnamed rivals say: Lucretius wants us to be believe that the theory

---

74   Giussani (1896–1898), Ernout-Robin (1925–1928), Furley (1966) 187–188, Brown (1984). Examples of 'illud' explained by an *AcI*-construction are found at II 184ff., 216ff., 581ff., 891ff., 934ff., III 319ff., V 146ff.

75   E.g. Bailey (1947). For an example of 'illud' with a relative clause Bailey refers to I 370ff. A better example is III 370–373 ('Illud in his rebus nequaquam sumere possis, / Democriti quod sancta viri sententia ponit, / corporis atque animi primordia singula privis / adposita alternis variare, ac nectere membra.'), where 'illud' governs not just a relative clause, but an *AcI* as well. A third option, proposed by Munro (1864), explaining 'quod dicunt' as parenthetical to the whole *AcI*-construction ('id quod dicunt'), corresponds to Bailey's interpretation in so far as it makes the *AcI*'s effectively dependent on 'dicunt'.

he describes and later rejects was not merely invented by him for the sake of argument, but was actually brought forward by the unnamed rivals themselves.

1054 mundi naturam: Brown (1984), following Bailey (1947), considers this to be a mere periphrasis for 'mundum'. Bailey concedes, however, that 'naturam' may have some force of its own, suggesting that—according to the unnamed opponents—the cosmos remains together *by nature* as opposed to the external blows of the Epicureans. It should be noted that here, as elsewhere in Lucretius, 'mundus' stands for the Greek κόσμος (cosmos), i.e. the structured whole consisting of heaven and earth and everything in between.[76] Cf. Lewis & Short 'mundus' II B, with refs. to Cic. *Tim.* 10.9–11 and Pliny *NH* II 8.

1054 stare: There is some ambiguity about the word 'stare', which, applied to the cosmos, could either refer to its 'staying together' or to its 'staying in one place'.[77] It is clear however that Lucretius is mainly thinking of the first possibility: compare lines 1081–1082, where the same thought is expressed as 'teneri in concilium' = 'to be held in union'.

1054–1055 sine ullis ictibus externis: This is best understood as a parenthetical remark of Lucretius' own, and need not imply that the unnamed rivals actually formulated their theory in opposition to Epicurus.

1056 summa atque ima: The comma could be placed either after or before 'summa atque ima'.[78] If after, 'summa atque ima' go with the preceding line, providing a new subject for 'resolvi'. If before, 'summa atque ima' will be the subject of 'sint nixa' in the same line, making 'omnia' an apposition. I slightly prefer the second option because in that way the words 'summa atque ima' prepare the way for Lucretius' parenthetical remark in the next line. The meaning of the lines is not, however, essentially effected either way.

1065–1067: As Brown (1984) observes, these lines do not imply the rejection of the astronomy involved. In V 650–704 Lucretius explicitly admits the possibility

---

76 See also Furley (1981) 4: "*Mundus* in Latin and *kosmos* in Greek meant a limited, organized system, bounded by the stars."
77 The same ambiguity also attaches to the Greek verb μένειν and its cognate noun μονή, which are used in statements of the corresponding Stoic theory. The relevant texts are Ar. Did. fr.23, quoted on p. 192 below (see also n. 84), and Achilles *Isagoge* 9, quoted in n. 87 on p. 193 below.
78 Furley (1966) 188.

that the sun passes below the earth each night. What he *does* object to is the belief that there are people down there (*illi*) to observe (*videant*) these phenomena.

1068–1075 & 1094–1101: The text of the passage as it has come down to us is damaged.[79] Part of a leaf of the archetype must have been torn off, partly destroying lines 1068–1075, and completely destroying lines 1094–1101, which would have been on the verso. Of the extant manuscripts, only the *Oblongus* has preserved whatever remained of lines 1068–1075. Fortunately, this is enough to gather the general sense. The second section, however, is lost beyond repair. As a result we do not know the precise relationship between what came before and what came after this lacuna.

4.3.1.4 Structure of the Passage

The passage consists of two main parts: a description of the rival theory (1053–1067) and a refutation of this theory (1068–1093 ff.), containing at least three objections (more may have been lost in the lacuna). In the course of the argument Lucretius presents eight propositions allegedly brought forward by the unnamed rivals: five as part of his description of the theory, and three more as part of his third objection. The structure of the passage can be set out as follows:

A. Description of the rival theory (1053–1067):
    1. Everything tends towards the centre of the universe (1053)
    2. In this way the coherence of the cosmos is explained—without external blows (1054–1057)
    3. Weights below the Earth press upward and stand upside down against the Earth (1058–1060)
    4. Animals living on the other side of the Earth walk upside down and do not fall down into the lower regions of the heavens (1061–1064)
    5. There are people living on the other side of the Earth who have day and night, and summer and winter reversed (1065–1067)
B. Refutation of the rival theory (1068–1093 ff.):
    i. There is no centre in the infinite universe (1070–1071)
    ii. Nothing would stop at this 'centre', because neither can a mere point in space offer resistance nor can a body suddenly lay down its weight (1071–1082)

---

79   See e.g. Bailey (1947) or Rouse-Smith (1982) ad loc.

iii. The anonymous rivals also claim (1083–1093 ff.) that:
   6. *not* everything tends towards the centre, but only earth and water, whereas air and fire move away from it (1083–1088)
   7. this is the reason why "the whole aether twinkles all around with stars, and the flame of the sun grazes through the blue of the sky" (1089–1090)
   8. and this is also why plants grow upwards (1091–1093)
      * * * * * * * * * * * (1094 ff.)

The eight propositions do not all enjoy the same status, but they serve different purposes in Lucretius' argument. Proposition 1 is the really important one. It is this view that makes the rival theory qualify as an alternative to Lucretius' own theory of external blows. Proposition 2 is important too, because it tells us explicitly what proposition 1 was meant to explain: viz. the coherence of the cosmos. Propositions 3–5 are merely corollaries of proposition 1. They are not important in themselves, nor are they objects of Lucretius' criticism, but they are used to ridicule the rivals' theory, and so prepare the reader's mind for the real criticism in part B of the passage. A similar point can be made with respect to propositions 6–8. Here too the first one, proposition 6, is the really important one. According to most commentators, Lucretius' third objection must have focused on this proposition, perhaps pointing out its inconsistency with proposition 1, or the disastrous consequences of proposition 6 alone: viz. the escape of air and fire into infinite space. By contrast, propositions 7 and 8 are relatively unimportant. Their only function is to tell us *why* the unnamed rivals felt the need to accept proposition 6 in the first place: viz. to account for the fact that the stars and the sun remain poised in the sky, and that plants grow upwards. Note that the facts contained in these two propositions are not themselves being disputed.

4.3.1.5 Identification of the Unnamed Opponents

It has long been debated who the unnamed opponents were. Most scholars believe that Lucretius is specifically targeting the Stoics, although some prefer to see Aristotle or early Platonists as the original targets.

The identification of the unnamed opponents seems to depend largely on the view one takes of Lucretius as a philosopher. At one end of the spectrum are those who see Lucretius as an Epicurean 'fundamentalist', whose work is mainly a Latin versification of the works of Epicurus himself. At the other end are those who wish to see Lucretius right in the middle of the philosophical debates of his time, when the Stoics had become the Epicureans' main antagonists.[80] In the

---

80   For a summary of the debate see Warren (2007) 22–24. Notable representatives of the first

first case, our passage must go back to Epicurus himself, who—it is believed—can only have had the Peripatetics or the dogmatic Academics for a target, since the Stoic school had only just come into existence.[81] In the second case, however, our passage may well represent the debates of Lucretius' own time in which the Stoics figured prominently.

In this subsection I will try to steer clear from both preconceived positions, and instead focus as much as possible on Lucretius' description of, and arguments against, the rival theory, in order to identify it. It may be thought that for the present purpose, i.e. the reconnaissance of Lucretius' own cosmological views, the identification of the anonymous rivals is not really necessary, but I will try to demonstrate that knowing them and their actual theories will enable us to better understand what Lucretius' argument is about.

As I said, the majority of commentators interpret the present passage as an attack on the Stoics.[82] In support of their identification they quote Arius Didymus fr.23 on Zeno:[83]

Ζήνωνος. Τῶν δ' ἐν τῷ κόσμῳ πάντων τῶν κατ' ἰδίαν ἕξιν συνεστώτων τὰ μέρη τὴν φορὰν ἔχειν εἰς τὸ τοῦ ὅλου μέσον, ὁμοίως δὲ καὶ αὐτοῦ τοῦ κόσμου, διόπερ ὀρθῶς λέγεσθαι πάντα τὰ μέρη τοῦ κόσμου ἐπὶ τὸ μέσον τοῦ κόσμου τὴν φορὰν ἔχειν, μάλιστα δὲ τὰ βάρος ἔχοντα· ταὐτὸν δ' αἴτιον εἶναι καὶ τῆς τοῦ κόσμου μονῆς ἐν ἀπείρῳ κενῷ καὶ τῆς γῆς παραπλησίως ἐν τῷ κόσμῳ περὶ τὸ τούτου κέντρον καθιδρυμένης ἰσοκρατῶς. οὐ πάντως δὲ σῶμα βάρος	Zeno's {tenet}: Of all the things in the cosmos which stand together with their own *hexis* the parts have their motion towards the middle of the whole, and the same applies also to the parts of the cosmos itself: therefore it is correct to say that all the parts of the cosmos have their motion towards the middle of the cosmos, especially those parts which have weight. The same fact explains both the immobility[34] of the cosmos

---

view are Furley (1966); id. (1978) 206–208 and Sedley (1998a), esp. 62–93, where Lucretius is called a 'fundamentalist'. The second view is championed by e.g. Schrijvers (1999) and Schmidt (1990).

81 On Epicurus' lack of engagement with Stoicism, see Kechagia (2010): I am grateful to David Sedley for providing me with this reference.

82 See Munro, Giussani, Ernout-Robin, Bailey ad loc.

83 Ar. Did. fr.23 Diels, apud Stob. *Ecl.* I 19,4 p. 166, 4 w. (= SVF I 99). On the validity of the ascription to Zeno, see Algra (2003) 15–19 = Algra (2002) 163–167.

84 The word μονή, lit. 'immobility', suggests that what is being explained is the cosmos

ἔχειν, ἀλλ' ἀβαρῆ εἶναι ἀέρα καὶ πῦρ· τείνεσθαι δὲ καὶ ταῦτά πως ἐπὶ τὸ τῆς ὅλης σφαίρας τοῦ κόσμου μέσον, τὴν δὲ σύστασιν πρὸς τὴν περιφέρειαν αὐτοῦ ποιεῖσθαι. φύσει γὰρ ἀνώφοιτα ταυτ' εἶναι διὰ τὸ μηδενὸς μετέχειν βάρους.

in infinite empty space and similarly of the earth in the cosmos, the earth being equably settled around its centre. Not in every respect, however, does body have weight, but air and fire are weightless: yet they too somehow tend towards the middle of the whole sphere of the cosmos, and they gather towards its periphery, for they are naturally *upward-moving* {i.e. *centrifugal*}[85] because they have no share in weight.[86]

The quotation seems highly apposite. All the major propositions of Lucretius' anonymous opponents are here: (1) the centripetal tendency of *all* matter (πάντα τὰ μέρη τοῦ κόσμου ἐπὶ τὸ μέσον τοῦ κόσμου τὴν φορὰν ἔχειν), (2) the explicit use of this theory to account for the coherence and immobility of the cosmos (ταὐτὸν δ' αἴτιον εἶναι καὶ τῆς τοῦ κόσμου μονῆς ἐν ἀπείρῳ κενῷ),[87] and (6) the paradoxical claim that air and fire, while tending towards the centre, are also naturally centrifugal (οὐ πάντως δὲ σῶμα βάρος ἔχειν, ἀλλ' ἀβαρῆ εἶναι ἀέρα καὶ πῦρ· [...] φύσει γὰρ ἀνώφοιτα ταυτ' εἶναι διὰ τὸ μηδενὸς μετέχειν βάρους),[88]—and

---

staying in one place. The terms συνεστώτων and ἕξιν, however, clearly show that it is the cosmos *staying together* that is at stake here. It is possible that the original Stoic argument combined both aspects: *staying in one place* and *staying together*; see Algra (1988). The same ambiguity is found in DRN I 1054 'stare', in the description of a rival theory: see p. 189 with n. 77 above.

85 'Upward' in Stoic cosmology equals 'away from the centre of the cosmos': see n. 61 above.
86 On the interpretation of this text and its parallels, see Samburksy (1959) 111–113, Furley (1966) 191–193, Hahm (1977) 107–126, Algra (1988) and Wolff (1989) 499–533 and 539–542.
87 Cf. Achilles *Isagoge* 9 (SVF II 554): Φασὶ μὲν οὖν μένειν τὸν κόσμον ἐν ἀπείρῳ κενῷ διὰ τὴν ἐπὶ τὸ μέσον φοράν, ἐπεὶ πάντα αὐτοῦ τὰ μέρη ἐπὶ τὸ μέσον νένευκε. Μέρη δέ ἐστιν αὐτοῦ γῆ ὕδωρ ἀὴρ πῦρ, ἃ πάντα νεύει ἐπὶ τὸ μέσον. Διὰ τοῦτο οὖν οὐδαμοῦ ῥέπει ὁ κόσμος. Cic. ND II 115 (SVF II 549): '... ita stabilis est mundus atque ita cohaeret, ad permanendum ut nihil ne excogitari quidem possit aptius. *Omnes* enim *partes* eius undique medium locum capessentes nituntur aequaliter'. Cleomedes I 1.91–92: Φήσομεν δὲ ὅτι ἀδύνατον αὐτὸν {sc. τὸν κόσμον} φέρεσθαι διὰ τοῦ κενοῦ· νένευκε γὰρ ἐπὶ τὸ ἑαυτοῦ μέσον καὶ τοῦτο ἔχει κάτω, ὅπου νένευκεν. Chrysippus apud Plut. *De Stoic. repugn.* 44, 1055a (SVF II 550): πιθανὸν πᾶσι τοῖς σώμασιν εἶναι τὴν πρώτην κατὰ φύσιν κίνησιν πρὸς τὸ τοῦ κόσμου μέσον.
88 Cf. Aët. I 12.4 (= SVF II 571): Οἱ Στωϊκοὶ δύο μὲν ἐκ τῶν τεσσάρων στοιχείων κοῦφα, πῦρ καὶ ἀέρα· δύο δὲ βαρέα, ὕδωρ καὶ γῆν. Κοῦφον γὰρ ὑπάρχει φύσει ὃ νεύει ἀπὸ τοῦ ἰδίου μέσου,

there are many other testimonies to confirm the Stoics' commitment to these views.[89]

The other propositions can be linked to the Stoics too. The theory of the antipodes (3–5), for instance, though not exclusively Stoic, seems to have been especially dear to them. It is given extensive treatment by the Stoic astronomer Cleomedes,[90] and the Stoics are mocked for holding such views by Plutarch.[91] Also (7) the theory that the sun and the stars require some kind of nourishment is well attested for the Stoics.[92] Even proposition 8, which probably stated that the upward growth of plants is somehow caused by the centrifugal tendency of air or fire, can be related to the Stoic tenet that attributes the growth of plants to the presence of an internal fire.[93]

βαρὺ δὲ τὸ εἰς μέσον. *Schol. Hes. Theog.* 134 Gaisf. *Gr. Poet. Min.* II 482 (= *SVF* I 100): φύσιν ἔχει πάντα τὰ κοῦφα ἀφιέμενα πίπτειν ἄνω. Cic. *ND* II 117: 'aer *natura* fertur ad caelum'. Plut. *De facie* 13, 927c: ἄνω γάρ, οὐ κύκλῳ τὰ κοῦφα καὶ πυροειδῆ κινεῖσθαι πέφυκεν. Plut. *De Stoic. repugn.* 42, 1053e (= *SVF* II 434): τό τε πῦρ, ἀβαρὲς ὄν, ἀνωφερὲς εἶναι λέγει, καὶ τούτῳ παραπλησίως τὸν ἀέρα. Achilles *Isagoge* 4 (partly quoted as *SVF* II 555b): τεσσάρων οὖν ὄντων τῶν στοιχείων συμβέβηκε τὸ πῦρ καὶ τὸν ἀέρα κουφότατα ὄντα ἐπὶ τὴν ἄνω φορὰν ἔχειν τὴν ὁρμὴν καὶ περιδινεῖσθαι. ὅτι δὲ πῦρ καὶ ἀὴρ κουφότατα καὶ ἀνωφερῆ, δῆλον μὲν καὶ ἐκ τῆς ὄψεως [...]. Cf. also Seneca *N.Q.* II 13.1–14.1; II 24.1; II 58.2; VII 23.1.

89   See nn. 87 and 88 above.

90   Cleomedes I 1.258–261: Πρὸς δὲ τοὺς ἀντίποδας οὐδὲν ἡμῖν κοινόν ἐστιν, ἀλλὰ πάντα ἀντέστραπται. Καὶ γὰρ τὰ ὑπὸ γῆν ἀλλήλων ἔχομεν κλίματα, καὶ τὰ κατὰ τὰς ὥρας ἡμῖν ἔμπαλιν ἔχει, καὶ τὰ κατὰ τὰς ἡμέρας καὶ νύκτας καὶ τὰ κατὰ τὰς αὐξήσεις τῶν ἡμερῶν καὶ μειώσεις.—"With the antipodeans we have nothing in common, but everything is reversed. For we occupy the regions that are 'down under' with respect to each other, and everything relating to the times is contrary between us, both with respect to days and nights and with respect to the lengthening and shortening of the days." [my translation] (cp. with Lucr. I 1065–1067).

91   Plut. *De facie* 7, 924a4–6: οὐκ ἀντίποδας οἰκεῖν ὥσπερ θρῖπας ἢ γαλεώτας τραπέντα ἄνω τὰ κάτω τῇ γῇ προσισχομένους [...];—"Do you not say that it is inhabited by antipodes, who cling to the earth like wood-worms or geckos turned upside-down [...]?" (cp. with Lucr. I 1058–1064).

92   See e.g. Cic. *ND* II 40 (= *SVF* I 504); 43; 83; 118 (= *SVF* II 593); III 37 (= *SVF* I 501b / II 421); Aët. II 17a.1 (*SVF* II 690); Aët. II 23.7 (*SVF* I 501d / II 658); Diog. Laërt. VII 145 (*SVF* II 650); Cleomedes I 8.79–82 (*SVF* II 572); Plut. *De Iside* 41, 367e (*SVF* II 663); Plut. *De facie* 25, 940c (*SVF* II 677).

93   See Ar. Did. fr.33 Diels, apud Stob. *Ecl.* I 25, 5 (*SVF* I 120): τὸ δὲ {sc. πῦρ} τεχνικὸν αὐξητικόν τε καὶ τηρητικόν, οἷον ἐν τοῖς φυτοῖς ἐστι καὶ ζῴοις, ὃ δὴ φύσις ἐστὶ καὶ ψυχή, and Cic. *ND* II 41 (*SVF* I 504): 'ille {sc. ignis} corporeus vitalis et salutaris omnia conservat alit auget sustinet sensuque adficit.'

Yet, the identification of the unnamed opponents with the Stoics is not universally agreed upon. In an article, published in 1966, David Furley critically reviews a number of passages in the *DRN* that had until then been interpreted as attacks on the Stoics. One of these passages is the one we are presently investigating.[94] Furley argues that this passage cannot have been aimed at the Stoics, but should be read as an attack on Aristotle instead. Furley's views were in turn contested by Jürgen Schmidt in 1990,[95] who tries to restore the Stoics as the prime targets of Lucretius' criticism. Unfortunately, Schmidt's work seems to have gone largely unnoticed, to the effect that in the 1992 Loeb-edition of Lucretius we can still read, without any qualification, that Lucretius' argument was probably aimed at Aristotle.[96] In 1998, the case has been reopened by David Sedley.[97] Dismissing Schmidt's refutation, Sedley endorses Furley's arguments against the Stoics and even adds one of his own. He does not, however, accept Furley's view that the intended target is Aristotle, but instead points to dogmatic Platonists under the leadership of Polemo.

Although I am not convinced by Schmidt's additional conclusion that Lucretius must have been influenced by Academic scepticism and therefore must have been working from neo-epicurean sources, I think that Schmidt has clearly demonstrated the inadequacy of Furley's arguments. However, in view of Sedley's recent resumption of the case and his introduction of a new possible target for Lucretius' criticism, I think that a new defence of the traditional interpretation of the passage is in order. On the following pages I intend to show that the Stoics are not only the most likely, but in fact the only possible targets of Lucretius' criticism in the present passage.

It is true that some of the unnamed rivals' views can be identified with Aristotelian and Platonic tenets as well. For instance, both Plato and Aristotle believed that earth and water have a natural tendency towards, and air and fire away from, the centre of the cosmos (proposition 6).[98] Both also had some conception of antipodeans, people living on the other side of the earth with their feet pressing towards ours (propositions 3–5).[99] Things become much

---

94   Furley (1966) 187–195.
95   Schmidt (1990) 212–222.
96   Rouse-Smith (1992) 86–87, note b.
97   Sedley (1998a), 78–82.
98   Plato *Ti.* 62c–63e (speaking of fire and earth); Arist. *Cael.* I 2, 269a18–19 and esp. *Cael.* IV.
99   Plato *Ti.* 63a (Cf. also Cic. *Luc.* 123, 7 against the adherents of Antiochus' Old Academy: "Vos etiam dicitis esse e regione nobis, e contraria parte terrae qui adversis vestigiis stent contra nostra vestigia, quos *antipodas* vocatis.") and Arist. *Cael.* IV 1, 308a20.

harder, however, if we want to reduce proposition 7, about the sun feeding itself, or 8, about plants growing due to the upward tendency of air and fire, to Platonic and Aristotelian views.[100] However, the most important and, to my mind, fatal objection against identifying the unnamed opponents' with Plato or Aristotle, is the absence in their works of any version of propositions 1 and 2. Proposition 1 is the claim that *all* things tend towards the centre of the universe. The closest match is Plato's and Aristotle's view that all *heavy* things (i.e. earth and water) tend towards the centre, but it is not clear how such a view could explain the integrity of the *whole* cosmos (i.e. including the *light* elements, air and fire), as the unnamed opponents say it does (prop. 2). It is true that Aristotle uses the centripetal tendency of *earth* to explain the immobility of the *earth*,[101] but this is not at stake here. If Lucretius' had wished to criticise *this* theory, he would have done so in book V, in connection with his own theory about the earth's stability (V 534–563). In the present passage, however, we are dealing with the integrity of the whole *cosmos*. Moreover, Furley and Sedley seem to ignore the fact that Lucretius' unnamed opponents *themselves* used the centripetal tendency of all matter to explain how 'the cosmos stays together' and 'cannot be dissolved in any direction' (prop. 2).[102] This implies that the unnamed rivals *themselves* assumed the existence of an external void into which the cosmos might otherwise be dissolved. Aristotle, on the other hand, explicitly denied the existence of an extracosmic void,[103] and Plato is generally believed to have held the same view.[104] Neither of them, therefore, had any need to explain the integrity of the cosmos as a whole. It is therefore extremely unlikely that Lucretius' criticism would have had either of them for a target.

This does not prove, of course, that the target must be Stoic. In fact, several arguments against the Stoics being Lucretius' main target can be conceived. Two powerful arguments have been brought forward by Furley, and a third has

---

100 Both theories are refuted by Aristotle, in *Mete.* II 2, 354b34–355a32 and *De an.* II 4, 415b28–416a9 respectively. His pupil Theophrastus, on the other hand, seems to acknowledge some relation between plant growth and the motion of fire, in *De igne* 56.
101 Arist. *Cael.* II 14, 296b28–297a2.
102 If, as I argued on p. 188 above, the AcI's in this passage are all dependent on 'dicunt' in line 1053, then it is clear that the opinion expressed in the present lines must also be attributed to the unnamed opponents themselves.
103 Arist. *Cael.* I 9, 279a12–17. Cf. *Phys.* IV 5, 212b14–17 and 8–9, 214b12–217b28, where Aristotle rejects the existence of any kind of void.
104 See e.g. Aët. II 9.4: Πλάτων Ἀριστοτέλης μήτ' ἐκτὸς τοῦ κόσμου μήτ' ἐντὸς μηδὲν εἶναι κενόν.

been devised by Sedley. Before we can conclude that the target is Stoic after all, these three arguments will have to be dealt with.

Furley's first argument turns on the apparent inconsistency between propositions 1 and 6. First, in line 1053, Lucretius has the unnamed opponents say that *all* things tend towards the centre, and then, in lines 1083–1088, that *not* all things tend towards the centre, but only earth and water, whereas air and fire move away from it. The inconsistency leaps to the eye, and most commentators think that this was precisely what Lucretius wished to point out. However, Furley claims that no such inconsistency is implied: in line 1053 Lucretius would have been thinking of *heavy* things *only*, and then, in lines 1083–1088, he would have made explicit the earlier, implicit, qualification. According to Furley, then, the unnamed opponents simply said that earth and water are centripetal, and air and fire centrifugal, and nothing more. This interpretation effectively paves the way for Furley's own identification of the unnamed opponents with Aristotle, and Sedley's with Polemo's Platonists. Furley does not explain why the obvious interpretation, which is accepted by every other commentator, should be rejected. It is clear, therefore, that line 1053 must be taken literally: according to Lucretius the unnamed opponents really said that *all* things (*both heavy and light*) converge to the centre. Only in this way the rival theory can provide a reasonable alternative to the Epicurean theory of *external blows*.

This brings us back to where we were: the unnamed opponents hold two inconsistent views: *all* things move towards the centre, and *not all* things move to the centre. As we saw above, both these views are in fact attested for the Stoics, and for no other school but the Stoics. According to Furley, however, 'the evidence for this Stoic theory is [...] confused'.[105] What the Stoics actually meant, he says, following Sambursky,[106] is that air and fire 'are only *relatively* centrifugal', i.e. 'only *in the presence of earth and water*' [my italics]. I do not believe this interpretation can be right. Arius Didymus fr.23 (see p. 192 above) informs us explicitly (as do several other reports) that according to the Stoics air and fire are *naturally* centrifugal (φύσει ἀνώφοιτα),[107] which suggests something much stronger and more basic than a mere *relative* lightness. It will not do to simply dismiss the evidence as being 'confused'. Besides, Sambursky and Furley seem to attribute to the Stoics a theory that begins to sound very much like the one held by Democritus and Strato, and, more importantly, by Epicu-

---

105   Furley is here referring to Ar. Did. fr.23 quoted on p. 192 above, and the texts quoted in nn. 87 and 88 above.
106   Sambursky (1959) 111.
107   See n. 88 above, where all explicit references to *'nature'* and *'naturally'* have been italicized.

rus and Lucretius themselves,[108] all of whom claimed that *all* things are heavy and move downwards *by nature*, while the lighter bodies are pressed upwards *against their nature* by the heavier ones. In the existing ancient reports, however, the Stoics are never included among those who held this theory. That the Stoics in fact felt themselves much more at home with the opposite view (that air and fire are *positively light*) can also be observed through a comparison between Seneca *N.Q.* II 13 & 24 and Lucretius *DRN* II 203–215. Lucretius views the downward movement of thunderbolts and shooting stars as manifestations of the natural tendency of fire, whereas the upward movement of flames is attributed to pressure from the heavier elements. For the *Stoic* Seneca it is the other way round: the natural movement of fire is upward, whereas the descent of thunderbolts and shooting stars is an exception brought about by pressure from above.[109] However, for the purpose of identifying Lucretius' unnamed opponents, it does not really matter how the Stoic theory is interpreted. It is enough that Lucretius' description of the rival theory closely resembles the existing reports on the Stoics. If all these reports are confused, as Furley maintains, Lucretius can hardly be blamed for entertaining the same confusion in his depiction of the same theory. From this point of view, then, there is no reason why Lucretius' unnamed opponents should not be identified with the Stoics, and there is every reason for not identifying them with Aristotle or with Platonists under the leadership of Polemo.

Furley's second argument concerns the expression 'in medium *summae*' in Lucretius I, 1053. In Bailey's commentary on I, 235 a useful list is given of the many different meanings which the word *summa* (lit. 'sum', 'sum total', 'totality') may have in the *DRN*, either by itself or in combination with other words. Three uses deserve our consideration in the present context:

(1) the totality of matter *in this world alone*,
(2) the totality of matter *everywhere*, and
(3) the totality of matter *and void*, usually translated as 'the universe'.

The first option can be easily discarded, even though 'summa' was used in precisely this sense only eight lines earlier (1045). According to Lucretius (as

---

108 For the evidence see p. 211 ff. below.
109 Seneca is here following the argument of Arist. *Mete.* I 4, 342a13–27 and II 9, 369a20–30: see n. 174 on p. 148 above.

well as most other philosophers), our world is *finite* and therefore the matter contained in it too. Yet, in lines 1070–1071 Lucretius criticises his opponents, because what is *infinite* cannot have a centre. In the present context, therefore, *summa* (whatever it is) must be something *infinite*, which the matter *in this world* is not. The second option is problematic too. According to Lucretius, the totality of matter is infinite, as he has stated just before (line 1051). In the present passage, however, Lucretius is examining a theory, which, if accepted, would cancel this provisional conclusion, because it would allow for the cosmos to remain intact without the need for an infinite amount of extra-cosmic matter. As a consequence Lucretius cannot here use the infinity of matter without committing a serious *petitio principii*.[110] That leaves us only the third option: *summa* is the totality of matter *and void*. For this to be infinite it suffices that only one of its two components be infinite. Although the infinity of *matter* has not yet been fully established, the infinity of the *void* has been proved, and therefore the infinity of their sum too. Consequently, for Lucretius' argument to be valid, *summa* must be read as the *totality of matter and void*, i.e. *the universe*. According to Lucretius, then, the unnamed opponents said that all things move towards the centre of *the universe*.

This is not, however, what the Stoics said, as Furley rightly points out. Arius Didymus fr.23 (see p. 192 above) and other sources clearly state that according to the Stoics all things move towards the centre of *the cosmos*,[111] which is finite and *can* have a centre. Furley concludes therefore that the Stoics cannot have been Lucretius' intended targets. This conclusion, however, seems to me to be too strong. As Schmidt observes,[112] Lucretius (or his source) may simply have misunderstood or misrepresented the Stoic position on this point—something not uncommon in ancient polemics. Moreover, the Stoic position may not have been as clear-cut as Furley presents it. As Plutarch testifies in ch. 44 of his *On Stoic self-contradictions* Chrysippus had explained the indestructability of the cosmos as a whole by its occupying 'the centre'. If we may trust Plutarch, Chrysippus actually said: "to the virtual indestructability of the cosmos a good deal is contributed even by the position that it has occupied in space, that is to say through its being in the centre, since, if it should be imagined to be elsewhere, destruction would most certainly attach to it" (transl. Cherniss,

---

110 Schmidt (1990), 217, does accept this solution with all the circularity it involves, making 'summa' Lucretius' latinization of Stoic 'τὸ ὅλον', the (finite) sum total of matter.
111 See n. 87 above, where I have italicized the explicit references to 'the centre *of the cosmos*'.
112 Schmidt (1990) 217; cf. Brown (1984) ad loc.

slightly modified).¹¹³ Whatever 'centre' Chrysippus had in mind here, it cannot be the centre of the *cosmos*, for how could the cosmos occupy its own centre? Not unreasonably therefore Plutarch identifies this 'centre' with the centre of *space*, and accuses Chrysippus of inconsistency, because, as Chrysippus says elsewhere, space, being infinite, cannot have a centre.¹¹⁴ If Plutarch could bring this criticism against the Stoics, why not Lucretius?

At any rate, even if it is found that on this point Lucretius' criticism does not really apply to the Stoics, it is far less appropriate against Furley's preferred candidate, Aristotle, who denied the infinity of the universe altogether.¹¹⁵ For him the universe coincides with the cosmos, which is finite and therefore *can* have a centre. As a result, Aristotle is even more invulnerable to Lucretius' criticism than the Stoics. The same appears to hold for Plato. It is true that in his own works Plato neither explicitly denies nor acknowledges the existence of extra-cosmic void. Yet, tradition associates him with Aristotle in denying the existence of empty space both inside and outside the cosmos.¹¹⁶ Therefore, although Furley certainly has a point here, his conclusion that Lucretius' target cannot be Stoic does not obtain.

A third argument against identifying the unnamed opponents with the Stoics is brought forward by David Sedley. His argument concerns lines 1089–1091, which he paraphrases as: 'the upward motion of fire from the earth feeds the heavenly bodies, which are themselves fiery'. Apparently then, Lucretius' unnamed opponents said that the heavenly bodies are nourished by *fire* rising from the earth. Sedley observes that this is very different from what the Stoics said, who held that the heavenly bodies were sustained by *moisture* evaporating from terrestrial waters.¹¹⁷ Sedley's argument can be countered in two ways. In

---

113 Chrysippus apud Plut. *De Stoic. repugn.* 44, 1054c7–10 (SVF II 551): Οἷά τε δ' εἰς τὴν ὥσπερ ἀφθαρσίαν πολύ τι αὐτῷ συνεργεῖν καὶ ἡ τῆς χώρας κατάληψις, οἷον διὰ τὸ ἐν μέσῳ εἶναι· ἐπεί, εἰ ἀλλαχῇ νοηθείη ὤν, καὶ παντελῶς ἂν αὐτῷ συνάπτοι ἡ φθορά. See also Plut. *De defectu oraculorum* 28, 425d–e.

114 On the interpretation of Plutarch *De Stoic. repugn.* 44, 1054b–1055a see Algra (1995) 282–307.

115 Arist. *Cael.* I 9, 279a7–17.

116 See n. 104 above. Sedley (1998a), 80–81 argues on the basis of a few passages that Plato could be and may have been interpreted by some as admitting the existence of extra-cosmic void. However, the evidence is very circumstantial and proves at best that *from this particular point of view* Plato cannot be entirely ruled out as a target for Lucretius' criticism.

117 See n. 92 above.

the first place, Lucretius does not actually say that the heavenly bodies are nourished by *fire*. His words are:

atque ideo totum circum tremere
    aethera signis
	and that therefore the whole aether
	    twinkles all around with stars and the

1090 et solis flammam per caeli caerula pasci,	flame of the sun grazes through the blue

quod calor a medio fugiens se ibi
    conligat omnis,
	of the sky, because all heat fleeing from
	    the centre gathers itself there,

What Lucretius says here is that the sun, being fiery, is grazing through the sky, because heat (i.e. fire) gathers itself there. When we are told that a certain animal always feeds in a certain place, this can mean two things: it feeds there because it is the natural place for its food to be, or because it is the natural place for the animal itself to be. So too in the case of the sun: the sun feeds in the sky either because that is the natural place for *its food*, or because it is the natural place for *the sun itself*. Sedley goes for the first option, but in fact Lucretius' language rather suggests the second, for, while saying nothing about the nature of the sun's food, he explicitly mentions the fiery nature of the sun itself ('solis *flammam*'). So, I would say, the *sun* is up in the sky, because the *sun is fiery* and it is natural for *fire* to collect in the sky. That the sun spends its time there *grazing* is simply a picturesque detail, with no bearing on the argument, just like the *twinkling* of the stars in the preceding line is not an object of Lucretius' criticism. The image of grazing heavenly bodies is used by Lucretius himself on two other occasions (I 231 & V 525), and appears to derive from Epicurus.[118]

If this is not enough, it may be noted that the Stoic theory *does* actually lend itself to the interpretation that the heavenly bodies are nourished by *fire*. According to the Stoics, the process of exhalation is nothing but the transformation of moisture into air,[119] and subsequently into fire.[120] By the time the

---

118 Cf. Epic. *Pyth.* 8 [93], writing about the motions of the heavenly bodies: ... κατά τινα ἐπινέμησιν τοῦ πυρὸς ἀεὶ ἐπὶ τοὺς ἑξῆς τόπους ἰόντος. "... due to a certain *grazing* of the fire which always moves towards the adjacent places." On the different translations proposed for the word ἐπινέμησις in this passage, see Mansfeld (1994), n. 35.

119 Plut. *De aud. poët.* 11, 31e (SVF I 535): τὸν ἐκ τῆς γῆς ἀναθυμιώμενον ἀέρα—"the *air* which is being exhaled from the earth."

120 Plut. *De Stoic. repugn.* 41, 1053a (SVF II 579.1–3): ἔμψυχον ἡγεῖται τὸν ἥλιον, πύρινον ὄντα καὶ γεγενημένον ἐκ τῆς ἀναθυμιάσεως εἰς πῦρ μεταβαλούσης.—"He (Chrysippus) believes the sun to be ensouled, being fiery and having been created from *the exhalation which has transformed itself into fire.*" Plut. *De comm. not.* cp. 46, 1084e (SVF II 806d): γεγονέναι δὲ καὶ

exhalation reaches the heavenly bodies it *is* fire, so that the stars can be said to be nourished by fire. Therefore, Sedley's conclusion that the unnamed opponents cannot have been the Stoics is not compelling.

Anyway, it requires a lot more work to connect the theory of the sun's nourishment with Sedley's and Furley's preferred candidates: the early Platonists and Aristotle, respectively. Sedley (1998a), p. 79 n. 85, suggests that the theory of the sun feeding on fire may be an early Platonist development from Plato's statement in the *Timaeus* that fire tends to the place (i.e. the heavens) which is most natural to it and where the largest portion of it (i.e. the sun and the stars) is collected.[121] However, without testimonies for such a development, Sedley's suggestion remains mere conjecture.

That Aristotle should be Lucretius' intended target, as Furley maintains, is even more unlikely, since we find him explicitly denying the fiery nature of the heavenly bodies[122] as well as their need for nourishment.[123] As a result Furley is forced to the dubious assumption that Aristotle in his younger years would have embraced theories which he later strongly rejected.[124]

In sum, allowing for a certain degree of misrepresentation on Lucretius' part, the unnamed opponents can only be plausibly identified with the Stoics. Only they, of all imaginable candidates, believe in the existence of an extra-cosmic void, only they envisage the possibility that the matter which constitutes our cosmos could be dispersed into this surrounding void, and only they counter this possibility by positing a centripetal tendency of all matter.

Some details of the rival theory may also be found in Plato and Aristotle, but the crucial items, viz. the existence of extra-cosmic void and the universal centripetal motion of all matter, can only be found in Stoicism. Therefore there can be no doubt that Lucretius' criticism is directed primarily against the Stoics.

4.3.1.6 Lucretius' Avoidance of Centrifocal Terminology

Before offering his formal arguments against the rival theory, Lucretius tries to set his reader's mind against it. This he does by carefully misrepresenting some of the theory's corollaries.

τὸν ἥλιον ἔμψυχον λέγουσι, τοῦ ὑγροῦ μεταβάλλοντος εἰς πῦρ νοερόν.—"They say that the sun too is created an ensouled being, *the moisture transforming itself into intelligent fire.*"

121 *Tim.* 63b1–3: ἐν τῷ τοῦ παντὸς τόπῳ καθ' ὃν ἡ τοῦ πυρὸς εἴληχε μάλιστα φύσις, οὗ καὶ πλεῖστον ἂν ἠθροισμένον εἴη πρὸς ὃ φέρεται.
122 Arist. *Cael.* II 7, 289a11–35; *Mete.* I 3, 339b16–340a18.
123 See n. 100 above.
124 See Schmidt (1990) 219–220.

The rival theory is, as we have seen above, a *centrifocal* theory: all motions are described as being directed towards, or away from, the centre. As a rule, such theories employ a *centrifocal spatial reference system*, defining 'downwards' as 'towards the centre', and 'upwards' as 'away from the centre'. This applies to the Stoics as well as to Aristotle. We can see this for instance in Arius Didymus fr.23 quoted above (p. 192), where 'being upward-moving' (εἶναι ἀνώφοιτα) is contrasted with 'moving towards the centre' (τείνεσθαι ἐπὶ τὸ μέσον),[125] as if they were logical opposites. Consequently, centrifocalists were able to say, just like 'parallelists', that heavy things tend downwards, and light things upwards.

In Lucretius' account of the rival theory we find nothing of the sort. Instead he consistently uses 'up' and 'down' and 'above' and 'below' and similar expressions in a *parallel* sense. As a consequence Lucretius is able to saddle his rivals with the paradoxical view (1058–1060) that on the other side of the earth weights 'tend *upwards* and rest *inversely* placed against the earth, like images of things we now see in the water', and (1061–1064) that 'animals wander *upside down*, and cannot fall back from the earth into the *lower* regions of the heavens', nothing of which the Stoics would have actually said, because in a centrifocal spatial reference system weights fall *down* on whatever side of the earth they happen to be.

4.3.1.7 The Lacuna and the Status of Lines 1102–1113

Before we go on to discuss Lucretius' formal argument against the rival theory, I want to briefly move forward, to the text following the eight-line lacuna, in order to decide whether or not it is part of the refutation of the rival theory. The text runs as follows:

* * * * * * * * * * *	* * * * * * * * * * * ⟨… there is a danger⟩
ne volucri ritu flammarum moenia mundi	lest, after the flying fashion of flames,
diffugiant subito magnum per inane soluta,	the walls of the cosmos suddenly flee
et ne cetera consimili ratione sequantur,	apart, being dissolved through the
1105 neve ruant caeli tonitralia templa superne,	great void, and lest the other parts
terraque se pedibus raptim subducat et omnis	in like manner follow, and lest the thundering quarters of the sky crash in
inter permixtas rerum caelique ruinas	from above, and the earth swiftly
corpora solventes abeat per inane profundum,	withdraw itself from under our feet, and as a whole amidst the mingled

---

125 See also n. 61 on p. 178 above.

temporis ut puncto nil extet reliquiarum	ruin of things and heaven, dissolving
1110 desertum praeter spatium et primordia	their bodies, depart through the
caeca.	profound void, so that in a moment of
nam quacumque prius de parti corpora	time nothing rests of the remains
desse	but deserted space and invisible first
constitues, haec rebus erit pars ianua leti,	beginnings. For from whichever part
hac se turba foras dabit omnis materiai.	you first assume the bodies to leave,
	this part will be the door of death to
	things, through this the whole mass of
	matter will remove itself abroad.

A lot has been written about the function of these lines and their connection with the preceding passage. Since Bailey (1947) it is generally assumed that it is a warning against the disastrous consequences that would ensue if air and fire would be centrifugal, as the Stoics claim they are. Yet, in fact only the first two lines (1102–1103) can be plausibly read in this manner: if fire is taken to be *centrifugal*, then the walls of the cosmos, being fiery,[126] might well be expected to flee apart and be dissolved into the great surrounding void. From line 1104 onwards, however, the account becomes purely Epicurean (as even Bailey and his followers admit): there is no reason why, on the assumption of *centrifugal* air and fire, the other elements should follow (1104), there is no reason why the sky should fall down (1105),[127] or why the earth should be withdrawn from under our feet, leaving nothing behind but empty space and scattered particles (1106–1113). These events can only be understood as ingredients of an Epicurean disaster scenario.[128] However, if so many details of

---

126 Cf. *DRN* I 73: 'flammantia moenia mundi'.
127 Bailey translates line 1105 as: 'lest the thundering quarters of the sky *rush upwards*' (rejecting Munro's '*tumble in from above*'), apparently understanding 'upwards' in the Stoic sense, as a synonym for 'away from the centre' (see n. 61 on p. 178 above). Yet, this will not do at all. Firstly, as Brown (1984) aptly points out, because the Latin will not allow it: *caelum ruere* was proverbial for the sky *falling down*, and *superne* in Lucretius' days always meant 'from above' or 'above', but never 'upwards'. Secondly, it would be against Lucretius' consistent practice, if, having managed throughout the passage to remain faithful to Epicurus' *parallel* conception of 'up' and 'down' (see previous section), he now suddenly were to conform to the *centrifocal* language of his rivals and use the word 'upwards' in the Stoic sense of 'away from the centre'. Besides, if 'superne' is to be understood in a *centrifocal* sense, so is 'se subducat' in the next line, but this is impossible, because in a centrifocal cosmology there is no lower place toward which the earth might withdraw.
128 Cf. *DRN* VI 605–607: ne pedibus raptim tellus subtracta feratur / in barathrum, rerumque

the description point to a purely Epicurean account, it seems more reasonable to interpret the first two lines accordingly as well. But would it be consistent with the Epicurean theory of atomic motion, if, on the assumption of an infinite extra-cosmic void, the walls of the world were to fly off in *every* direction? I believe it would. In *DRN* II 1133–1135, for example, Lucretius describes how every compound thing, including the cosmos itself, having reached the limits of its growth, and the greater its extension, will scatter and emit *in all directions* (in cunctas undique partis) more atoms than is due.[129] Apparently then, even with an infinite amount of matter all around, a cosmos loses atoms *on all sides*, and it will do so even more when there is only empty space around: there is no reason at all why the first two lines of the present passage could not have been framed from a purely Epicurean point of view. Therefore I move to reject Bailey's interpretation and to return to the one proposed by Munro, who thought that in the lacuna Lucretius would first have formally concluded his criticism of the anonymous opponents, and then reasserted his *own* position, viz. that for our finite cosmos to remain stable in infinite space, an infinite supply of extra-cosmic matter is needed. The twelve lines following the lacuna would then be a description, on purely Epicurean lines, of what might happen to the cosmos in the absence of this extra-cosmic matter. These lines, therefore, are not part of the anti-Stoic passage, which would have ended somewhere in the lacuna.

4.3.1.8 Lucretius' Criticism of the Rival Position

It is now time to have a closer look at Lucretius' criticism of the rival theory and see what this may tell us about Lucretius' and Epicurus' own position. I will therefore review the three points of Lucretius' criticism, and see how valid and cogent they are, first, against Lucretius' most likely targets, i.e. the Stoics, and, second, against any form of centrifocal theory, and also to what extent Lucretius was justified in rejecting the theory.

As we saw above (§ 4.3.1.4 on p. 190), Lucretius' criticism of the rival position consists of (at least) three points:

---

sequatur prodita summa / funditus, et fiat mundi confusa ruina.—'{fear} that the earth may be withdrawn from under our feet and fall into the great abyss, and that the sum of things, now compromised, may follow, and a confused ruin of the world come about.'

129   *DRN* II 1133–1135: quippe etenim quanto est res amplior, augmine adempto, / et quo latior est, *in cunctas undique partis* / plura modo dispargit et a se corpora mittit.

i. There is no centre in the infinite universe (1070–1071)
ii. Nothing would stop at this 'centre', because neither can a mere point in space offer resistance nor can a body suddenly lay down its weight (1071–1082)
iii. The anonymous rivals also claim (1083–1093 ff.) that:
   6. *not* everything tends towards the centre, but only earth and water, whereas air and fire move away from it (1083–1088)
   7. this is the reason why "the whole aether twinkles all around with stars, and the flame of the sun grazes through the blue of the sky" (1089–1090)
   8. and this is also why plants grow upwards (1091–1093 ff.)

(i.) The first point of Lucretius' criticism is aimed at the unnamed opponents' claim that all things tend towards the centre of the *universe*. According to Lucretius, this is an impossible notion, because the *universe*, being *infinite*, cannot have a centre. As Furley points out, however, this criticism does *not* seem to apply to the Stoics, who are reported to have said that all things tend towards the centre of the *cosmos*, which is *finite*. Either Lucretius does not have the Stoics in mind after all (so Furley), or, if he does, he is misunderstanding or misrepresenting their position (so Schmidt and Brown).

If Lucretius had been aware of the actual Stoic position, could he have accepted it? Probably not. If every part of the cosmos moves towards the centre of the cosmos *qua whole*, as the Stoics say, this means that the motion of each part is determined by the whole, and hence by every other part of this whole, whether adjoining or not. Yet, in Epicurean physics objects can only influence each other by direct physical contact.[130] Therefore, if there is to be a focus of universal attraction in the cosmos, the location of this focus should be independent of the location of each of the attracted bodies, but the only thing in Epicurean physics that exists independent of all bodies, is *space*. Accordingly, if there is to be a focus, it can only be defined with respect to space, which brings us back to Lucretius' criticism that space, being infinite, can have no centre. One wonders, however, what would happen if Lucretius' adversaries would simply concede the point and say: "All right, let's not call it 'centre', then; it is simply a special point in space."

(ii.) The second point of Lucretius' criticism concerns the incorporeal nature of space. This thing which the opponents call the 'centre' (whether of space or of the cosmos), being spatially defined, must be a spatial and therefore incorpo-

---

130 See *DRN* III 166 and Epic. *Hdt.* 67, with Furley (1981) 8, 12 and id. (1999) 420.

real point. However, as Lucretius has stated before, space can offer no resistance but must by its very nature yield to any movement.[131] There is no reason, therefore, why anything should stop at this 'centre', rather than continue to move in whatever direction it was moving.[132] Lucretius also mentions an alternative version of the theory, according to which bodies upon reaching the 'centre' would somehow 'lay down the force of their weight' and stop moving. Here Lucretius seems to make a mistake: if, as the unnamed opponents say, all things have a natural tendency to move towards the 'centre' (and no further), it is not the object stopping at the 'centre' that needs explaining, but rather its continued motion beyond the 'centre'. If the object stops this is simply the fulfilment of this natural tendency, not the result of some kind of resistance. Nor does it 'lay down the force of its weight', but stopping at the 'centre' is precisely what its weight compels it to do. The origin of Lucretius' mistake seems to lie in his incapability or unwillingness to set aside for the moment his own (not yet proved!) Epicurean conception that all things fall down to infinity (see p. 214 ff. below), a tendency which can only be resisted by force. However, behind the mistake lies a valid observation: the rival theory seems to posit an *anomaly*: one point in space that behaves differently from any other point in infinite space. And apparently the Epicureans found such an anomaly hard to swallow. In fact,

---

131   See *DRN* I 437–439: 'sine intactile erit, nulla de parte quod ullam / rem prohibere queat per se transire meantem, / scilicet hoc id erit, vacuum quod inane vocamus' and II 235–237: 'at contra nulli de nulla parte neque ullo / tempore inane potest vacuum subsistere rei, / quin, sua quod natura petit, concedere pergat'. This was sound Epicurean doctrine: see esp. Sextus Emp. *Adv. Math.* X 221–222: ἀχώριστα μὲν οὖν ἐστι τῶν οἷς συμβέβηκεν ὥσπερ ἡ ἀντιτυπία μὲν τοῦ σώματος, εἶξις δὲ τοῦ κενοῦ· οὔτε γὰρ σῶμα δυνατόν ἐστί ποτε νοῆσαι χωρὶς τῆς ἀντιτυπίας οὔτε τὸ κενὸν χωρὶς εἴξεως, ἀλλ' ἀίδιον ἑκατέρου συμβεβηκός, τοῦ μὲν τὸ ἀντιτυπεῖν, τοῦ δὲ τὸ εἴκειν.—'unalienable from the things of which they are properties, are, for instance, resistance from body and yielding from void: for it is neither possible to imagine body without resistance, nor void without yielding, but they are the eternal property of each, resisting of the one, and yielding of the other.' Cf. also the scholion ad Epic. *Herod.* 43: ... τοῦ κενοῦ τὴν εἶξιν ὁμοίαν παρεχομένου καὶ τῇ κουφοτάτῃ καὶ τῇ βαρυτάτῃ— '... the void providing equal yielding to the lightest and the heaviest of atoms'.

132   A very similar criticism against the Stoic position is brought forward by Plutarch, in his *On the face in the moon*. Having earlier (7, 924b5–6) reminded the reader that the Stoics considered such things as a 'centre' *incorporeal*, he later states that "there is no body that is 'down' towards which the heavy bodies are in motion and it is neither likely nor in accordance with the intention of these men {i.e. the Stoics} that the incorporeal should have so much influence as to attract all these objects and keep them together around itself" (11, 926b4–7, transl. Cherniss, slightly modified). See also Cherniss' useful remarks on p. 65, note d.

however, the anomaly is only apparent: if every atom is thought of as having an *in-built* centripetal tendency, then this tendency would be a property of matter, not of space.

(iii.) We do not know how Lucretius' third argument continued, since its final part has been lost in the eight-line lacuna following on line 1093. Bailey suggests that Lucretius may first have pointed out the inconsistency of centrifugal air and fire with the earlier claim that *all* things move towards the centre, and then argued that this theory could not be true anyway, because, if fire and air are centrifugal, there is nothing to stop them from continuing their centrifugal motion *ad infinitum* and being dispersed and lost in the infinite void.

Several sources confirm that the Stoics did in fact make these two (seemingly) inconsistent claims, and so the Stoics seem to be the perfect targets for Lucretius' criticism. According to Furley, however, the Stoics *meant* to say that the *primary* movement of *all* bodies is towards the centre, and that air and fire are only *relatively* centrifugal, in so far as they are squeezed upwards by the heavier elements earth and water. If this interpretation were right (which I do not believe) the Stoic position would in fact answer both parts of Lucretius' criticism. In the first place the observed inconsistency would turn out to be only apparent, and in the second place the danger of air and fire escaping into infinite space would be removed, as their centrifugal tendency would automatically stop as soon as they had risen above the heavier elements, whereupon their natural centripetal tendency would prevail. Such a theory, then, would be invulnerable to this part of Lucretius' criticism, and if he had really wanted to exclude every kind of centrifocal theory, he should have considered it. It is not as if he could not have conceived such a theory: in II 184 ff. he himself argues for a single downward tendency of all matter ('downward' in this case, of course, in a *parallel* sense). Moreover, we know of at least one philosopher who combined a single downward tendency of all matter with a *centrifocal* conception of 'up' and 'down': Strato, Aristotle's second successor as head of the Lyceum.[133]

If Lucretius had been aware of the possibility of such a theory, how would he have responded? There is a passage in his work that might suggest that he would still have rejected it. In *DRN* I 984–997, Lucretius offers (*inter alia*) the following argument for the infinity of space (my paraphrase):[134] 'If space were

---

[133] Aët. I 12.7 (Strato fr.51 Wehrli); Simpl. *In Arist. De cael.* 267.30–34 Heiberg (fr.52) [quoted on p. 212 below]; ibid. 269.4–6 (fr.50) [quoted in n. 137 on p. 211 below]. It must be noted, however, that Strato followed Aristotle in denying the existence of extra-cosmic void: Aët. I 18.4 (fr.55 Wehrli); Theodoret. *Graec. affect. cur.* IV 15.2–3 (fr.54).

[134] On this argument see p. 182 above.

finite, all matter would since long by force of its weight have been heaped up at the bottom, and nothing would ever be done below the vault of heaven, nor would there even be a heaven or a sun. Yet, in fact the atoms are forever in constant motion. Therefore, space cannot be finite.' Lucretius seems to think that the existence of an absolute *bottom* in the universe would cause all matter to be packed together into an inert mass. In the theory we are presently investigating the 'centre' provides just such a 'bottom' towards which all matter converges, so that the likely result in this case would also be an inert mass, quite unlike the world we actually see around us. So, even a general centripetal tendency of all matter, as maintained by Strato, would probably have been unacceptable to an Epicurean.

I will not leave it at this, however. Even though the parallel suggests that Lucretius would probably have rejected a general centripetal tendency of all matter, I do not think he would have been entirely justified in doing so. According to Epicurus, all atoms are forever moving; when their motion is checked in one direction they rebound, and continue moving in another direction.[135] This, I think, would be enough to prevent matter from being packed together into an inert mass, even when its natural motion would be impeded by an absolute 'bottom' or 'centre'.

It is now time to sum up my conclusions. In lines 1070–1093 ff. Lucretius rejects the centrifocal theory of the Stoics. His reason for doing so at this point is that their theory is the only remaining obstacle to his own Epicurean view that the infinite universe contains an infinite amount of matter. Other existing centrifocal theories, like Plato's, Aristotle's and Strato's, would not qualify at this point as they all assume the universe to be finite, a view which Lucretius has already rejected (before line 1014). The Stoic theory, then, was the only *existing* centrifocal theory that still needed to be refuted. However, if Lucretius had really wanted to dissipate any possible obstacle, he should not have contented himself with attacking the specifics of the Stoic theory, but also have explored conceivable alternative theories. Above I have tried to show that he would probably have rejected such theories as well, but I have also tried to show that he would perhaps not have been entirely justified in doing so.

4.3.1.9   Conclusion

Although Lucretius presents his rejection of centrifocal cosmology as the inevitable outcome of a logical argument, on closer scrutiny his arguments do

---

135   Epic. *Hdt.* 43; Lucr. *DRN* II 62–111.

not seem to be as strong as they appear. It is quite possible that Lucretius had other motives for rejecting centrifocal cosmology. For one thing, this rejection allows Lucretius to finally conclude (as he will probably have done explicitly in the lacuna after line 1093) that *infinite space* contains an *infinite amount of matter*. In the following books of the DRN the infinity of matter is simply taken for granted. It plays an especially important role in II 1023–1147, where the infinity of matter leads Lucretius to assume the existence of an infinite number of worlds, and this in turn provides the basis for the *principle of plenitude* (see § 2.3.2.1.6 on p. 21 above), which is the corner-stone of Epicurus' anti-teleological cosmology, and upon which the simultaneous truth of all alternative explanations in astronomy and meteorology rests (V 526–533: see § 2.3.2 on p. 13 ff. above). If on the other hand Lucretius had accepted some kind of centrifocal system this would not just leave the infinity of matter unproved, since it would remove the need for 'external blows' to keep the world together, but it would actually rule out the very existence of extra-cosmic matter (let alone an *infinite* amount of it), because this external matter, being subject to the same centripetal tendency as internal matter, would either have to be considered part of the cosmos already, and therefore not be 'external' to it, or else to be forever added to it, producing a forever increasing cosmos, which seems to be contradicted by the senses. Accepting a centrifocal cosmology would therefore exclude the infinity of matter, the infinite number of worlds, and the simultaneous truth of all alternative explanations.

Justifiably or not, Lucretius rejects centrifocal cosmology. Whether or not this rejection automatically entails the acceptance of a parallel-linear cosmology will be investigated in the next section.

4.3.2 *Downward Motion* (DRN II 62–250)

4.3.2.1 Introduction

In the previous section we saw Lucretius rejecting a theory in which all things move naturally towards the centre of the universe. We also saw—although Lucretius tries to obscure this by imposing his own parallel linear terminology—that the most likely proponents of this theory, the Stoics, identified this universal centripetal tendency with the downward motion that seems to be characteristic of heavy bodies. Theoretically, Lucretius' rejection of this theory can mean one of two things: either the downward motion of heavy bodies is *not natural*, or, if it is, it is *not centripetal*.

A version of the first option is often ascribed to Democritus, who is supposed to have claimed that heavy objects only acquire their downward tendency when caught in a cosmic whirl or vortex. If this ascription is correct, it is remarkable that Epicurus and Lucretius, who in many respects may be considered Democritus' philosophical heirs, entirely ignore this option. Like most ancient philosophers they did not hesitate to make the downward tendency of heavy bodies a natural and unalienable property of the primary constituents of matter.

However, in contrast to most other philosophers of their time, Epicurus and Lucretius reject the notion of *centripetal* downward motion. Instead, as we shall see, they assume that all heavy bodies move downward along *parallel* trajectories.

Another point of contention concerns the status of *upward* motion. While it was generally agreed that downward motion was a fundamental property of the first elements, it was not so obvious whether upward motion was so too. Whereas Plato and Aristotle, and to a certain extent the Stoics as well,[136] attributed both tendencies to the elements—downward motion to the heavy elements earth and water, and upward to the light elements air and fire—, the Epicureans maintained that in reality only heaviness and natural downward motion existed, while the upward tendency of lighter bodies was ascribed to expulsion and displacement by heavier ones.

Below I want to examine the Epicurean stance on each of these three questions, viz. whether downward motion is natural or forced, whether it is centripetal or parallel, and how it relates to upward motion. To facilitate the flow of my argument I will discuss them in reverse order.

4.3.2.2   Upward versus Downward Motion

In *DRN* II 184–215 Lucretius argues that nothing moves upwards of its own accord, but only downwards, and that those things which are observed to move upwards do so only because of the outward pressure exerted by the surrounding medium. Just as a plank of wood, when forcibly held under water and then released, is ejected violently by the surrounding water, so a flame must be thought to be forced upwards (204 *expressa*) by the surrounding air.

Epicurus' views on this subject are reported by Simplicius in three separate passages which Hermann Usener included as fragment 276 of his *Epicurea*. The relevant parts of the first two passages are quoted below; the third one provides no additional information and will be omitted.[137]

---

136   *Pace* Sambursky (1959), 111, and Furley (1966), 191–193: see p. 197f. above.
137   The third reference is in Simpl. *In Arist. De caelo* I 8, 277a33, 269.4–6 Heiberg (*Epicurea* 276 (3) / Strato 50 Wehrli): Ἰστέον δέ, ὅτι οὐ Στράτων μόνος οὐδὲ Ἐπίκουρος πάντα ἔλεγον εἶναι τὰ σώματα βαρέα καὶ φύσει μὲν ἐπὶ τὸ κάτω φερόμενα, παρὰ φύσιν δὲ ἐπὶ τὸ ἄνω […]

(1) Simplic. *In Arist. De caelo* III 1, 299a25 = 569.5–9 Heiberg:[138]

| Οἱ γὰρ περὶ Δημόκριτον καὶ ὕστερον Ἐπίκουρος τὰς ἀτόμους πάσας ὁμοφυεῖς οὔσας βάρος ἔχειν φασί, τῷ δὲ εἶναί τινα βαρύτερα ἐξωθούμενα τὰ κουφότερα ὑπ' αὐτῶν ὑφιζανόντων ἐπὶ τὸ ἄνω φέρεσθαι, καὶ οὕτω λέγουσιν οὗτοι δοκεῖν τὰ μὲν κοῦφα εἶναι τὰ δὲ βαρέα. | The followers of Democritus and later Epicurus say that the atoms, being all of the same nature, have weight, but that due to the fact that some things are heavier, the lighter ones are expelled by the heavier ones, when these settle down, and so move upwards, and in this way, they say, some bodies seem light and others heavy. |

(2) Simplic. *In Arist. De caelo* I 8, 277b1 = 267.30–268.4 Heiberg:[139]

| Ταύτης δὲ γεγόνασι τῆς δόξης μετ' αὐτὸν Στράτων τε καὶ Ἐπίκουρος πᾶν σῶμα βαρύτητα ἔχειν νομίζοντες καὶ πρὸς τὸ μέσον φέρεσθαι, τῷ δὲ τὰ βαρύτερα ὑφιζάνειν τὰ ἧττον βαρέα ὑπ' ἐκείνων ἐκθλίβεσθαι βίᾳ πρὸς τὸ ἄνω, ὥστε, εἴ τις ὑφεῖλε τὴν γῆν, ἐλθεῖν ἂν τὸ ὕδωρ εἰς τὸ κέντρον, καὶ εἴ τις τὸ ὕδωρ, τὸν ἀέρα, καὶ εἰ τὸν ἀέρα, τὸ πῦρ. | This opinion was later adopted by Strato and Epicurus, who assumed that every body has weight and moves *towards the centre*, but that, due to the fact that the heavier ones settle down, the less heavy are extruded upwards by force, so that, if one were to remove the earth from below, the water would reach to the centre, and if one removed the water, the air would, and if one removed the air, the fire. |

These two reports confirm and supplement Lucretius' account in several respects:

(a) We now learn that the same theory had been held by Democritus before (see below),[140] and by Strato afterwards. Yet, the passage in which the

---

138 Also included among the fragments of Democritus as A61a D–K.
139 Also included among the fragments of Strato as fr.52 Wehrli.
140 Cf. Simpl. *In Arist. De caelo* IV 4 311b13, 712.27–29 Heiberg (Democrit. A61b D–K): οἱ περὶ Δημόκριτον οἴονται πάντα μὲν ἔχειν βάρος, τῷ δὲ ἔλαττον ἔχειν βάρος τὸ πῦρ ἐκθλιβόμενον ὑπὸ τῶν προλαμβανόντων ἄνω φέρεσθαι καὶ διὰ τοῦτο κοῦφον δοκεῖν.

theory is also claimed for Strato is somewhat misleading. It says that according to Strato and Epicurus every body has weight and moves *towards the centre*. This was certainly not Epicurus' view, as we have seen above and shall further explore below. The centrifocal language of Simplicius' report is probably due to a conflation of Epicurus' views with those of Strato, who—as an Aristotelian—*did* equate 'downwards' with 'towards the centre'.

(b) We also learn that weight and lightness of (*compound*) *bodies* are somehow linked to the weight of the composing *atoms*. The details of this relation are provided by Lucretius in *DRN* I 358–369, where the relative weight and lightness of bodies are attributed to the admixture of smaller and greater quantities of void.

(c) Just as in Lucretius' account, the upward motion of lighter bodies is said to be caused by their being *expelled* (ἐξωθούμενα) or *extruded* (ἐκθλίβεσθαι; cf. Lucr. 204 *expressa*) upwards by the surrounding heavier bodies.

(d) Not just the *upward* motion of *lighter* bodies, but also the *downward* motion of *heavier* bodies is affected by the surrounding medium. As we saw above, Lucretius likens the *upward* motion of fire through the surrounding air to the upward thrust of a plank of wood submerged in water. In the two accounts of Simplicius the *downward* motion of *heavier* bodies is also compared to motion through a fluid: the verb ὑφιζάνειν (*sinking* or *settling down*) is typically used to denote the retarded downward motion of heavier bodies in a somewhat lighter liquid medium, e.g. earth and mud in water, and dregs in olive-oil or wine.[141]

Why did Epicurus posit only one natural elementary motion, viz. *downwards*, instead of two, viz. *upwards* and *downwards*? Although our sources are silent on the matter, Epicurus may have reasoned something like the following. If both heaviness and lightness are supposed to be fundamental properties of certain elements, then the transformation of something heavy (e.g. a log) into something light (e.g. fire and smoke) would require for the composing elements

---

141 See e.g. Nemesius *De nat. hom.* 5.151–153 (Einarson): ἐὰν γὰρ εἰς ὕδωρ γῆν ὀλίγην βαλὼν ταράξῃς, διαλύεται εἰς ὕδωρ ἡ γῆ· ἐὰν δὲ παύσῃ ταράττων, στάσιν λαβόντος τοῦ ὕδατος ὑφιζάνει; Philoponus *In Arist. Mete.* 33.38 (Hayduck) βαρυτέρα τοῦ ὕδατος οὖσα ἡ γῆ ὑποχωρεῖ τούτῳ καὶ ὑφιζάνει; Joh. Chrysostom. *Homiliae* 59.147.27–28 (MPG) τῆς ἰλύος ὑφιζανούσης; Galen *De methodo medendi* 10.973.13–14 (Kühn) τῇ τοῖς οἴνοις ὑφιζανούσῃ τρυγί; id. *De simplicium medicamentorum temp.* 11.414.13–14 (Kühn) ὃ δὴ καὶ ὑφιζάνει τῷ χρόνῳ, τρὺξ μὲν ἐπὶ τῶν οἴνων, ἀμόργη δ' ἐπ' ἐλαίου καλούμενον.

to be transformed themselves. Yet, as Lucretius says again and again, elements, if they are to be real elements, cannot change.¹⁴² Therefore, all the elements must have the same fundamental properties, and be all 'of the same nature' (ὁμοφυεῖς).

But why should this primary property they all share be heaviness, and not lightness? On this point neither Epicurus nor Lucretius express themselves clearly, but they may have thought something along the following lines. In our experience heavier objects are generally more densely packed, and better able to offer resistance than lighter bodies.¹⁴³ Therefore, if weight and downward motion are considered fundamental, it is not hard to imagine how the heavier bodies (e.g. earth, stone, water) would be able to press the lighter ones (e.g. air, fire, smoke) upwards. If, on the other hand, lightness and upward motion are taken as fundamental, it is very difficult to conceive how the lighter bodies (air, fire, smoke), being rarefied and volatile, would be able to press the heavier ones (earth, stone, water) downwards. For this reason weight and downward motion seem the better candidates for primacy.

4.3.2.3 Downward Motion and the Atomic Swerve

Above we have seen that Lucretius rejected the anonymous rivals' assumption of a centripetal tendency of all matter. We have also seen that in describing their *centrifocal* theory he consistently uses *parallel-linear* terminology. Is this enough to conclude that Lucretius' own theory was parallel-linear? Perhaps. Nevertheless, it may be interesting to try and find some more positive proof for attributing a parallel-linear theory to the Epicureans.

In *DRN* II 83–85 two kinds of motion are ascribed to the atoms: they move either because of their own *weight* ('gravitate sua') or because of collisions with other atoms.¹⁴⁴ In II 190 and 205 we are told that *weights*, insofar as they are weights, all tend to move *downwards* ('deorsum').¹⁴⁵ Yet, this still does not tell us whether this downward motion is *parallel* or *centripetal*.

The clearest statement is found in the next section of Lucretius' account. In II 216–250 Lucretius presents a third kind of atomic motion, the *swerve* or

---

142   See esp. *DRN* I 665–674 and 782–797.
143   See e.g. *DRN* I 565–576 and II 100–108. That this observation was not restricted to Epicureans is shown by Arist. *Phys.* IV 9, 217b11–12: ἔστι δὲ τὸ μὲν πυκνὸν βαρύ, τὸ δὲ μανὸν κοῦφον.
144   *DRN* II 83–85: cuncta necessest / aut *gravitate sua* ferri primordia rerum / aut ictu forte alterius. Cf. Epic. *Hdt.* 61: οὔθ' ἡ ἄνω οὔθ' ἡ εἰς τὸ πλάγιον διὰ τῶν κρούσεων φορά, οὔθ' ἡ κάτω διὰ τῶν ἰδίων βαρῶν.
145   *DRN* II 190: pondera quantum in se est cum deorsum cuncta ferantur; *DRN* II 205: pondera quantum in se est deorsum deducere pugnent.

declination ('clinamen').¹⁴⁶ This swerve is introduced with the express purpose of allowing the atoms to meet and collide (221–224):¹⁴⁷

Quod nisi *declinare* solerent, omnia deorsum imbris uti guttae caderent per inane profundum, nec foret offensus natus nec plaga creata principiis: ita nil umquam natura creasset.	For if they were not used to *swerve*, all things would, like drops of rain, fall down through the profound void, and no collision would be born, nor blow created among the atoms: nature thus would never have created aught.

Without the swerve, i.e. if their motion were determined by weight alone, the atoms would never meet. This means that their downward trajectories do *not intersect* at any point; in other words: they are *parallel*. The same theory is also explicitly ascribed to Epicurus himself in a number of passages collected as fr.281 in Hermann Usener's *Epicurea*. The conclusion that downward motion must be parallel can also be inferred from a famous passage, § 60, of Epicurus' *Letter to Herodotus*:

Καὶ μὴν καὶ τοῦ ἀπείρου ὡς μὲν ἀνωτάτω καὶ κατωτάτω οὐ δεῖ κατηγορεῖν τὸ ἄνω ἢ κάτω. εἰς μέντοι τὸ ὑπὲρ κεφαλῆς, ὅθεν ἂν στῶμεν, εἰς ἄπειρον ἄγειν ὄν, μηδέποτε φανεῖσθαι τοῦτο ἡμῖν, ἢ τὸ ὑποκάτω τοῦ νοηθέντος εἰς ἄπειρον ἅμα ἄνω τε εἶναι καὶ κάτω πρὸς τὸ αὐτό· τοῦτο γὰρ ἀδύνατον διανοηθῆναι.	Furthermore, of the infinite it is necessary that one not use the expressions "up" or "down" in the sense of "highest" and "lowest." For certainly, while it is possible to produce [a line] to infinity in the direction overhead from wherever we may be standing, [it is necessary] that this [view] never seem right to us, or that the lower part of the [line], imagined to infinity, be at the same time up and down with respect to the same thing. For this is impossible to conceive.

---

146   Cf. Cic. *De fato* 22 (Epic. fr.281b Us.): 'Itaque tertius quidam motus oritur extra *pondus* et *plagam*, cum *declinat* atomus intervallo minimo', Aët. I 12.5 (Epic. fr.280a Us.): κινεῖσθαι δὲ τὰ ἄτομα ποτὲ μὲν κατὰ στάθμην ποτὲ δὲ κατὰ παρέγκλισιν, τὰ δ' ἄνω κινούμενα κατὰ πληγήν, κατὰ ⟨ἀπο⟩παλμόν.

147   See also Cic. *De fin.* I 19 (Epic. fr.281a Us.): "... si omnia deorsus e regione ferrentur et, ut dixi, ad lineam, numquam fore ut atomus altera alteram posset attingere itaque attulit rem commenticiam: *declinare* dixit atomum perpaulum, quo nihil posset fieri minus."

ὥστε ἔστι μίαν λαβεῖν φορὰν τὴν ἄνω νοουμένην εἰς ἄπειρον καὶ μίαν τὴν κάτω, ἂν καὶ μυριάκις πρὸς τοὺς πόδας τῶν ἐπάνω τὸ παρ' ἡμῶν φερόμενον ⟨εἰς⟩ τοὺς ὑπὲρ κεφαλῆς ἡμῶν τόπους ἀφικνῆται ἢ ἐπὶ τὴν κεφαλὴν τῶν ὑποκάτω τὸ παρ' ἡμῶν κάτω φερόμενον· ἡ γὰρ ὅλη φορὰ οὐθὲν ἧττον ἑκατέρα ἑκατέρᾳ ἀντικειμένη ἐπ' ἄπειρον νοεῖται.

Therefore one may assume one upward course imagined to infinity and one downward, even if something moving from us toward the feet of those above us should arrive ten thousand times at the places over our heads, or something moving downward from us at the heads of those below. For the whole course is nonetheless imagined to infinity as one [direction] opposed to the other.[148]

According to Konstan (1972), the argument of this passage is directed against Aristotle's *centrifocal* view, which defines 'up' with reference to the periphery, and 'down' with reference to the centre of the cosmic sphere. Epicurus objects to this view (which is also the Stoic one) because it would entail that a continued downward motion through the centre would be both 'downwards' and 'upwards', which he finds inconceivable. Instead he posits a single upward and single downward direction, both extending to infinity. In other words: he evinces a *parallel-linear* conception of 'up' and 'down'.

Apparently, for Epicurus as for Lucretius, the parallel-linear model was the only conceivable alternative to the centrifocal system of their adversaries, and rejecting the latter therefore automatically implied the former. In addition they may have thought that they had the evidence of observation on their side: Lucretius compares the downward motion of the atoms to the fall of raindrops, which certainly looks parallel.

4.3.2.4 Downward Motion: Natural or Forced?
For all their differences Lucretius and his centrifocalist opponents agreed on one thing: they all assumed that downward motion was something natural to the first elements. By contrast, Democritus is commonly claimed to have held that matter has *no* primary form of motion at all, but only acquires its (undeniable) upward and downward tendencies when caught in a *cosmic whirl*. This claim was first made by Adolf Brieger in 1884 and, independently, by Hugo Liepmann in 1885, and has been embraced by most subsequent scholars.[149]

---

148 Text and translation: Konstan (1972), who also provides a thorough analysis of this passage.
149 Furley (1976), 80 n. 16, and (1983), 94, cites Dyroff (1899); Burnet (1892) 343–345; Kirk &

The reason for ascribing such a view to Democritus is as follows. Aristotle repeatedly accuses Democritus of failing to assign a natural motion to the atoms. (Brieger mentions *Metaph.* XII 6, 1071b32; *Phys.* VIII 1, 252a34; *GA* II 6, 742b17 and *De caelo* III 2, 300b8.) This is often taken to mean that Democritus' atoms do *not have a natural downward tendency*. In addition it is stated by Aëtius (I, 3, 18 and I, 12, 6) that Democritus denied the atoms the property of *weight*. This claim is contradicted, however, by Aristotle (*GC* I, 8, 326a8) and Theophrastus (*De sens.* 61), who state that Democritus' atoms are *heavier* in proportion to their *size*.

This contradiction is resolved by most scholars in the following manner. Alone in the extra- or pre-cosmic void the atoms have no natural tendency to move in a particular direction. The *cosmic whirl*, however, has the effect of driving larger and bulkier bodies towards the centre, and *this* is the downward tendency we generally associate with *weight*.

David Furley[150] strongly objects to this solution of the problem, because it does not take into account the dynamics of whirls, nor the views on weight that *can* be plausibly and positively attributed to the early atomists. Furley offers the following arguments:

1. In our every-day experience, the downward tendency of heavy objects is a necessary *condition* for the sorting effect of a whirl. Heavy objects sink down and because of the friction with the bottom they collect where the motion of the whirl is least: in the centre *of the bottom*. On the other hand, objects that are light enough to remain suspended or even to float to the surface are driven away from the centre. The upward and downward tendencies of heavy and light bodies must therefore precede, and exist independently from, the whirl.[151] I must admit that I do not entirely agree with this particular argument. It is unlikely that the early cosmologists were aware of the precise mechanics of a whirl. They do not seem to have realised, for instance, that *bottom friction* was a necessary factor, for, during the formation of the cosmos, when, according to their theories, the cosmic whirl would have been most active, nothing yet was formed that might serve as a 'bottom' to the whirl.

Raven (1957) 415 f.; Guthrie (1965) 400–404; Alfieri (1953) 88 ff.; and O'Brien (1984). More recently, a variant of this theory was defended by Chalmers (1997).
150 Furley (1976) and (1983).
151 Furley (1983) 95–96.

2. The cosmic whirl is in fact also used by certain early cosmologists to explain why some heavy bodies, viz. the celestial bodies, *do not fall down, but remain poised in heaven*.[152] Therefore, the general *downward* tendency of heavy bodies cannot be attributed to the whirl.[153]

3. According to the doxographical reports, the cosmic whirl causes heavy objects to move *towards the centre*.[154] This sounds exactly like the behaviour attributed to heavy objects in *centrifocal* cosmologies like Aristotle's and the Stoics', where 'towards the centre' equals 'downward'. In reality, however, as Furley observes, whirls only draw heavy objects *horizontally* towards the central *axis* of their rotation. Yet, even if we admit—against Furley—that the cosmic whirl is somehow capable of drawing heavy objects *from all directions* towards its central *point*, as Aristotle's testimony suggests it does,[155] this is still far removed from the typical behaviour Democritus assigns to heavy objects. In Aristotle's *De caelo* II 13, 294b13–30, Democritus is explicitly numbered among those who believe that the earth is kept from falling down by the resistance of the air underneath. This means that for Democritus the centre, where the earth is, is not the focal point of gravity, and the downward motion of heavy objects is not *centrifocal* but *parallel*.[156]

To these three arguments we may perhaps add the following:

4. The early cosmologists attributed to the cosmic whirl the property of driving dense matter to the centre and tenuous substances to the periphery.[157] This means that neither tendency, the inward and the outward, is prior to the other, but both are subordinate to the rotating movement of the whirl. Simplicius, however, ascribes to Democritus the view that *all* atoms are heavy and tend to move downwards, but that some lighter bodies are observed to move upwards, because they are *extruded* by the heavier bodies

---

152 See Leucippus fr. A1.23–26 D–K; Anaxagoras frs. A12, A42, A71 D–K. That Democritus used the cosmic whirl in much the same way is suggested by the fact that he too, like Anaxagoras, believed the celestial bodies to be heavy and stone-like: cf. Democritus frs. A85, A87, A90, A39 D–K.
153 Furley (1983) 96–97.
154 Arist. *Cael.* II 13, 295a13–14; Anaxagoras fr. A42; Leucippus fr. A1.21 D–K.
155 Arist. *Cael.* II 14, 297a12–19.
156 Furley (1983) 98–100; id. (1981) 11–12, and id. (1976) 80–81.
157 Anaxagoras frs. B15, A42; Leucippus A1.11–26.

around them.[158] In other words: upward motion is secondary to downward motion. Therefore, if both reports are right, the downward and upward tendencies of heavy and light bodies cannot be identified with the inward and outward motions caused by the cosmic whirl.

Furley concludes (rightly, in my opinion) that the cosmic whirl cannot serve to account for the downward motion of heavy bodies, and was in fact never used so by Democritus (or anyone else, for that matter). Democritus, just as Epicurus and Lucretius afterwards, considered *parallel* downward motion a *natural* and *fundamental* property of all the atoms.[159] Perhaps Aristotle, when he accused Democritus of *failing* to assign a natural motion to the atoms, was simply thinking of his own theory of natural motion as motion *towards a natural place*.[160] What is important to us is the realisation that probably no one in antiquity denied the fundamental character of downward motion, whether *parallel* or *centrifocal*, and that Lucretius was therefore justified in believing that by refuting *centrifocal* downward motion he had by the same token proved the only conceivable alternative, i.e. *parallel* downward motion.

4.3.2.5   Summary
At the beginning of this subsection I set out to answer three questions, viz. (1) whether the Epicureans considered downward motion natural or forced, (2) whether they conceived of downward motion as centripetal or parallel, and (3) how they explained upward motion in relation to downward motion. To these questions we have found the following answers:

Ad 1:   Like almost every other philosophical school the Epicureans considered downward motion a natural concomitant of weight. If Democritus claimed otherwise (which is doubtful), the Epicureans appear to have been ignorant of the fact, and did not feel any need to justify their position.

Ad 2:   As a consequence they could safely assume that by rejecting the centripetal model of their adversaries they had by the same token proved the only available alternative, viz. parallel downward motion. That they

---

158   Simpl. *In Arist. De caelo* 569.5–9 Heiberg (Democr. fr.A61a D–K; see text 1 on p. 212 above) and ibid. 712.27–29 (Democr. fr.A61b; see text in n. 140 on p. 212 above).
159   This is, as Furley (1976), 81 n. 20, (1983), 100, himself acknowledges, a return to the interpretation of Zeller (1879).
160   See Furley (1976) and (1983).

conceived of downward motion as parallel is testified by their introduction of the so-called *clinamen* to ensure that atoms can collide.

Ad 3: In contrast to most other philosophers of their time the Epicureans considered only weight and downward motion fundamental to bodies, attributing the upward motion of lighter bodies to expulsion by heavier ones.

For our investigation the second point is, of course, the most important, since parallel downward motion is generally assumed to imply a flat earth. The third point will be useful in our discussion of Lucretius' cosmogony on pp. 223 ff. below.

4.3.3 *The Apparent Proximity of the Sun* (DRN IV 404–413)

We have now established that Epicurus and Lucretius assumed a parallel downward motion. In §4.2.4 on pp. 177 ff. above we have also seen that in such cosmologies the earth is most often and most conveniently considered *flat*. In the absence of more specific statements about the shape of the earth it would be interesting to have a passage that at least *suggests* a more specific shape. One such passage may be the following. In DRN IV 404–413 Lucretius describes the well-known illusion of the sun appearing to *touch* the mountains above which it is seen to rise, although in fact (IV 410–413) …

inter eos solemque iacent immania ponti aequora substrata aetheriis ingentibus oris, interiectaque sunt terrarum milia multa, quae variae retinent gentes et saecla ferarum.	between the mountains and the sun lie vast plains of sea spread below the great aetherial regions, and thrown in between are many thousands of lands which various peoples and breeds of wild beasts inhabit.

Although the purpose of these lines is to demonstrate the enormous distance that separates us from the sun,[161] Lucretius' way of expressing this distance at the same time suggests that, on a cosmic scale, the sun is really *not so far away*, but lies closely to the edge of a *flat* earth. Only in this way would it be possible to measure the entire distance by the intervening lands and seas.

---

161 Cf. Cleomedes II 1.136–139: 'Again, often when setting or rising on a mountain peak, the Sun sends out to us the appearance of its touching the peak, although its distance from every part of the earth is as vast as is to be expected when the earth has the ratio of a point in relation to its height.' (transl. Bowen-Todd).

4.3.4 *Climatic* Zones? (DRN V 204–205)

Another passage where the shape of the earth might seem to play a role is *DRN* V 204–205. Here, in the course of an argument against divine providence Lucretius writes:

Inde duas porro prope partis fervidus ardor adsiduusque geli casus mortalibus aufert.	Further, almost two parts {of the earth} the boiling heat and continuous fall of frost take away from mortal men.

Most commentators see these lines as a reference to the theory of the five zones.[162] According to this theory the earth is divided into *five* climatic *zones* or *belts*, three of which are uninhabitable to man: the arctic and antarctic zones due to extreme cold, and the equatorial zone due to extreme heat, leaving only the two intervening temperate zones for human habitation. An interesting parallel to these lines is found in the speech of the Epicurean spokesman Velleius in Cicero's *De natura deorum* I, 24:

… terrae maxumas regiones inhabitabilis atque incultas videmus, quod pars earum adpulsu solis exarserit, pars obriguerit nive pruinaque longinquo solis abscessu.	… we observe that enormous regions of the earth are uninhabitable and uncultivated, because part of them is burned by the approach of the sun, and part is stiffened by snow and hoar-frost and the distant retreat of the sun.

Here too the commentaries refer us to the theory of the five zones.[163] An interesting detail, which most commentators fail to mention, is that this the-

---

162   Merrill (1907): "The ancients thought the torrid **zone** uninhabitable" with a ref. to Ov. M. I 49–50 '[**zonae**] quarum quae media est, non est habitabilis aestu; nix tegit alta duas'. Bailey (1947): "*duas partis*: {…} the 'two-thirds' are the tropic and arctic **zones**." Costa (1984): "*duas partis*: 'two-thirds': the equatorial and arctic **zones**. This refers to a belief that there were torrid, temperate and arctic belts or **zones** in the heavens which caused corresponding ones on the earth, only the temperate one [sic!] being habitable by man". Schmidt (1990) 197: "Kalte **Zone** [sic!] um die Pole, verbrannte **Zone** am Äquator, gemäßigte **Zonen** dazwischen." [my emphasis]. Similarly Abel (1974) 1036.50–1037.59.

163   See e.g. Goethe (1887) 37, and Pease (1955) vol. 1, 202–203. Similarly Abel (1974) 1037.58–59.

ory presupposes a *spherical earth*.¹⁶⁴ So, do these two passages prove that the Epicureans knew and accepted the earth's sphericity after all? Hardly. In neither passage the technical term 'zone' or any of its known translations is used, nor is the number *five* mentioned or implied. On the contrary: Lucretius and Velleius seem to be thinking of *three* rather than *five* 'parts'. If Lucretius had wished to refer specifically to the theory of the *five* zones, *three* of which are uninhabitable to man, he would not have spoken of *two* parts but of *three*, as the elder Pliny does, more than a century later (*NH* II 172): "... terrae *tris partis* abstulit caelum ...", "... *three parts* of the earth (out of five) the sky has taken away ...". There is no reason, therefore, to assume that either Lucretius or Velleius refers specifically to the five-zone-*theory*. Instead, the two references apply just as well, if not better, to the ancient *observation* that regions to the extreme north and south of our part of the earth were uninhabitable to man because of the freezing cold and the scorching heat respectively. Herodotus, for instance, writes that the country south of the Aethiopian 'deserters' is desolate (ἔρημος) because of the heat (*Hist.* II 31.1), and that because of the continuous winter the northernmost regions of our continent are uninhabited (ἀνοίκητα) (*Hist.* IV 31.2). These observations were also accepted by flat-earth philosophers, and accommodated to their worldview¹⁶⁵ (see figure 4.2 below).¹⁶⁶

In later times, the recognition of the earth's sphericity provided such observations with a new theoretical basis, transforming the three climatic divisions into *zones* encircling the earth: the frigid northern region was now defined by the arctic circle and became the *arctic zone*, and the torrid southern region was defined by the two tropic circles and became the *tropic or equatorial zone*, while the region in between became the *temperate zone*. Symmetry demanded the introduction of two further zones south of the winter tropic: a southern temperate zone, also known as the *counter-temperate* zone, and a southern frigid zone,

---

164   Schmidt (1990) 215, referring to the same passage, does mention the earth's sphericity: "Kennzeichnend für die Argumentation des Lukrez ist in diesem Zusammenhang, daß er ebenfalls an einer anderen Stelle, nämlich in V 204f., die **Zonenlehre**, deren Kenntnis die **Kugelgeographie** voraussetzt, für einen Beweis heranzieht" [my emphasis]. Similarly Abel (1974) 1037.18–35.
165   See e.g. Anaxagoras fr. A67 / Diogenes of Apollonia fr. A11 D–K (quoted on p. 248 below) and Leucippus fr. A27 D–K (quoted in n. 207 on p. 249 below).
166   On the ancient three-part theory as opposed to the later five-zone theory see Abel (1974) 1012.55–1013.24.

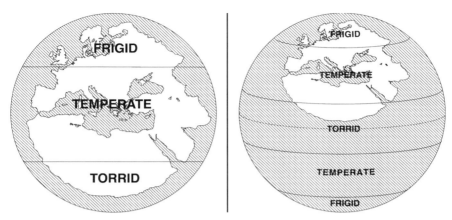

FIGURE 4.2 *Three climatic regions versus five climatic zones*

also called the *counter-arctic* or *ant-arctic*. Thus the theory of the *five zones* was born.[167] The older theory of the three climatic regions, and the newer theory of the five zones are illustrated in figure 4.2.

If the two Epicurean passages must be related to one of these two world-views, it is clear that they should be thought of as referring to the older theory, which presupposes a flat earth. It is also possible, however, to interpret the two passages as neutral with respect to the two world-views and hence to the shape of the earth: one could argue that both passages are simply stating the existence of three types of climate, frigid, temperate and torrid, without reference to their geographic division or the shape of the earth. Either way, the passages should not be interpreted as specific references to the five-zone theory and hence to a spherical earth.

4.3.5   Lucretius' Cosmogony (DRN V 449–508)
4.3.5.1   Introduction

Until now the picture arising from our investigation seems unambiguous: Lucretius rejects centrifocal natural motion, which is commonly thought to imply a spherical earth, and instead, like Epicurus before him, assumes that natural motion is parallel, an assumption which—in turn—is thought to imply a flat earth. In addition we have found two passages that may not strictly exclude the earth's sphericity, but sit more easily with the notion of a flat earth.

The next passage I want to discuss, Lucretius' cosmogonical account in book V of the *DRN*, will—if my interpretation is correct—throw all this into

---

167   On the five-zone theory in general see Abel (1974).

confusion. The entire passage runs from line 416 to line 508, but the relevant portion for our purposes begins only at line 449. In these lines (449–508) Lucretius describes the formation of the major parts of the cosmos, resulting in a four-layered structure with earth at the bottom, then water, then air and finally aether. The reason for including this passage in the present argument is that it can—as I will try to prove—only plausibly be understood in a *centrifocal* sense, which flatly contradicts Epicurus' and Lucretius' otherwise *parallel-linear* cosmology.

4.3.5.2   Origins and Parallels

Although the extant works and fragments of Epicurus do contain some general remarks concerning cosmogony (see esp. the *Letter to Pythocles* 4 [89–90]), the relevant portion of Lucretius' account has no parallel in any work by any known Epicurean. The closest parallel is a curious chapter in Aëtius' *Placita*. A typical chapter in Aëtius' work reports by name the views of several philosophers concerning one or a number of related topics, which are presented in indirect speech (*AcI*-constructions). In chapter I 4, however, which deals with the coming-into-being of the cosmos, only one view is given, which is presented in direct speech and not attributed to anyone. The only certain clue to its provenance is the fact that it mentions atoms.[168] It is this account that furnishes the closest parallel to Lucretius' cosmogony, which it matches in all essential details (see below). For ease of reference the text and a literal translation are printed below; the division into sections is my own.

Πῶς συνέστηκεν ὁ κόσμος	How the cosmos was constituted.
1 Ὁ τοίνυν κόσμος συνέστη περικεκλασμένῳ σχήματι ἐσχηματισμένος τὸν τρόπον τοῦτον.	The cosmos, then, was constituted and shaped with a bent shape in the following manner.
2 Τῶν ἀτόμων σωμάτων ἀπρονόητον καὶ τυχαίαν ἐχόντων τὴν κίνησιν συνεχῶς τε καὶ τάχιστα κινουμένων εἰς τὸ αὐτό, πολλὰ σώματα συνηθροίσθη, [καὶ] διὰ τοῦτο ποικιλίαν ἔχοντα καὶ σχημάτων καὶ μεγεθῶν.	As the atomic bodies have an unguided and haphazard motion and are constantly and most swiftly moving, many bodies were gathered together in the same place, and thereby had a variety of shapes and sizes.

---

168   For the various ascriptions of this account see e.g. Spoerri (1959) 7–8.

3 Ἀθροιζομένων δ' ἐν ταὐτῷ τούτων, τὰ μέν, ὅσα μείζονα ἦν καὶ βαρύτατα, πάντως ὑπεκάθιζεν· ὅσα δὲ μικρὰ καὶ περιφερῆ καὶ λεῖα καὶ εὐόλισθα, ταῦτα καὶ ἐξεθλίβετο κατὰ τὴν σύνοδον τῶν σωμάτων εἴς τε τὸ μετέωρον ἀνεφέρετο.

As they gathered in the same place, those that were larger and heaviest settled down completely, while those that were small and round and smooth and slippery were extruded during the concourse of atoms and carried up into the upper region.

4 Ὡς δ' οὖν ἐξέλιπε μὲν ἡ πληκτικὴ δύναμις μετεωρίζουσα, οὐκέτι δ' ἦγεν ἡ πληγὴ πρὸς τὸ μετέωρον, ἐκωλύετο δὲ ταῦτα κάτω φέρεσθαι, ἐπιέζετο πρὸς τοὺς τόπους τοὺς δυναμένους δέξασθαι· οὗτοι δ' ἦσαν οἱ πέριξ, καὶ πρὸς τούτοις τὸ πλῆθος τῶν σωμάτων περιεκλᾶτο, περιπλεκόμενα δ' ἀλλήλοις κατὰ τὴν περίκλασιν τὸν οὐρανὸν ἐγέννησεν.

When the force of the blows stopped lifting them up, and the blows no longer drove them into the upper region, they were prevented from being carried down ⟨and⟩ were squeezed into those places that were able to receive them: these were the places all around, and to these the majority of the bodies were bent round, and as they became entangled with each other during the bending they generated the sky.

5 Τῆς δ' αὐτῆς ἐχόμεναι φύσεως ⟨αἱ⟩ ἄτομοι, ποικίλαι οὖσαι, καθὼς εἴρηται, πρὸς τὸ μετέωρον ἐξωθούμεναι τὴν τῶν ἀστέρων φύσιν ἀπετέλουν· τὸ δὲ πλῆθος τῶν ἀναθυμιωμένων σωμάτων ἔπληττε τὸν ἀέρα καὶ τοῦτον ἐξέθλιβε· πνευματούμενος δ' οὗτος κατὰ τὴν κίνησιν καὶ συμπεριλαμβάνων τὰ ἄστρα συμπεριῆγε ταῦτα καὶ τὴν νῦν περιφορὰν αὐτῶν μετέωρον ἐφύλαττε.

Having the same nature and being varied, as was said, the atoms that were expelled to the upper region produced the nature of the heavenly bodies; the majority of the bodies that were being exhaled struck the air and extruded it: and the air, turned into wind during its movement and embracing the heavenly bodies, drove them round and preserved their present revolution in the upper region.

6 Κἄπειτα ἐκ μὲν τῶν ὑποκαθιζόντων ἐγεννήθη ἡ γῆ, ἐκ δὲ τῶν μετεω–ριζομένων οὐρανὸς πῦρ ἀήρ.

And then, from the bodies which settled down, the earth was generated, and, from the bodies which were lifted up, the sky, fire and air, were generated.

7 Πολλῆς δ' ὕλης ἔτι περιειλημμένης ἐν τῇ γῇ, πυκνουμένης τε ταύτης κατὰ τὰς ἀπὸ τῶν πνευμάτων πληγὰς καὶ τὰς ἀπὸ τῶν ἀστέρων αὔρας, προσεθλίβετο

Since a lot of matter was still contained in the earth and this was compressed by the blows of the winds and the breezes of the

πᾶς ὁ μικρομερὴς σχηματισμὸς ταύτης καὶ τὴν ὑγρὰν φύσιν ἐγέννα· ῥευστικῶς δ' αὕτη διακειμένη κατεφέρετο πρὸς τοὺς κοίλους τόπους καὶ δυναμένους χωρῆσαί τε καὶ στέξαι, ἢ καθ' αὑτὸ τὸ ὕδωρ ὑποστὰν ἐκοίλανε τοὺς ὑποκειμένους τόπους.

8 Τὰ μὲν οὖν κυριώτατα μέρη τοῦ κόσμου τὸν τρόπον τοῦτον ἐγεννήθη.

heavenly bodies, the earth's entire configuration, which was made up of small particles, was squeezed together and generated the moist nature: and since this nature was disposed to flow, it was carried down into the hollow places and those able to hold and contain it; or (else) the water by itself hollowed out the underlying places by settling there.

The most important parts of the cosmos, then, were generated in this way.

In the absence of explicit Epicurean parallels it is reasonable to wish to compare the cosmogony represented by Lucretius and Aëtius on the one hand with the cosmogonical views of Epicurus' philosophical ancestors, esp. Leucippus and Democritus, on the other hand. It turns out, however, that in spite of certain similarities, Lucretius' and Aëtius' cosmogonies are very different from these.[169] In Presocratic cosmogonies, including those of Leucippus and Democritus, the separation of the major cosmic parts is caused by the operation of a whirl or vortex (δίνη/δῖνος), which drives finer matter towards the periphery and coarser matter towards the centre, thus producing a spherical cosmos with the earth at its centre.[170] Paradoxically the whirl also causes certain heavy bodies *not* to move towards the centre but to remain poised at a certain distance, as is the case with the sun, the moon and the stars, which are composed of heavy (stony or earthlike) matter and yet do not fall down towards the earth.[171] In Lucretius' and Aëtius' cosmogonies, on the other hand, we find no trace of the whirl as the moving principle for the formation of the cosmos, but instead the major cosmic parts are separated as a result of their own upward and downward tendencies in proportion to their relative weights (this shall be further elaborated in § 4.3.5.3 below). This applies to the sun and moon as well, which—according to Lucretius—occupy the region half way between earth and aether, because of their intermediate weight. In this respect Lucretius' and Aëtius' cosmogonies correspond to Epicurus': in *Pyth.* 5 [90–91] Epicurus explicitly denies the

---

169 Similar observations in Spoerri (1959) 8–29.
170 See p. 218 with n. 157 above.
171 See p. 218 with n. 152 above.

δίνη such a crucial role, and also makes the sun, moon and stars consist of light substances like air and fire, rather than earth and stone.[172]

It is clear, therefore, that the type of cosmogony represented by Lucretius and Aëtius is essentially different from the one ascribed to Leucippus and Democritus, and therefore needs to be investigated and interpreted in its own right. Before we go on I will first try to establish more clearly that in the cosmogonical accounts of Lucretius and Aëtius the moving principle is weight, or rather weight-difference.

4.3.5.3 Weight as the Moving Principle of Cosmogony

Neither Lucretius nor Aëtius is very explicit initially about the grounds for the upward and downward motions that cause the separation of the major cosmic parts. In Aëtius' account it is the 'larger and heaviest' (3: μείζονα καὶ βαρύτατα) bodies that sank down, while the 'small, round, smooth, and slippery' ones moved upwards. According to Lucretius, the earthy particles sank down 'because they were heavy and entangled' (450: propterea quod erant gravia et perplexa), while those bodies whose particles were 'smoother', 'rounder' and 'smaller' moved upwards. Later on in Lucretius' account it becomes clear that from this array of material properties the real determining factor is *weight*. In lines 471–475, for instance, Lucretius explains that the first-beginnings of sun and moon took up a position half-way between earth and aether, 'because neither were they so *heavy* as to be pressed down and settle, / nor so *light* as to be able to glide through the highest regions' (474–475). The predominant role of weight in cosmogony is confirmed in the concluding section of Lucretius' account (495–501), where Lucretius describes the resulting layered structure of the cosmos. Earth, which is *heaviest*, sank down; then came the sea, then air, then aether, each *lighter* than the one below, with aether, *lightest* of all, on top. It may be concluded, then, that the separation of the main cosmic masses is brought about by their respective upward and downward tendencies, which in turn result from their different relative weights.

It is also interesting to see *how* these upward and downward tendencies depend on weight. In line 450 (already quoted above) we are told that first earthy particles sank down '*because* they were *heavy* and entangled'. Apparently being heavy (and entangled) is all that is needed to produce this downward motion. Then, as these earthy particles converged and became more and

---

172 But see Aëtius II 13.15 where Epicurus is said not to have committed himself to any particular substance of the stars, and II 20.14 where Epicurus is said to have called the sun an 'earthlike condensation'.

more entangled, they *squeezed out* (453: expressere; cf. Aët. I 4, 3: ἐξεθλίβετο) and forced upwards every lighter substance. So, whereas downward motion is a *natural* and inevitable consequence of weight, upward motion is not, but rather comes about by *force*. This does not mean that downward motion is entirely free from external forces. Lucretius uses various expressions to describe this downward motion, the recurrent ones being *sedere* and *subsidere*, which are typically used to denote a retarded natural downward motion through a liquid medium, like mud in water or dregs in wine.[173] This image is made explicit in lines 495–497, where Lucretius writes that 'the weight of the earth settled down, and being as it were the *mud* of all the world, {...} *sank deep down like dregs*' (*subsedit funditus ut faex*). In the same sense Aëtius I 4 uses the rare compound ὑποκαθίζειν.[174]

The way in which downward and upward motion are spoken of here corresponds exactly to the Epicurean theory of weight and upward and downward motion as explained by Lucretius in DRN II 184–215, and by Epicurus himself in fr.276 Us. (see p. 211 ff. above). There is only one difference: in those passages upward and downward motion were conceived of as *parallel*, the present context, however, requires them to be *centrifocal*.

4.3.5.4   Main Clues for a Centrifocal Interpretation

The clearest indication that Lucretius' cosmogony must be interpreted centrifocally is found right at the beginning of the relevant section (V 449–451):

Quippe etenim primum terrai corpora quaeque,	For, first of all, in fact, all bodies of earth,
propterea quod erant gravia et perplexa, coibant	because they were heavy and entangled, came together in the *middle* and all took the *lowest* seats.
in *medio* atque *imas* capiebant omnia sedes.	

---

173   Cf. e.g. Vitr. 8.6.15 'limus enim cum habuerit quo subsidat'; Col. 12.52.14 'si quae faeces aut amurcae in fundis vasorum subsederint'; Plin. NH 31.21 'in Averno etiam folia subsidere'; ibid. 33.103 'faex quae subsedit'; ibid. 34.135 'faece subsidente'; Macr. *Saturnalia* 12.7 'faex in imo subsidit'.

174   For other instances of this verb in this sense see: Galen, *De compositione medicamentorum secundum locos* 13.285.2–4 (Kühn): ἅπαντα μίξαντες εἰς ἀγγεῖον κεραμοῦν, θερμαίνομεν ἐπ' ἀνθρακιᾶς, κινοῦντες ἐπιμελῶς, ἵνα μηδὲν ὑποκαθίσῃ τοῦ φαρμάκου; idem, *De compositione medicamentorum per genera*, 13.788.11–12 (Kühn): ἕψε δὲ ἐπὶ μαλακοῦ πυρός, κινῶν σπάθῃ δᾳδίνῃ ἀδιαλείπτως, ὥστε μὴ ὑποκαθίσαι, τάχιστα γὰρ κατακαίεται.

The most natural interpretation of these lines is to take 'came together in the middle' and 'took the lowest seats' as two equivalent statements, expressing one and the same process.[175] This would make the present lines just another of those many instances in the *DRN* where two statements are conjoined, by means of *et, ac, atque* or *-que*, to describe one and the same state of affairs.[176] If this is true, 'coming together in the middle' and 'taking the lowest seats' amount to the same thing: 'the lowest seats' *are* 'the middle'.

If the bottom of the cosmos is to be identified with its centre, so the periphery must be identical with the top. This too can be understood from Lucretius' text. In lines 457 ff. we are told that aether rose up ('se sustulit') in the same way as we see exhalations rising from the surface of lakes and rivers gather high above ('sursum in alto') as clouds. The concluding sentence runs as follows (467–470):

Sic igitur tum se levis ac diffusilis aether corpore concreto circumdatus undique ⟨flexit⟩ et late diffusus in omnis undique partis omnia sic avido complexu cetera saepsit.	Thus, then, at that time, the light & spreading aether, having placed itself, with compacted body, all around, ⟨curved⟩ in all directions, and spreading wide in all directions everywhere, thus fenced in all else with a greedy embrace.

The first word, 'sic' (thus), can only possibly refer to the process just described of aether *rising up*, like clouds gathering *high above*. But how can *rising up* result in an embrace *all around?* The only possible way is if the upper region *is* all around, i.e. if 'up' equals 'towards the periphery' or 'away from the centre'.

This view is endorsed by Lück (1932) 30, who claims that in antiquity the terms 'above' and 'below' were *generally* used to denote the periphery and the centre of the cosmos. In applying this qualification to Lucretius he is, of course, mistaken: in the preceding sections we have clearly seen that the Epicureans did *not* normally conceive of 'up' and 'down' in this manner. A centrifocal interpretation of Lucretius' cosmogony is therefore strongly rejected by, among

---

175   Cf. Plin. *N.H.* 2.11.6 '*imam* atque *mediam* in toto esse terram'; Manil. *Astron.* 1.167 '*ima*que de cunctis {sc. tellus} *mediam* tenet undique sedem'; et ibid. 170 'ne caderet *medium* totius et *imum*'.

176   See e.g. I 170: inde enascitur *atque* oras in luminis exit; or I 514: corpore inane suo celare *atque* intus habere. On this practice see e.g. Bailey (1947) 145–146, Kenney (1971) 25, 74, 122 (n. ad III 346) and Montarese (2012) 233.

others, Giussani (1896–1898), and Bailey (1947). Below I will first investigate their respective alternative interpretations and then return to the centrifocal interpretation they both reject.

4.3.5.5   Giussani's interpretation

Having rejected the centrifocal interpretation of Lucretius' cosmogony, Giussani (1896–1898) opts for a purely parallel-linear view, where 'the earthy elements condense horizontally at the bottom, and the light and celestial elements extend more or less horizontally on high'.[177] Applying this interpretation to the final section of Lucretius' account (lines 495–508), as Giussani seems to want us to, we see before our eyes a structure consisting of four superimposed, horizontal layers, with earth at the bottom, then sea, then air, and on top of all the aether. This is a nice picture and one fully consistent with a parallel-linear interpretation of the passage.

There are several problems to this interpretation. In the first place, if nothing yet was formed, what was this 'bottom' on which the earthy particles came to rest? Secondly, we can be pretty sure that this is *not* what the cosmos looked like according to the Epicureans. Only a couple of lines below, in V 535 ff., Lucretius clearly expresses the view that the earth is situated in the *centre*, not at the *bottom*, of the cosmos, and the same passage also tells us that air is found not only above, but also below the earth. Elsewhere (II 1066) we learn that aether 'holds [our cosmos] with a greedy embrace'.[178] Even within the cosmogonical passage itself Lucretius intimates as much, telling us (449–451) that particles of earth came together *in the middle*, and (467–470) that aether 'curved in all directions' and (echoing II 1066) 'thus fenced in all else with a greedy embrace.'

Giussani recognises the problem, which he tries to solve by assuming a *second stage* to Lucretius' cosmogony.[179] The concluding section of Lucretius' account (495–508) does not—according to Giussani—describe the definitive position of the four elements, just their position at the end of the *first stage*. Not until the *second stage*, which—Giussani claims—is described in lines 467–470, does aether bend itself so as to *surround* the rest of the cosmos *from all sides*, and only then the earth acquires its final *central* position.

This interpretation will not work, however, for a number of reasons. In the first place, if the earth first settled at the lowest point, and only during the

---

177   Giussani (1896–1898), note ad V 449–494, 1st observation.
178   Cp. also I, 1062–1063 'loca caeli ... inferiora', which suggests that part of the heavens is below the earth.
179   Giussani (1896–1898), note ad V 449–494, 1st observation; & note ad V 496.

second stage took up its present central position, why does Lucretius in 449–451 mention these two motions in tandem and even in reversed chronological order? Secondly, if the earth's downward and centripetal motions are *not* identical, then the latter remains unexplained. In lines 449–450 Lucretius told us that the particles of earth moved towards the centre and towards the lowest places, 'because they were heavy and entangled'. Now, nobody would object to heaviness causing downward motion, but why should heaviness also produce a centripetal tendency, unless 'towards the centre' and 'downwards' are the same thing? Thirdly, if Lucretius' cosmogony is assumed to consist of two stages, it is not logical that the second and final stage should be awarded just two casual remarks in the course of the account (449–451 and 467–470), while the conclusion (495–508) is reserved for a detailed description of what is only the outcome of the first stage. Such a conclusion would also make for a very clumsy transition to the next, astronomical, passage, which presupposes a fully, not a half, developed cosmos. Besides, if the state of affairs described in the conclusion does *not* represent the final and *present* situation, why is it written largely (501–505) in *present* tenses (influit, commiscet, sinit, fert)? Finally, as we have seen above, lines 467–470 clearly describe the *logical outcome* (cf. 467 'sic') of the process described in the preceding lines, not a subsequent development. Giussani's second cosmogonical stage is a fiction.

4.3.5.6 Bailey's interpretation

Giussani's interpretation is also rejected by Bailey, on the rather vague ground that he does 'not fully understand' it. Yet Bailey's own interpretation is hardly less problematic, as we shall see. Bailey starts, like Giussani, by emphatically rejecting a centrifocal reading of the passage. ''Top' and 'bottom'', Bailey says, 'are for him [i.e. Lucretius] absolute terms in relation to ourselves.' Lucretius' reference (V 449–451) to the particles of earth 'coming together in the centre and taking the lowest seats' Bailey tries to resolve on the assumption that Lucretius 'is thinking only of the parts visible from Earth and known to our sensation, the upper hemisphere, as one might call it.' 'To this', Bailey continues, 'the horizon forms the bottom, and the Earth is in the centre of it ...'

At first sight this might seem like a good solution to the problem, but on closer inspection it turns out to be rather meaningless: if the 'bottom' of the cosmos is defined by the present horizon, which in turn depends on the earth's present position, and if the earth is said to have been formed at the 'bottom', then Lucretius is telling us nothing more than that the earth was formed where it is now—wherever this may be. It also seems rather odd that Lucretius should limit his account of the coming-into-being of the cosmos to its visible upper portion only, especially since elsewhere (e.g. V 534–536) he does not hesitate at all to think and speak of what is below the earth. More importantly, if the 'bottom' is defined in this way, there is no reason why the earthy particles should have stopped there instead of moving further downwards towards some other more secure and pre-existing 'bottom'. However all this may be, by defining 'bottom' and 'downward' in this manner Bailey is clearly advocating, like Giussani, a *parallel-linear* conception of the passage.

Immediately afterwards, however, he contradicts himself, when he allows that 'the light particles *rise up to the circumference*' [my Italics], for if upward motion is motion towards the circumference, downward motion—its natural opposite—must be motion towards the centre. Far from rejecting this conclusion, Bailey actually embraces it, approvingly quoting Lück who said 'that in antiquity 'below' in reference to the world means the middle and 'above' the periphery.' This is a purely *centrifocal* conception of the passage. Bailey ends up endorsing the very view he set out to reject! The interpretations offered by Costa (1984) and Gale (2009) are essentially the same.

4.3.5.7 A Centrifocal Interpretation of the Passage

As two attempts to explain Lucretius' cosmogony in a parallel-linear sense have proved unsuccessful, it is now time to re-examine the alternative, a centrifocal interpretation of the account. I will do this in the form of a brief commentary, focussing on those passages and those aspects where the choice between the two alternatives is relevant for our understanding of these passages. Some of these have already been discussed above, but for the sake of completeness the relevant points will be repeated below.

449–451: The particles of earth came together in the centre and occupied the lowest seats. In a centrifocal model these are two different ways to express one and the same motion. As there is only one motion, a single explanation suffices: the particles of earth moved towards the centre / the lowest seats, because they were heavy and entangled. Nobody will object to heaviness causing *downward* motion, which—in a *centrifocal* model—equals *centripetal* motion.

457–470: In these lines the formation of the 'fire-bearing aether' is described. In line 458 aether is said to have 'lifted itself' ('se sustulit'). In lines 460–466 aether's upward motion is compared to the way in which in the early morning mist is seen to rise from the waters and land, collecting up in the sky to form the texture of clouds. In line 467 Lucretius commences his conclusion with the words: 'sic igitur'—'**in this way** then'. One would expect the conclusion to be that aether *rises up* in the same way as mists do. Instead Lucretius concludes (467–470) that in this way aether 'placed itself ... *all around* and *curved in all* *directions*, and spreading wide *in all directions everywhere*, thus *fenced in* all else with a greedy *embrace*.' This conclusion only follows if *upward* motion is identified with motion *away from the centre* and *towards the periphery*, i.e. if we explain the passage *centrifocally*.

471–472: 'These (earth and aether) were followed by the first-beginnings of sun and moon, whose globes revolve through the air between the two (= between earth and aether).' If we follow Giussani's parallel-linear interpretation, 'between the two' would have to mean: in a horizontal plane sandwiched between the (equally horizontal) planes of aether and earth. It would follow that the sun and the moon would never set. (It is interesting to note that this image comes actually very close to the view attributed to Xenophanes,[180] which is vehemently rejected by the Epicurean Diogenes of Oenoanda, fr.66). In a centrifocal system however, 'between the two' would mean: in a hollow sphere intermediate between the centre (the earth) and the periphery (the aether). This would at least allow for the sun and the moon to set and continue their courses below the plane of the horizon, an option which Lucretius explicitly admits (see *DRN* V 650–655).

495–508: In these lines Lucretius describes the layered structure of the cosmos that results at the end of the whole cosmogonical process, with earth at the bottom, above this the sea, above this the air, and on top of all fire-bearing aether. If we interpret the passage centrifocally, as I think we should, these layers must be conceived of as (hollow) *spheres* embracing one another. This would allow for the (*spherical*) earth to be at once at the bottom and in the centre (as was stated in 449–450; the earth's central position is stated again in line 534), for air to be not just on our side but also on the other side of the earth (where it serves to support the earth—as Lucretius in lines 534–563 claims it does), and for aether to hold the rest of the cosmos 'with a greedy embrace' (as Lucretius said in 467–470, repeating what he wrote in II 1066).

In contrast to Giussani's and Bailey's *parallel* interpretation, a *centrifocal* reading of the passage turns out to produce an entirely logical and *internally* coherent account of the cosmogony.

4.3.5.8   Incompatibility of Lucretius' Cosmogony with Epicurean Physics
This conclusion, however, presents us with a serious problem. While two of the passages we investigated, viz. Lucretius' rejection of centrifocalism in I 1052–

---

180   Xenophanes fr.A41a D–K = Aëtius II 24.9.

1093 and the account of atomic motion in II 62–250, pointed clearly to a *parallel-linear* conception of downward motion, the present passage cannot be interpreted in any other way than *centrifocal*. Now, it may be observed that the passages which imply a parallel downward motion are concerned either with the behaviour of atoms in a pre- or extra-cosmic state (Lucr. II 216–250; Epic. fr.281 Us.) or with the behaviour of compound bodies inside our fully formed cosmos (Lucr. I 1052–1093; id. II 184–215; Epic. fr.276 Us.), whereas the present passage deals exclusively with the behaviour of atoms and compound bodies *inside a cosmos in the process of being formed*. Someone might argue that there is no real contradiction if under special circumstances, such as the formation of a cosmos, atoms and compound bodies behave differently from the way they normally do. Such an argument might actually work if the motion of atoms and compound bodies during the formation of the cosmos had been described as forced motion, as it is in most Presocratic cosmogonies, where heavy bodies are driven to the centre by the vortex. In Lucretius' cosmogony, however, the centripetal motion of heavy bodies is equated with *downward* motion, which is the natural concomitant of weight. In other words, downward motion within a cosmos during its formation should be identical to downward motion outside the cosmos as well as inside a fully formed one, but it is not. Lucretius' cosmogony turns out to be incompatible with other parts of his and Epicurus' system.

Now, as we have seen, those passages which point to a *parallel* downward motion appear to be well integrated into the overall argument: Lucretius' rejection of *centrifocal* cosmology is presented as a logical consequence of the infinity of matter and space, and Lucretius' subsequent assumption of a *parallel* downward motion may be interpreted as a necessary implication of this rejection. In addition, this *parallel* downward motion is also implied by several passages in works and fragments of Epicurus himself. By contrast, Lucretius' *centrifocal* cosmogony is not. Although many details of this cosmogony correspond beautifully to other parts of Lucretius' and Epicurus' cosmology, the crucial detail that downward motion is *centrifocal* seems to be unique to this passage and not founded on anything that was stated before.

As a consequence, if the cosmogony cannot be reconciled with other passages, and seeing that those passages are firmly integrated into Epicurean physics, while the cosmogony is not, it is hard to avoid the conclusion that Lucretius' cosmogony is a *Fremdkörper*, that somehow found its place among the writings whose content Lucretius chose to work up into his poem. If so, what might be its origin and how did Lucretius end up including it? It is remarkable that, except for its being *centrifocal* rather than *parallel-linear*, it contains some unmistakable Epicurean echoes. Like the cosmogonies of Leucippus and

Democritus it is atomistic (explicitly so in Aëtius' version) and stresses the lack of design,[181] but unlike its Presocratic precursors, and in agreement with the few details provided by Epicurus himself (*Pyth.* 4–5 [89–91]), it makes do without a cosmic vortex and it creates the heavenly bodies out of light rather than heavy substances. So, even though it is incompatible with orthodox *parallel-linear* Epicureanism, it still seems to be rooted in Epicureanism somehow. For now, however, it suffices to say that Lucretius' cosmogony seems to be an anomaly that does not fit the general *parallel-linear* picture that arises from other passages. Before drawing any firmer conclusions we had better move on and see what other passages may tell us about the direction of downward motion and the shape of the earth.

4.3.6 Stability of the Earth (DRN V 534–563)

In V 534–563 Lucretius argues that the earth remains at rest in the middle of the cosmos by being supported by air. The same view is also expressed in a fragment of Epicurus' *On nature*,[182] and is explicitly attributed to Epicurus in a scholion to his *Letter to Herodotus* 73 (τὴν γῆν τῷ ἀέρι ἐποχεῖσθαι—'that the earth rides on air'). This alone is enough to conclude that Epicurus' and Lucretius' cosmology is *parallel-linear*, for in a *centrifocal* cosmology the earth would need no underprop: see p. 177 ff. and especially the pictures on p. 179 above. It is remarkable that Lucretius should return to a *parallel* cosmology so soon after his *centrifocal* cosmogony. This rather confirms our suspicion that the cosmogony is somehow anomalous among the rest of Lucretius' and Epicurus' views.

Lucretius' and Epicurus' theory about the stability of the earth is reminiscent of the view ascribed to Anaximenes, Anaxagoras and Democritus. These three, according to Aristotle, *Cael.* II 13, 294b13–30, thought that the earth could float on the air because of its *flatness*, which prevents the air from escaping. The explanation offered by Epicurus and Lucretius is somewhat different. In their version the air is said to be able to support the earth because together they form an organic unity, with the substance of the earth gradually blending into the air underneath. In this way, just like the neck does not feel the weight of the head, the air would not feel the weight of the earth. It is interesting to note that in contrast to its Presocratic precursor this version of the theory does not specify the shape of the earth.

---

[181] Aëtius I 4, 2: Τῶν ἀτόμων σωμάτων ἀπρονόητον καὶ τυχαίαν ἐχόντων τὴν κίνησιν ... Cf. DRN V 419–420.

[182] Epicurus *On nature* XI, fr.42 Arr. See also frs.22 and 23 Arr.

4.3.7 The Size of the Sun (DRN V 564–591)

In *DRN* V 564–591 Lucretius argues that the sun, like the other heavenly bodies, is more or less the size it appears to be. A shorter version of the argument, by Epicurus himself, is preserved in the *Letter to Pythocles* (6 [91]), where a scholion informs us that Epicurus had also discussed the matter in book XI of his *On nature*.

The argument in both passages may be summarized as follows (passing by some difficult details and minor differences): if a terrestrial fire is near enough for its light and heat (Lucretius 564–573), or its colour (Epicurus according to the scholion), or its outline (Lucretius 575–584), or its flicker and glow (Lucretius 585–591), to be observed, then its size does not diminish with increasing distance; but the sun's (and the moon's and the stars') light and heat (etc.) *are* observed: therefore the sun's (etc.) size does *not* diminish with increasing distance, and therefore the sun must be the size it appears to be. Ancient critics took this to mean that the sun according to Epicurus was actually very small, about the size of a human foot,[183] and this interpretation is followed by most modern commentators.[184] It also receives some support from Lucretius who follows up the present passage with an account of how 'such a tiny sun' (tantulus ille sol) can shed so much light and warmth.

Several scholars have suspected a relation between this theory and Epicurus' views about the structure and size of the cosmos. Cyril Bailey,[185] for instance, connects the theory with Epicurus'—unattested—view that our 'world was comparatively small and the sun not very distant'. '[T]his', Bailey continues, 'would lead naturally to the belief that it [i.e. the sun] was not very large.' David Sedley,[186] too, suspects such a connection. He claims (without argument) that for Epicurus the earth was flat and that the sun set somewhere not too far past its westernmost edge (cf. my remarks on Lucr. IV 410–413 on p. 220 above). As a result, Sedley argues, it would have caused Epicurus no small embarrassment to discover that even in lands far to the west the sun, though much nearer, does not appear any bigger than at home. Orthodox astronomers could simply ascribe this to the negligible dimensions of the earth as compared to the radius

---

183   See n. 53 on p. 176 above.
184   An alternative interpretation is offered by Keimpe Algra (2001), esp. 17–19, who suggests that Epicurus and Lucretius are referring to the *relative* size of the sun, i.e. the portion of our field of view that is occupied by the sun, which is proportional to the ratio of the sun's size and its distance. According to this interpretation, E. and L. did not commit themselves to a very small sun, but rather refrained from assigning a specific size.
185   Bailey (1947) III 1408. See also Giussani (1896–1898) ad vv. 564–611.
186   Sedley (1976) 48–54.

of the solar orbit, but for Epicurus this was not possible. Instead he explained the observed invariability of the sun's size on the assumption that the sun was somehow exempt from the laws of optics, an exemption that seemed to be supported by the analogy of terrestrial fires seen from a distance. Whereas the image of other objects in our experience shrinks with increasing distance, the sun's—just like distant terrestrial fires—does not. It follows that the sun's image arrives at our eyes having more or less the same size it set out with, i.e. its actual size. Therefore the sun is about the same size it appears to be, which is approximately the size of a human foot. In this way Epicurus' theory about the size of the sun is linked to his (supposed) view that the sun is relatively close-by.

Neither Bailey nor Sedley explains why Epicurus and Lucretius wanted the sun to be so close to earth in the first place. In two recent publications David Furley attempts to supply the missing argument.[187] He suggests that Epicurus' commitment to a nearby, and hence very small, sun may have been partly motivated by his wish to harmonize certain astronomical observations with his own flat-earth cosmology. Before Epicurus' time, Aristotle had already reported the observation that as we travel north or south the stars change their position, while some stars that never set in the north do so in the south, and others that are seen in the south are invisible in the north.[188] Aristotle and others explained these observations on the assumption that the earth is spherical. Furley points out that such observations could in fact be reconciled with a flat earth, if the heavenly bodies are assumed to be relatively close by: he compares this to the effect of walking under a painted dome. (In fact, as we observed on p. 172 above, this model can only explain part of the reported observations: while it may account for the fact that observers at different latitudes see the same stars at different positions, it cannot explain how certain stars completely disappear from sight.)

In order to give an indication of how close-by the heavenly bodies would have to be, Furley focuses on a special case of these observations. It was observed that the sun's position in the sky at noon on a certain day, for instance the day of the summer solstice, is not constant but depends on the observer's latitude, as could be demonstrated by the relative length of shadows.[189] Orthodox astronomers of course explained these observations with reference to the earth's curvature, but, as Furley demonstrates, the same facts might also be accommodated to a flat earth on the assumption that the sun is relatively near.

---

187   Furley (1996), id. (1999) 421, 428–429.
188   Arist. *Cael.* II 14, 297b24–298a10. See also p. 171 'third proof' above.
189   See p. 172 'fifth proof' above.

A famous instance of these observations is ascribed to Eratosthenes, who noted that at noon during the summer solstice in Syene (modern Aswan in southern Egypt) the sun is vertically overhead, but in Alexandria, which is situated some 5000 stades or 788 km to the north, appears 7.2° out of plumb.[190] Eratosthenes used these data to calculate the circumference of the (spherical) earth, but Furley suggests that these same data might be used to calculate the sun's distance if the earth is assumed to be flat. He does not provide the actual calculation, only its results, but it is not hard to reconstruct. Assuming the earth to be flat, we can construct a right-angled triangle Sun-Syene-Alexandria, with a top angle of 7.2° and a base of 5000 stades (see the figure below). The distance between an observer in Syene and the sun is given as 5000/sin(7.2°), which yields 39,579 stades or 6,238 km. To an observer in Alexandria the sun is slightly further away.[191]

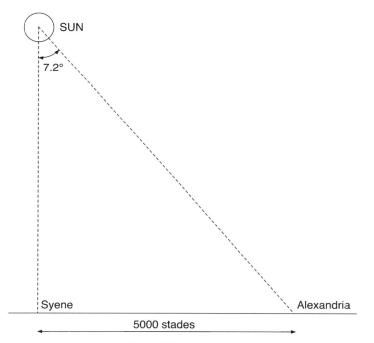

FIGURE 4.3 *Furley's calculation of the sun's distance on the assumption of a flat earth*

---

190 See Cleomedes I 7.49–110.
191 To be precise: 5000/tan(7.2°) = 39,894 stades or 6,287 km.

According to his argument, then, the earth's flatness requires the sun to be relatively close-by, which in turn implies that it cannot be very large. And this, according to Furley, may explain why Epicurus was so keen on proving its small size. In this way Epicurus' argument for the sun's size can be interpreted as a symptom of his commitment to a flat-earth theory, and his wish to uphold this theory against the pretensions of contemporary astronomy.

4.3.8 Centrifocal Terminology (DRN V 621–636)

In *DRN* V 614–649, Lucretius discusses the turnings of the sun and the moon. The first theory presented (621–636) is the view held by Democritus, who proposed that the turning speed of the heavenly bodies decreases with decreasing distance from the earth:

Nam fieri vel cum primis id posse videtur,	For, in the first place it seems that this
Democriti quod sancta viri sententia ponit:	may be the case, what the sacred opinion of the man Democritus states: that in
quanto quaeque *magis* sint *terram* sidera *propter*,	proportion as each heavenly body is *nearer the earth*, so much the less can it
tanto posse minus cum caeli turbine ferri;	be moved with the whirling of the sky,
625 evanescere enim rapidas illius et acris	since its swift and violent forces vanish
imminui *super* viris, ideoque relinqui	and grow less *below*, and therefore the
paulatim solem cum posterioribu' signis,	sun is slowly left behind with the signs
inferior multo quod sit quam fervida signa.	which come behind it, because the sun is much *lower* than the burning signs.
Et magis hoc lunam: quanto *demissior* eius	And more so than the sun the moon: in proportion as its course is *lower*, being
630 cursus abest *procul a caelo terrisque propinquat*,	*far from the sky*, and *approaches the earth*, so much the less can it keep up its course
tanto posse minus cum signis tendere cursum;	with the signs; and in proportion as it is moved with an even fainter whirling,
flaccidiore etiam quanto iam turbine fertur	being *lower* than the sun, so much the more all signs catch it up all around and
inferior quam sol, tanto magis omnia signa	move past it. Therefore it so happens that the moon appears to return to every sign
hanc adipiscuntur circum praeterque feruntur.	more quickly, because the signs return to it.
635 Propterea fit ut haec ad signum quodque reverti	
mobilius videatur, ad hanc quia signa revisunt.	

The theory *as such* does not allow us to draw any conclusion about the shape of the earth or the direction of falling objects: there is no intrinsic reason why the heavenly bodies could not move in this special way in either kind of cosmology. The problem lies in the terminology employed. Throughout the passage Lucretius uses words like 'lower' and 'below' side by side with expressions like 'nearer the earth' and 'far from the sky', as if they were synonyms (instances of both kinds have been italicized in the text and translation above). This is *centrifocal* language, which has no place in a *parallel* cosmology.

Another instance of centrifocal terminology occurs in V 714, where the moon is said 'to keep the path of her course *below the sun*' ('cursusque viam *sub sole* tenere'), which can only mean that the moon is 'nearer the earth than the sun'.[192]

I have found no traces of centrifocal language in Epicurus' *Letters*, nor in the cosmological and astronomical fragments of book XI of his *On nature*. A clear instance, however, occurs in a fragment of the later Epicurean Diogenes of Oenoanda (fr.13 I.11–13):

Ἔτι δ' οἱ μὲν ὑψηλὴν ζώνην φέρονται, οἱ δ' αὖ ταπεινήν.	Moreover, some (of the heavenly bodies) move in a high orbit, others however in a low one.
	tr. SMITH

Here, as in the two passages of Lucretius, 'high' and 'low' must be understood as 'far from the earth' and 'near the earth' respectively.

The use of centrifocal terminology in these passages may call to mind Lucretius' cosmogonical account in V 449–508, which I argued can only be understood in a *centrifocal* sense (see p. 223 ff. above), making it incompatible with Lucretius' and Epicurus' otherwise *parallel-linear* cosmology. The passages we are now investigating, on the other hand, may not warrant such a dramatic conclusion. In the cosmogony terms like 'up' and 'down' are explicitly linked to the natural motion of heavy bodies, so that the use of *centrifocal* terminology directly affects our conception of this motion. However, in the passages currently under investigation the natural motion of heavy bodies is not at stake. Save for the use of *centrifocal language* the theories that are being described do not necessarily exclude a *parallel-linear* cosmology. The use of centrifocal terminology in these passages, however ill-suited to Epicurean cosmology, may therefore be nothing but a slip into conformity with the accepted language of astronomy.

---

192 Bailey (1947) ad loc.

4.3.9 Sunrise and Sunset (DRN V 650–679)

In V 650–679 Lucretius discusses the possible causes of sunset and sunrise. Two alternative theories are recognized: either the same sun passes unaltered below the earth and emerges again the next morning, or the sun is extinguished every night, to be rekindled the following day. The same theories of rising and setting are discussed by Epicurus in his *Letter to Pythocles* 7 [92], where they are applied to other heavenly bodies as well. The second of these two theories is the object of a fierce attack by the Stoic astronomer Cleomedes (II 1.426–466). Cleomedes points out that, since the earth is spherical, times of rising and setting differ with latitude as well as longitude. To illustrate the latitudinal variation he produces a list of actually reported minimum night times for a number of places at different latitudes. (This is an application of 'proof 6' of the earth's sphericity: see p. 173 above). The longitudinal variation of the times of sunset and sunrise ('proof 7': see p. 173 above) Cleomedes had already discussed in his first book (I 5.30–44). There he argued that the times of sunset and sunrise depend on longitude, and that this time-difference can be measured by simultaneous observations of eclipses at different longitudes. Since, then, the times of setting and rising are different for every place on earth it follows that the sun would have to be lit and extinguished at one and the same time incalculably many times. The obvious conclusion, which Cleomedes fails to draw, seems to be that Epicurus did *not* believe the earth to be *spherical*, but flat. Only in this way, Cleomedes argues elsewhere (I 5.30–37), without mentioning Epicurus, would it be possible for the sun to rise and set for everyone at the same time. Yet, since in reality the times of sunrise and sunset are different at different places, Epicurus must have been either ignorant of the facts, or have wilfully ignored them.

Although both conclusions—that Epicurus believed the earth to be flat, and that therefore he must have been ignorant of the facts—seem inevitable, I think both can be avoided. Against the first conclusion one could argue that the theory does *not necessarily* mean that the sun's extinction and rekindling *coincide* with its setting and rising: even on the assumption of a flat earth Epicurus would have had to admit that times of setting and rising vary due to the unevenness of the earth's surface, while obviously the sun's extinction and rekindling each occur at one specific time; and in the same way one could assume that on a spherical earth the sun is extinguished only after the last person has seen it set, and rekindled before the first person observes its rise. Moreover, the theory of the sun's extinction and rekindling is only one of two possible theories, so that the conclusion that the earth is flat would be only hypothetically true anyway. In defence against the charge of ignorance about the facts, one might point out the inaccuracy of the observations themselves: timekeeping in antiquity was notoriously unreliable, and comparing times at different places even more so.

An interesting change of attitude towards the explanation of sunset and sunrise can be observed in the work of the 2nd century AD Epicurean Diogenes of Oenoanda (see p. 37 ff. above). In fragment 13 of his Epicurean inscription, Diogenes promises to deal with the question of risings and settings. Before embarking on this subject he explains that with problems such as these one should not confine oneself to a single explanation when several options present themselves. Thus far he is speaking like a true Epicurean. Then he adds something that has no counterpart in any other Epicurean writing and seems to be against the spirit of Epicurean multiple explanations: "It is correct, however, to say that, while all explanations are possible, *this one is more plausible than that*." (tr. Smith, my emphasis). It seems that Diogenes reserves to himself the right to prefer certain explanations above others. Unfortunately his actual discussion of settings and risings has not been preserved. There is, however, another fragment, that touches upon the same problem. In fragment 66 he rebukes certain adversaries for "dismissing the unanimous opinion of all men, both laymen and philosophers, that the heavenly bodies pursue their courses round the earth both above *and below* ..." (tr. Smith, my emphasis). It is clear that Diogenes too subscribes to this 'unanimous opinion of all men', thus silently passing by Epicurus' alternative explanation that the heavenly bodies are extinguished at night. Diogenes may have justified his preference for the commonly accepted view with an appeal to plausibility, and he may well have done so in response to arguments such as those of Cleomedes.

4.3.10   *The Earth's Conical Shadow* (DRN V 762–770)

In DRN V 762–770 Lucretius discusses the subject of lunar eclipses. Among a number of alternative explanations he also mentions the theory of orthodox astronomy, that the moon, which on this theory receives its light from the sun, is eclipsed when it falls into the shadow of the earth (762–764):

Et cur terra queat lunam spoliare vicissim lumine et oppressum solem super ipsa tenere, menstrua dum rigidas coni perlabitur umbras, ...?	And why should the earth in turn be able to rob the moon of light, and keep the sun oppressed, itself being above, while in its monthly course the moon glides through the rigid shadows of the cone, ...?

The same theory was also mentioned, as a scholion informs us, in book XII of Epicurus' *On nature*,[193] and is alluded to in ch. 13 [96] of Epicurus' *Letter to Pythocles*,[194] but only Lucretius provides the essential detail that the earth's shadow is shaped like a *cone*.

That the earth's shadow must of necessity have this shape was already known to Aristotle,[195] and is aptly demonstrated by Cleomedes and the elder Pliny.[196] Their argument runs as follows: supposing the sun and the earth to be spherical the earth's shadow will have one of three possible shapes: (1) if the sun is smaller than the earth it will produce a funnel-like shadow; (2) if the sun is equal to the earth a cylindrical shadow will result; and (3) if the sun is larger than the earth the shadow will be conical. Now, if the shadow were funnel-like it would extend over a large part of the night sky and obscure the moon almost continuously, which is not observed to happen. If, on the other hand, the shadow were cylindrical it would still be so large that the moon could not fail to be obscured every full moon, which goes against the evidence as well. Only a conical shadow can account for the observed fact, that the moon is eclipsed only during full moon and then only rarely and for a relatively short time. Therefore the earth's shadow must be conical, and the sun larger than the earth.

There are two implications to this theory that need looking into: (a) it seems to presuppose a *spherical* earth, and (b) it requires for the sun to be *larger* than the earth. I will deal with each of these implications below.

Ad a: A perfect cone has a circular base and the only figure that will at all times present a circular outline to the sun is a sphere. A perfectly conical shadow therefore implies a spherical earth. Moreover, the conical shape of the earth's shadow also implies that, when during a lunar eclipse the moon crosses the earth's shadow, the obscured segment of the moon is always convex, as is in fact observed. In Aristotle's *De caelo* II 14, 297b24–31 this observation is presented as a proof of the earth's sphericity ('proof 1' on p. 171 above). Therefore, by thus referring to the earth's conical shadow Lucretius might seem not only to presuppose a spherical earth, but even to acknowledge and accept Aristotle's proof of the earth's sphericity. However, I do not think this inference is necessary. In the first place a sphere is not the only shape that would produce

---

193   Scholion ad Epic. *Pyth.* 13 [96]: σελήνην δὲ (ἐκλείπειν) τοῦ τῆς γῆς σκιάσματος (ἐπισκοτοῦντος)—'the moon is eclipsed when the earth's shadow obscures it.'
194   Epic. Pyth. 13 [96]: ἐπιπροσθέτησιν ... γῆς—'interposition of the earth'.
195   Arist. *Mete.* I 8, 345b1–9.
196   Cleomedes II 2.19–30 & II 6.60–108; Pliny *N.H.* II 51.

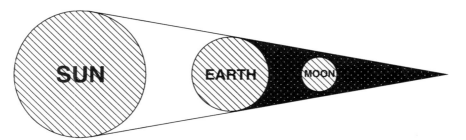

FIGURE 4.4 *The moon in the conical shadow of the earth*

a conical shadow (or a convex obscuration). A disk, for instance, will produce a perfectly conical shadow if the sun's rays hit it perpendicularly. And so will an ellipse when hit by the sun's rays at an appropriate angle. Moreover, the word 'cone' may be used here in a loose sense only, to describe any three-dimensional figure with a roughly circular base whose outline contracts with increasing distance. There is no real need therefore to suppose that the reference to a conical shadow implies a spherical earth.

Ad b: This brings us to the other implication: if the earth's shadow is shaped like a cone—and even if we interpret 'cone' in the loose sense indicated above—, the sun must be *larger* than the earth, which contradicts Epicurus' and Lucretius' supposed claim (see p. 236 above) that the sun is very small. Three possible solutions present themselves: either that supposition was wrong and the Epicureans did not hold the sun to be very small,[197] or Epicurus and Lucretius failed to appreciate the geometrical implications of the present theory, or this is just another of those cases where, as Wasserstein observed,[198] the Epicureans failed to harmonize the explanation of one astronomical phenomenon with those of another. All three solutions are damaging to Furley's thesis that the small size of the sun was meant to reconcile certain astronomical observations with a flat earth. If the Epicureans did not believe the sun to be very small after all, Furley's thesis must be rejected forthwith. If, on the other hand, the Epicureans were unable to grasp the geometrical implications of the present theory, they can hardly be expected to have understood the geometry involved in proving the relative proximity of the sun and hence its small size, in the way Furley suggests they did. Finally, if the Epicureans did not care to harmonize the explanations of different phenomena, why should they have cared to

---

197 See n. 184 on p. 236 above.
198 See p. 32 ff. above.

harmonize their flat-earth theory with one particular set of observations, as Furley suggests they did? In sum, the Epicureans' inclusion of the present theory among a number of possible alternatives throws serious doubt on Furley's thesis that their theory about the size of the sun had anything to do with their supposed commitment to a flat earth.

4.3.11 *The 'Limp' of the Cosmic Axis* (DRN VI 1107)

There is an intriguing passage near the end of the *DRN*, where Lucretius discusses the relation between local climates, racial characteristics and endemic diseases.[199] Five examples are given (VI 1106–1109):

Nam quid Brittannis caelum differre putamus,	For in what way do we suppose the climate of the Britons to differ, and that
et quod in Aegypto est *qua mundi claudicat axis*,	which is in Egypt, *where the cosmic axis limps*, or in what way that which is in
quidve quod in Ponto est differre, et Gadibus atque	Pontus to differ, and in Gades,[200] and all the way to the black tribes of men
usque ad nigra virum percocto saecla colore?	with their scorched colour?

It is the second of these lines I wish to focus on: "Egypt, where the cosmic axis limps." Many commentators have done their best to make some sense of these words, which has resulted in a great diversity of different interpretations. As some of these interpretations seem to imply or presuppose a specific shape of the earth, it seems worthwhile to devote some time and space to a critical survey of these interpretations.

Before I start I will first briefly discuss two of the terms involved:

- **mundi axis:** According to most commentators (e.g. Bailey), *mundi axis* is the imaginary axis through the earth around which (the rest of the) cosmos was observed to revolve in just under 24 hours.[201] Some commentators (Robin, Leonard & Smith) speak, anachronistically, of the *earth's* axis, which is how the cosmic axis came to be known after Copernicus. Others (Lambinus,

---

199 For a brief analysis of the context of this passage see p. 112 ff. above.
200 Modern Cádiz in Southern Spain.
201 The same expression, in Greek, is found in Eudoxus fr.124.84–85 Lasserre (= Simpl. *In Arist. De caelo* 495.22–23 Heiberg): σφαῖρα περὶ τὸν ἄξονα τοῦ κόσμου στρεφομένη. For the meaning *mundus* = *cosmos* see n. 76 on p. 189 above and text thereto.

Munro, Merrill) take *mundi axis* as referring to the visible end-point of the cosmic axis, i.e. *the celestial north pole*. This meaning is also suggested by the only other instance of the word 'axis' in the *DRN*, in VI 720: '{flabra} quae gelidis ab stellis *axis* aguntur'—'{winds} which are driven from the ice-cold stars of the *north pole*'.
- **claudicat**: The most puzzling word in Lucretius' line is *claudicat*. According to the dictionaries, *claudicare* can have two different meanings:
  i. (literal:) *to limp, to walk with a limp*
  ii. (metaphorical:) *to be defective, to malfunction, to fail, to falter*
  The commentators, on the other hand, want *claudicat* to mean:
  a. (Lambinus, Munro, Merrill, thinking of the celestial north pole:) *is depressed, lies low*
  b. (Robin, Bailey and most others, thinking of the cosmic axis) *slants, slopes, tilts, inclines*

The problem is how to get from meaning i or ii to meaning a or b. Most commentators simply state that meaning i implies a or b. Sometimes they point out internal parallels in support of their view: III 453 '*claudicat* ingenium', IV 436–437 '*clauda* videntur / navigia', IV 515 'libella ... *claudicat*' (for some reason VI 834 '*claudicat* ... pinnarum nisus' is never mentioned), but they fail to explain how each parallel works, for this is hardly obvious: what have 'the mind' or 'a ship' (or 'the support of feathers') in common with the cosmic axis or the celestial north pole? The only obvious parallel seems to be that of the level (libella) in IV 515, adduced by Bailey, where 'claudicat' seems to mean 'is inclined'. However, this parallel is misleading: *claudicat* here only **seems** to mean 'is inclined', because a properly functioning level should be horizontal, and therefore its 'malfunction' is 'being inclined', but surely it is not the 'proper function' of the cosmic axis to be horizontal. Munro's explanation of *claudicare* as 'leaning over like a limping man' does not convince either: the cosmic axis and the north pole do not seem to have much in common with a limping man, nor do limping men necessarily lean over. Although I am convinced that in the present context *claudicare* must mean something like (a) *to be inclined*, or (b) *to lie low*, a good explanation of how this meaning can be derived from one of the verb's original meanings has yet to be produced. For the moment I will, therefore, simply assume each of the two proposed interpretations and see where it leads us.

This having been said it is time to turn to the interpretation of the entire line. The problem, in my view, is a triple one:

I. what does it mean to say that the cosmic axis (or celestial north pole) *limps*?
II. what is the *relevance* of this 'limp' in the present context?
III. why does it do so in *Egypt* of all places?

In the following discussion I will use these three questions to judge the validity of each explanation. The explanations that have been proposed can be reduced to three different interpretations:

1. the inclination of the cosmic (or terrestrial) axis with respect to the ecliptic or zodiac
2. the inclination of the cosmic axis with respect to the plane of the flat earth
3. the relatively small elevation of the celestial north pole in Egypt

The fullest and yet most confused commentary is that of Robin, who touches upon all three interpretations with ample parallels, yet fails to observe the fact that they are different and mutually exclusive. I will use his commentary as a guide to the three interpretations.

1. Robin sees in line 1107 an allusion to the *inclination of the earth's axis*, which he links to the *obliquity of the zodiac* (λόξωσις τοῦ ζῳδιακοῦ / *obliquitas (orbis/circuli) signiferi*). This latter term was used in ancient astronomy to denote the fact that the sun's *annual* path through the sky—often called *zodiac* after the belt of constellations by which it is marked—is inclined by about 23.5° with respect to the direction of the fixed stars' *daily* rotation. Because of this obliquity the sun does not stay on the equator, but wanders to and fro between the two tropics, and in doing so produces the four seasons. Both Epicurus and Lucretius were in fact familiar with this *obliquity of the zodiac*, which is mentioned by Epicurus among a number of possible explanations for the wanderings of the sun,[202] and by Lucretius as one possible way to account for the seasonal variation of day-lengths.[203] In modern, post-Copernican, astronomy the daily westward rotation of the fixed stars has been replaced by a daily eastward rotation of the earth around its own axis, and the apparent yearly motion of the sun through the signs of the zodiac is now attributed to a yearly rotation of the earth around the sun. The phenomenon which the ancients commonly referred to as the

---

202 Epic. *Pyth.* 9 [93].
203 Lucr. *DRN* V 691–693.

obliquity of the zodiac is now referred to as the *inclination of the earth's axis*. There is no indication, however, that anyone before the time of Copernicus ever referred to this phenomenon in terms of the terrestrial or celestial axis. Robin is mixing up ancient and modern terminology. Moreover, the *obliquity of the zodiac* is not even relevant to the subject of Lucretius' line: while it may explain the *annual* variation of the seasons, it has nothing to do with the *geographic* diversity of climates, which is what the present passage is about. Finally, there is no specific link between the *obliquity of the zodiac* and Egypt: it is a 'global' constant that is the same for Egypt and Britain and any other place on earth. In short, whatever Lucretius is saying in VI 1107, we can be confident that he is *not* referring to the *obliquity of the zodiac*.

2. Robin goes on to cite a number of Presocratic fragments dealing with the *inclination of the cosmos* (Anaxagoras, Diogenes of Apollonia, Empedocles) or, alternatively, *of the earth* (Leucippus and Democritus). He does not seem to view these as anything but illustrations of his own, erroneous, interpretation. In fact however, as Bailey has rightly observed, they provide the key to a different and far more promising interpretation. The most telling texts are Diog. Laërt. II 9.1–3 (Anaxag. fr.A1.31–32 D–K) (ignored by Bailey):

Τὰ δ' ἄστρα κατ' ἀρχὰς μὲν θολοειδῶς ἐνεχθῆναι, ὥστε κατὰ κορυφὴν τῆς γῆς τὸν ἀεὶ φαινόμενον εἶναι πόλον, ὕστερον δὲ τὴν ἔγκλισιν λαβεῖν.	The heavenly bodies were originally carried along like a dome, so that the ever-visible pole was vertically above the earth, but later they (the stars / the pole?) acquired the inclination.

and Aëtius II 8.1 (Anaxag. fr.A67 D–K = Diog. Apoll. fr.A11 D–K):

Διογένης καὶ Ἀναξαγόρας ἔφησαν μετὰ τὸ συστῆναι τὸν κόσμον καὶ τὰ ζῶια ἐκ τῆς γῆς ἐξαγαγεῖν ἐγκλιθῆναί πως τὸν κόσμον ἐκ τοῦ αὐτομάτου εἰς τὸ μεσημβρινὸν αὐτοῦ μέρος, ἴσως ὑπὸ προνοίας, ἵνα ἃ μὲν ἀοίκητα γένηται ἃ δὲ οἰκητὰ μέρη τοῦ κόσμου κατὰ ψύξιν καὶ ἐκπύρωσιν καὶ εὐκρασίαν.	Diogenes (of Apollonia) and Anaxagoras said that after the formation of the cosmos and the creation of the animals out of the earth, the cosmos somehow spontaneously inclined towards its southern part, perhaps by providence, in order that some parts of the cosmos might become uninhabitable, others inhabitable, according to the freezing and scorching and temperation.

These texts are concerned with the problem of why the celestial north pole is not vertically above the plane of the (flat) earth, but appears at a certain angle above the northern horizon (38° to an observer in Athens). Anaxagoras and Diogenes assumed that after its coming-into-being the whole *cosmos*, including the fixed stars and the celestial north pole, had somehow tilted, causing the *celestial axis* to become *inclined* towards the south.[204] This *inclination* was also somehow responsible for the *latitudinal*[205] variation of climates. A similar view is attributed to Empedocles,[206] and a variant to Leucippus[207] and Democritus,[208] who supposed that not the cosmos, but the (flat) earth was inclined.

It is to such views, according to Bailey, that Lucretius must be alluding in line 1107. For some reason, however, Bailey only gives Leucippus' and Democritus' version of the theory (which he wrongly ascribes to Anaxagoras and Diogenes too), making Lucretius say that "the *earth* was tilted upwards to the

---

204 Verbs like (ἐγ-/ἐπι-)κλίνομαι (incline, slope) always indicate a deviation from the *horizontal* plane. The direction of the slope, i.e. the direction of its *descent*, is indicated by εἰς/πρὸς/ἐπὶ + acc.

205 I use 'latitudinal' and 'latitude' here simply to refer to a place's position with respect to the north and the south, irrespective of the assumed shape of the earth.

206 Aët. II 8.2 (Emp. A58 D–K): Ἐμπεδοκλῆς τοῦ ἀέρος εἴξαντος τῇ τοῦ ἡλίου ὁρμῇ ἐπικλιθῆναι τὰς ἄρκτους, καὶ τὰ μὲν βόρεια ὑψωθῆναι, τὰ δὲ νότια ταπεινωθῆναι, καθ' ὃ καὶ τὸν ὅλον κόσμον.— "Empedocles says that when the air had yielded before the force of the sun, the Bears tilted, and the northern regions (of the earth/the cosmos?) were lifted, and the southern depressed, and accordingly the whole cosmos." Personally, I believe that the attribution of this view to Empedocles is wrong. While Anaxagoras, Leucippus and Democritus, and probably Diogenes too, were *flat*-earthers, Empedocles may well have believed the earth to be *spherical*; see p. 166 with n. 22 above.

207 Aët. III 12.1 (Leuc. A27 D–K): Λεύκιππος παρεκπεσεῖν τὴν γῆν εἰς τὰ μεσημβρινὰ μέρη διὰ τὴν ἐν τοῖς μεσημ¬βρινοῖς ἀραιότητα, ἅτε δὴ πεπηγότων τῶν βορείων διὰ τὸ κατεψῦχθαι τοῖς κρυμοῖς, τῶν δὲ ἀντιθέτων πεπυρωμένων.—"Leucippus said that the earth inclined toward its southern parts because of the drought in the southern parts, since the northern parts are rigid due to cooling-down by the frost, while the opposite parts are scorched." See also Diog. Laërt. IX 33.6–8 (Leuc. A1.31–33 D–K): ... κεκλίσθαι τὴν γῆν πρὸς μεσημβρίαν· τὰ δὲ πρὸς ἄρκτῳ ἀεί τε νίφεσθαι καὶ κατάψυχρα εἶναι καὶ πήγνυσθαι.—"... the earth is inclined towards the south, and the regions lying to the north are always snowy and cold and frozen."

208 Aët. III 12.2 (Democr. A96 D–K): Δημόκριτος διὰ τὸ ἀσθενέστερον εἶναι τὸ μεσημβρινὸν τοῦ περιέχοντος αὐξομένην τὴν γῆν κατὰ τοῦτο ἐγκλιθῆναι· τὰ γὰρ βόρεια ἄκρατα, τὰ δὲ μεσημβρινὰ κέκραται· ὅθεν κατὰ τοῦτο βεβάρηται, ὅπου περισσή ἐστι τοῖς καρποῖς καὶ τῇ αὔξῃ.— "Democritus says that, because the southern portion of the cosmic envelope is weaker, the earth during its growth started to tilt in that direction; for the northern regions are intemperate, but the southern regions temperate. Hence it is weighed down, where it abounds with crops and growth."

north and downwards to the south" [my italics]. It is hard to see, however, how '*mundi axis*' could be made to stand for '*the earth*'. It seems much more obvious to regard Lucretius' line as an allusion to the Anaxagorean view that (paraphrasing Bailey:) "the *cosmic axis* was tilted upwards to the north and downwards to the south".

Contrary to Robin's, this interpretation is very relevant to the context of the passage. Not only Anaxagoras and Diogenes, but also Leucippus and Democritus explicitly linked the *inclination* to the latitudinal variation of climates.

It is not clear, however, and Bailey fails to explain, how this inclination can be connected to *Egypt*. On a *flat* earth the *inclination of the cosmos* is a *constant*: the celestial axis is no more inclined in Egypt than among the Britons or in any other location. So, although this interpretation offers an interesting parallel with the theories of acknowledged flat-earthers Anaxagoras, Leucippus and Democritus, it may not be what Lucretius had in mind.

3. This brings us to the third and last interpretation, which is in fact the oldest: according to Lambinus (1563), quoted with approval by Merrill, Lucretius is alluding to the fact that for an observer in Egypt the celestial pole appears *low* in the sky ('axis, sive polus arcticus, qui nobis sublimis semper appâret, illis est depressus'). Munro thinks along the same lines, and he aptly quotes Cleomedes I 5.47–48, who also contrasts Egypt and Britain in this respect:

... παρὰ μὲν Συηνίταις καὶ Αἰθίοψιν ἐλάχιστον φαίνεται τὸ τοῦ πόλου ὕψος, μέγιστον δὲ ἐν Βρεττανοῖς, ἐν δὲ τοῖς διὰ μέσου κλίμασιν ἀναλόγως.	... among the Syenites[209] and Aethiopians the elevation of the pole appears least, but among the Britons greatest, and proportionately at the intervening latitudes.

Munro's quotation is repeated by Robin, who fails to observe its incompatibility with his own interpretation and with his other quotations. However, although the present interpretation is incompatible with the previous one, it is not unrelated. While Anaxagoras, Diogenes, Leucippus and Democritus had been content with the observation that the north pole is not vertically overhead, but appears at a certain angle above the northern horizon (38° in Athens), the progress of knowledge and the expansion of the Greek 'horizon'

---

209 The inhabitants of Syene (modern Aswan), in southern Egypt, reputed to be exactly below the summer tropic circle (see p. 254 ff. below). In reality Syene is situated about 0° 39', i.e. 72 km, north of the tropic.

made it clear that this angle is by no means fixed, but depends on the location, or more precisely the latitude, from which the observation is made (varying from about 24° in Syene to about 54° in Britain). Accordingly the elevation of the pole could be and was actually used as a measure for the geographical latitude of a place[210]—as it still is. It had also been known for a long time that the temperature of a place somehow depends (among other things) on its latitude,[211] which could now be expressed in terms of the polar height. It would, therefore, make excellent sense for Lucretius, speaking of the variation of climates, to characterise Egypt by its relatively small polar elevation, which would then be a measure for its southern position and hence its hot climate.

At this point we encounter a problem: the small polar elevation in Egypt may be apt to distinguish it from Britain, or from 'the intervening latitudes' of Pontus and Gades, but what about 'the black tribes of men with their scorched colour', i.e. the Aethiopians,[212] who live even farther to the south, and for whom the north pole lies even lower? The key to this problem may be found in the next couple of verses (1110–1113):

Quae cum *quattuor* inter se diversa videmus	And as we see these *four* (regions)
quattuor a ventis et caeli partibus esse,	to be diverse among each other,
tum color et facies hominum distare videntur	according to the *four* winds and quarters of the sky, so the colour and aspect of men are seen to differ greatly,
largiter et morbi generatim saecla tenere.	and diseases to possess the nations race by race.

Although in the previous lines Lucretius clearly mentioned *five* places or peoples,—the Britons, Egypt, Pontus, Gades, and the Aethiopians—, he now

---

210 By Hipparchus, for instance: see p. 172 with n. 44 above.

211 See p. 222 ff. and nn. 207 and 208 with the text thereto above. Cf. also Arist. *Mete.* II, 5, 362b16–18: οὐ γὰρ ὑπερβάλλει τὰ καύματα καὶ τὸ ψῦχος κατὰ μῆκος, ἀλλ' ἐπὶ πλάτος—"for excesses of heat and cold take place, not according to longitude, but according to latitude."

212 That the Aethiopians are meant is clear from the nearly identical line in VI 722, where these same 'black tribes of men' are situated in the hot country south of Egypt where the Nile originates. The theory that the Aethiopians were burnt black by the sun was first mentioned by Herodotus II 22 and later became commonplace: see e.g. [Arist.] *Probl.* x 66, 898b; Strabo XV 1, 24 (quoting the IV BC tragedian Theodectes); Ovid. *Met.* II 235–236; Manilius IV, 758–759; Sen. *NQ* IVa 2, 18.1–2; Lucan X 221–222; Pliny *NH* II 189, 2–3; Hyg. *Astron.* I 8, 3, 12; Ptol. *Tetr.* II 2, 56; *Etymologicum Magnum* s.v. Αἰθίοψ.

refers to them as if there were only *four* of them, which should roughly correspond to the *four* cardinal winds. The problem is illustrated in the map below:

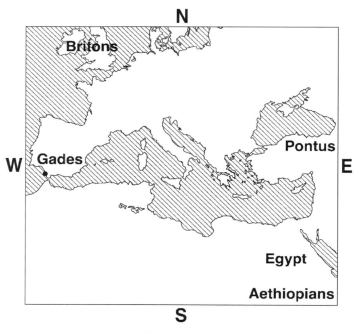

FIGURE 4.5 *Lucr.* VI 1106–1113: *five places, four winds*

The Britons obviously represent the north, and Egypt, where the celestial north pole lies low, the south. Pontus and Gades, which lie at roughly equal distances to the east and west of Rome, must represent these two directions. This leaves the Aethiopians, the southernmost people known to the ancients: they too must represent the south, as they do in Cleomedes' work (see the quotation above) and in many other scientific writings.[213] It is clear then that Lucretius did not mean the Aethiopians and Egyptians to be contrasted: they both represent the south, where the north pole lies low. Still, this double representation of the south, with Pontus and Gades squeezed in between, when, moreover, the subsequent reference to the four winds presupposes a single representative for each wind, is a bit awkward. Perhaps, if he had lived to revise his work, Lucretius

---

213   See e.g. Strabo I 2, 24–28, Pliny *NH* II 189, 2–3; Ptol. *Tetr.* II 2, 56 and Asclepiades of Bithynia according to Aëtius V 30, 6.

would have removed one of the two references to southern peoples. As for Pontus and Gades: it is clear that these places were not chosen for their difference in *latitude* (which is minimal), but for their difference in *longitude* (which is considerable).[214] This means, however, that the supposed reference in line 1107 to the polar elevation does not apply to their climatic difference. A strange discrepancy results: while in lines 1110–1113 Lucretius suggests that climates vary with latitude *and longitude*, the theory he seems to allude to in line 1107 only accounts for the *latitudinal variation*.[215]

Of the three interpretations we have just considered, only the third (Lambinus' and Munro's) answers all three questions we posed at the beginning: it explains the enigmatic words 'qua mundi claudicat axis' in a way that is both meaningful and relevant to the context and accounts for the fact that the phenomenon is situated in Egypt of all places.

Now it is time to come to the crux of this section: the Cleomedes-quote, which Munro uses to support his interpretation of line 1107, is part of an argument for the *sphericity* of the earth: *"If the earth's shape were plane and flat,"* says Cleomedes,[216] *"the pole would be seen by everyone at an equal distance from the horizon, and the arctic circle[217] would be the same. Yet, nothing like this is present in the phenomena, but instead, among the Syenites and Aethiopians the elevation of the pole appears least, but among the Britons greatest, and proportionately at the intervening latitudes."* This means that, if our interpretation of *DRN* VI 1107 is right, Lucretius is here alluding to an observation which in antiquity was used by others as evidence of the earth's sphericity ('proof 4' on p. 172 above). There is no indication, however, that Lucretius himself linked these observations to any particular shape of the earth. Besides, it is possible, as we have seen above, to harmonize such observations with a flat earth, if the sun is assumed to be relatively close by.

---

214 In Cleomedes' work (I 5.37–44) the extreme west and east are represented by Iberia (Spain) and Persia respectively. Strabo and others prefer India to represent the east: see Bowen-Todd (2004) p. 66, n. 11. The only other work where Pontus is used to represent the east is Vitr. VI 1.1 (see next note).

215 The same discrepancy can be observed in Vitruvius VI 1.1, where the assertion that climates vary with the inclination of the cosmos (i.e. with latitude), is followed by a list of places that includes representatives of the east (Pontus) and the west (Spain); the south is represented by Egypt, and the centre by Rome, while the north is not represented.

216 Cleomedes I 5.44–49.

217 The circle that encompasses the ever-visible portion of the heavens.

4.3.12 *Philodemus and the Gnomon (Phil. De sign. 47.3–8)*

A similar instance is found in a fragment of Lucretius' contemporary and fellow Epicurean Philodemus. In the *De signis* 47.3–8 Philodemus gives us three examples of inferences that go wrong, because they are based on too limited a sample:

[ἐάν γ]ἐ τις λέγηι [πάντας] ἀνθρώ‖[π]ους εἶναι λευ[κοὺς ἀ]πὸ τῶν | παρ' ἡμῖν ὁρμώμε[νος ἢ] τοὐ|ναντί[ο]ν ἀπὸ τῶν Αἰθιόπων, | ἢ πανταχοῦ τοὺς ὀρθοὺς γν[ώ]‖[μ[ο]νας περὶ μεσημβρίαν ἐν | ταῖς θεριναῖς τροπαῖς [ἀνε]λεῖν | σκιάν, ἆρ' οὐ μάταιος ἔ[σ]ται;	If a person says that *all* men are white, starting from those among us, or the opposite, starting from the Aethiopians, or that vertical gnomons *everywhere* wipe out their shadows around midday at the summer solstice, isn't he talking nonsense? tr. DE LACY, modified

In the first two examples the nature of the limitation is clearly stated: the investigation has been limited to 'those among us', or, alternatively, to the Aethiopians, resulting in the wrong conclusion that *all* men are white or that *all* men are black. In the third example the nature of the limitation is suppressed, but perhaps we are meant to supply it from the previous example: *if one were to limit one's investigation to the country of the Aethiopians,* one might come to the wrong conclusion that vertical gnomons *everywhere* are shadowless around midday at the summer solstice. In fact this was only observed to happen at the indicated time in Syene, an Egyptian town bordering on Aethiopia, and therefore sometimes considered Aethiopian: see e.g. Alexander *In Aristotelis Mete.* p. 103.30–31 Hayduck, reporting the same observation, with people instead of gnomons:

Λέγονται μέντοι οἱ περὶ Συήνην οἰκοῦντες τῆς Αἰθιοπίας ἐν τῇ μεσημβρίᾳ ὄντος τοῦ ἡλίου κατὰ τὰς θερινὰς τροπὰς ἄσκιοι γίνεσθαι.	It is said that when the sun is at its midday culmination at the time of the summer solstice those living around Syene in Aethiopia become shadowless.

Philodemus does not explain where each of his three examples goes wrong, or what the true facts are in each case, but his argument requires that both he and his audience are aware of the true facts: not everywhere men are white but only among us, not everywhere men are black but only in Aethiopia, and not everywhere vertical gnomons are shadowless around midday at the summer solstice but only in Syene. This last statement may be compared with a passage in Cleomedes (I, 7.71–78):

Φησὶ τοίνυν, καὶ ἔχει οὕτως, τὴν Συήνην ὑπὸ τῷ θερινῷ τροπικῷ κεῖσθαι κύκλῳ. ὁπόταν οὖν ἐν Καρκίνῳ γενόμενος ὁ ἥλιος καὶ θερινὰς ποιῶν τροπὰς ἀκριβῶς μεσουρανήσῃ, ἄσκιοι γίνονται οἱ τῶν ὡρολογίων γνώμονες ἀναγκαίως, κατὰ κάθετον ἀκριβῆ τοῦ ἡλίου ὑπερκειμένου. {...} Ἐν Ἀλεξανδρείᾳ δὲ τῇ αὐτῇ ὥρᾳ ἀποβάλλουσιν οἱ τῶν ὡρολογίων γνώμονες σκιάν, ἅτε πρὸς τῇ ἄρκτῳ μᾶλλον τῆς Συήνης ταύτης τῆς πόλεως κειμένης.	Eratosthenes says, and it is the case, that Syene is located below the summer tropical circle. So when the sun, as it enters Cancer and produces the summer solstice, is precisely in mid-heaven, gnomons of sundials are necessarily shadowless, since the sun is located vertically above them. {...} But in Alexandria at the same hour gnomons of sundials do cast a shadow, since this city is located further north than Syene.

tr. BOWEN-TODD, modified

Philodemus' reference matches not just the content but even the vocabulary of Cleomedes' account, which suggests that Philodemus was aware of such observations: he knew and accepted that the length of the shadow of a gnomon at noon at the summer solstice depends on the observer's latitude.

Now the quotation from Cleomedes is part of his report of Eratosthenes' famous calculation of the circumference of the earth,[218] which was based on the assumption that *the earth is a sphere*. Moreover, the same and similar observations are reported by the elder Pliny (*NH* II 183) as part of one of his formal proofs for the *sphericity of the earth* ('proof 5' on p. 172 ff. above). This means that Philodemus too was familiar with observations that could be used as evidence of the earth's sphericity. Yet, again there is no indication that Philodemus was aware of this implication, but had he been, these observations too could be reconciled with a flat earth on the assumption that the sun is relatively close by.

4.4 Conclusions

Having completed our review of relevant passages we may now be in a position to answer the six questions formulated at the outset of this chapter (see p. 164 above):

---

218 See also p. 169 and p. 238 above.

1   Did the Epicureans Posit a Parallel Downward Motion of All Bodies?

The ascription to the Epicureans of a flat-earth cosmology is usually supported with reference to their assumption of a parallel downward motion. The evidence for this assumption, however, now turns out to be more ambiguous than has hitherto been supposed. It is true that certain passages clearly imply a parallel downward motion: Lucretius' rejection of the centrifocal alternative in I 1053–1093, his theory of the 'swerve' in II 216–250 and his view that the earth rests on air in V 534–563 all point to his acceptance of a parallel downward motion, and corresponding passages and fragments of Epicurus confirm that this was Epicurus' view. Other passages, however, contradict this conclusion: both Lucretius, in V 621–636 and 713–714, and Diogenes of Oenoanda, in fr.13 I.11–13, use centrifocal language, which conflicts with their presumed parallel-linear cosmology, and Lucretius' cosmogonical account in V 449–508 even requires the assumption of a centripetal downward motion of all bodies, which is absolutely incompatible with an otherwise parallel-linear system.

In view of this evidence it may seem that Epicurean cosmology, or at least Lucretius' version of it, is a hopeless mixture of two incompatible worldviews, but perhaps we need not be so pessimistic. All those passages in Lucretius which point to a parallel cosmology are firmly rooted in, and interconnected with, sound Epicurean doctrine: the rejection of centrifocalism follows only after a thorough criticism of that alternative, and allows Lucretius to finally conclude that the infinite universe contains an infinite amount of matter (cf. Epic. *Hdt.* 41–42), and consequently an infinite number of worlds (Lucr. *DRN* II 1048–1089, Epic. *Hdt.* 45); the acceptance of a parallel-linear cosmology, implied in II 184–250, appears to follow automatically from the rejection of the centrifocal alternative, and has a clear counterpart in Epicurus' own writings (e.g. Epic. *Hdt.* 60), while the passage about the earth being supported underneath (cf. Epic. *Phys.* XI fr.42 Arr. and *Hdt.* 73 scholion) in turn follows logically from the acceptance of parallel downward motion. Lucretius' centrifocal cosmogony, on the other hand, appears to be an isolated passage, unconnected (as far as its *centrifocal* character is concerned) with what comes before or after and unparalleled in the remaining works and fragments of any other Epicurean. Lucretius' use of centrifocal terminology in two astronomical passages does not constitute a real parallel, because—except for the terminology—the theories they expound in no way require or imply a centrifocal downward motion (as Lucretius' *cosmogony* does). The aberrant terminology in these passages must probably be attributed to the sources from which Epicurus and Lucretius borrowed these astronomical theories,[219] and therefore presents no real obstacle to

---

219   See section 2.4 on pp. 58 ff. above.

the view that Epicurus' and Lucretius' cosmology was parallel-linear. The only real obstacle to this view is Lucretius' cosmogony, which may represent a later development (within or outside the circle of Epicurus' followers), which somehow made its way into Lucretius' otherwise orthodox account of Epicurean physics.

2 Does Parallel Downward Motion Imply a Flat Earth?

Parallel downward motion seems to imply a flat earth, and centripetal motion a spherical one. Underlying both implications is the assumption that heavy objects everywhere fall at right angles to the earth's surface (see the illustrations on p. 179 above). Although the assumption seems obvious, there is no cogent logical reason for accepting it, and there is no evidence that the Epicureans felt bound by it. It cannot, therefore, be taken for granted that the Epicureans' commitment to parallel downward motion implies belief in a flat earth.

3 Do Epicurean Astronomical Views Presuppose a Flat Earth?

Since the Epicurean views on the shape of the earth cannot be simply inferred from their conception of downward motion, we had to look for other clues that are independent of this assumption. Several such clues were found in passages dealing with astronomical matters. Here too, however, the evidence is ambiguous. The following passages seem to imply or presuppose a flat earth:

a) IV 404–413: the way the sun's distance is described in terms of intervening lands and seas suggests that Lucretius was thinking of a flat earth, with the sun rising not too far past its easternmost extremity. However, as independent evidence of the Epicureans' position this passage does not count for much: most likely it is simply the poet's picturesque way of conveying to the reader the enormous distance of the Sun, without implying any doctrinal stance concerning the shape of the earth.

b) V 204–205: Lucretius' mention of three climatic regions must be considered either a reference to the antiquated theory that the earth's *disk* is divided into three climatic strips, or to the untheoretical observation which underlies both this old theory and the newer five-zone theory. In the first case these lines would imply a belief in a flat earth, in the second case they are neutral with respect to the earth's shape. Accordingly, the passage cannot be used as evidence for the Epicureans' commitment to either shape of the earth.

c) V 564–613 & Epic. *Pyth*. 6 [91]: according to Furley, the Epicureans' preoccupation with the (small) size of the sun and the other heavenly bodies was partly motivated by their wish to bring the observation that the aspect of the sky changes with the observer's latitude into line with their own flat-earth cosmology. Although Furley's suggestion is very attractive, I do not think it can be maintained. In the first place it is not certain that the Epicureans really claimed the sun and the other heavenly bodies to be very small,[220] secondly, if the sun's small size had been so important to the Epicureans one would have expected them to take heed of it in other contexts as well, which they did not (see below), and, thirdly, the small size of the sun and the other heavenly bodies only accounts for some aspects of the mentioned phenomena (notably those underlying 'proofs 4' and '5': see below), while ignoring the rest. However, with Furley's interpretation out of the way there is no further reason to connect Epicurus' theory about the size of the heavenly bodies with his views concerning the shape of the earth.

d) v 650–704 & Epic. *Pyth.* 7 [92]: the theory that the sun is extinguished at sunset and rekindled at sunrise seems to imply that sunset and sunrise are simultaneous for everyone on earth, which in turn would imply that the earth is flat. However, since this is only one of two possible explanations for the sun's setting and rising, the only valid conclusion would be that the earth *may* be flat, not that it definitely *is* flat. Besides, Epicurus and Lucretius do not actually state that the sun's extinction and rekindling occur simultaneously with its setting and rising: one could assume that the sun is extinguished only after the last observed sunset and rekindled before the first observed sunrise. Interpreted in this way the theory could be compatible with a spherical earth as well.

By contrast, the following passages seem to presuppose or imply a spherical earth:

e) v 762–770 (cf. Epic. *Pyth.* 13 [96–97]): the theory that the moon is eclipsed by falling into to the earth's *conical* shadow seems to imply both that the earth is spherical and that the sun is larger than the earth. As with the previous case it must be noted that this theory is only one of several possible explanations, and that therefore the only warranted conclusion is that the earth *may* be spherical and the sun *may* be larger than the earth, not that they definitely are. Yet, even if we assume this explanation of lunar eclipses to be correct, the earth's sphericity does not necessarily follow: a conical shadow can be produced by any shape that presents a circular outline to the sun, even a flat disk or an oval, and the number of possible, non-spherical shapes increases dramatically if 'cone' is interpreted in a loose sense. The

---

220 See n. 184 on p. 236 above.

second implication, that the sun may be larger than the earth, concerns us only in so far as it contradicts the Epicureans' supposed claim that the sun is very small, and hence Furley's theory that the small size of the sun played a role in reconciling certain astronomical observations with their own flat-earth cosmology (see item c above). In short, the present passage has no real implications for the shape of the earth as such, but it provides further confirmation that, whatever shape the Epicureans had in mind, astronomical observations played no part in it (see also below).

f) VI 1107: this line may be an allusion to the fact that the celestial north pole stands lower in the sky for an observer in Egypt than for observers in more northerly countries, a fact that was used by others as a proof ('proof 4': see below) of the sphericity of the earth. It must be noted, however, that Lucretius himself does not draw this conclusion.

g) Philodemus *Sign.* 47.3–8: this is an unmistakable reference to the observation that the length of shadows is not the same everywhere, but depends on the observer's latitude. This observation, too, was used by others as a proof ('proof 5': see below) of the earth's sphericity. It must be noted, however, that Philodemus himself does not link this observation to the shape of the earth.

In short, the evidence of these seven passages appears to be conflicting, with some passages suggesting a flat, and others a spherical earth. On closer scrutiny, however, most passages turn out to be compatible with either shape. Only one passage, explaining the sun's setting and rising as possibly due to its extinction and rekindling, seems to point unambiguously to a flat earth. Yet, even here the implication for the earth's shape is dissipated if one allows that the sun is extinguished only when the sun has set for every observer, and rekindled before the first observer sees it rise. Moreover, since the theory is only one of two possible theories, its implications for the shape of the earth are hypothetical at best.

In sum, of the astronomical passages we investigated *none* has definite and incontrovertible implications for the shape of the earth.

4 Why Did the Epicureans Hold On to the Claim That the Earth is Flat?

If neither their theories concerning the direction of downward motion nor their astronomical views allow us to ascribe to the Epicureans the firm conviction that the earth is flat, this question—at least in its present form—becomes meaningless. A better question would be: why did the Epicureans not reject (as did most other philosophers) the possibility that the earth is flat? This question I will try to answer in the final paragraph of this section.

5 Were They Familiar with Contemporary Astronomy?

This question has been sufficiently answered by David Sedley and David Furley,[221] who point to Epicurus' dealings with contemporary mathematical astronomers and their theories in book XI of his *On nature*. To this may be added the many unmistakable references to individual astronomical theories in Epicurus' *Letter to Pythocles* and book V of Lucretius' *DRN*.[222] It is clear, then, that the Epicureans were familiar, at least to a certain extent, with the theories of contemporary astronomy.

6 Did They Know of the Astronomical Arguments for the Earth's Sphericity and Put Up a Reasoned Defence of Their Own Position?

If the Epicureans indeed refused to endorse the earth's sphericity, it seems apposite to ask to what extent they were aware of, and dealt with, the astronomical proofs for the earth's sphericity. Below I will discuss one by one the seven proofs distinguished on p. 169 ff. above:

Proof 1. The convexity of the earth's shadow during a lunar eclipse.
As Furley points out, this observation could be accounted for by many other shapes besides a sphere, including a flat disk. It is possible that the ancients realized this too, for after Aristotle no one else adduced this proof. Consequently, Epicurus and Lucretius can hardly be blamed for ignoring it.

Proof 2. Observations of, and from, departing and approaching ships.
There are no records of this proof before the time of Strabo (64 BC – 24 AD), whose work postdates both Epicurus and Lucretius. Therefore, Epicurus and Lucretius may have been unaware of it. Yet, even if they had known of such observations, they could have plausibly dismissed them as mirages such as are often observed at sea.

Proof 3. The aspect of the sky varying with latitude.
This proof was first reported by Aristotle and remained popular ever since. If Epicurus and Lucretius were serious about their refusal to commit to the earth's sphericity, they could hardly have afforded to ignore this proof. Furley remarks that observations of this kind could in fact be reconciled with a flat earth if the heavenly bodies were assumed to be relatively close-by, like paintings on a ceiling, and he suggests that Epicurus' concern to prove that the

---

[221] Sedley (1976) esp. 26–43 and 48–54; Furley (1996) 120–121.
[222] See p. 42 ff. above.

sun and the other heavenly bodies are very small was partly motivated by his wish to uphold his own flat-earth theory in the face of such observations. However, while Furley's model can explain why from different latitudes heavenly bodies are observed at different angles, it fails to account for the fact that at certain latitudes certain stars are completely blocked from view. The only other way to explain this fact, without recourse to the earth's sphericity, would be to attribute it to the unevenness of the earth's surface. Moreover, Furley's theory is contradicted by the Epicureans' acceptance of the possibility that the moon is eclipsed by falling into the earth's conical shadow—a theory which presupposes that the sun is *larger* than the earth. Anyway, there is no independent evidence that Epicurus or Lucretius or any other Epicurean was aware of the observations underlying this proof (except for the special cases singled out as 'proof 4' and 'proof 5') and of their possible implications for the shape of the earth.

Proof 4. The elevation of the celestial north pole varying with latitude.
Although this proof cannot be dated with any certainty before the time of Hipparchus (ca. 190–120 BC), there is every chance that it is much older, being a special case of 'proof 3' and analogous in structure to 'proof 5'. The observations underlying this proof are probably alluded to by Lucretius in VI 1107, although there is no indication that he was aware of their possible implications for the shape of the earth. In this case Furley's painted-ceiling model shows how such observations could be reconciled with a flat earth, but again there is no evidence that the Epicureans were committed to such a model in any way.

Proof 5. The length of shadows varying with latitude.
This proof can be dated to at least as early as Pytheas (ca. 325 BC). The observations that underlie this proof are alluded to by the later Epicurean Philodemus (ca. 110 – ca. 40 BC), in a way that suggests that he as well as his audience were familiar with them, although he makes no reference to their possible implications for the shape of the earth. In this case, too, the observations could be made to agree with a flat earth using Furley's painted ceiling model.

Proof 6. The maximum day-length varying with latitude.
This proof, too, must have been known to Pytheas (ca. 325 BC), and so could have been known to Epicurus. Therefore, if Epicurus and Lucretius refused to endorse one specific shape of the earth, and wanted to defend this position, one would expect them to have taken note of this proof. Instead, by accepting the possibility that the sun is extinguished at night, Epicurus and Lucretius seem to ignore the observations underlying this proof. However, given the inaccuracy

of time measurement and the impracticability of comparing times at different locations in antiquity, there seems to be some justification for disregarding the evidence.

Proof 7. Longitudinal time difference established by eclipses.
This proof cannot be dated with any certainty before the time of Hipparchus (ca. 190–120 BC), and so Epicurus may not have known it. Anyway, as with the previous proof, the theory of solar extinction and rekindling clearly shows that both Epicurus and Lucretius must have ignored the observations relating to this proof. In this case, too, their disregard for the evidence is somewhat justified by the unreliability of ancient time measurement in general, and by the scarcity of pertaining observations (viz. eclipses).

In short, at least some of the observations underlying the ancient proofs (notably proofs 4 and 5) of the earth's sphericity were known to some Epicureans. There is no indication however, that they acknowledged their evidential value, let alone subjected them to a reasoned criticism. Yet, the Epicureans' indifference towards the evidence may not be wholly unjustified. Indeed, each of these 'proofs' only counts as proof, when viewed in the light of certain assumptions, which the Epicureans did not share.

Despite the strong claims in modern studies about the Epicureans' commitment to a flat earth, this flat earth of theirs turns out to be rather elusive. Epicurus and his followers never said that the earth is flat, and even their most ardent ancient critics never accused them of saying so. It is true that the Epicureans assumed a parallel downward motion (with the curious exception of Lucretius' cosmogony), but they did not infer from this, as their predecessors had, that the earth is flat. Nor do their astronomical theories suggest one specific shape of the earth. It would seem, then, that the Epicureans had no firm conviction as to the shape of the earth at all: to them, as Woltjer (1877) already suggested, the earth like the cosmos (or the sun), might have any shape.[223]

This conclusion agrees well with the Epicureans' general attitude to astronomical problems, where any theory that explains a given problem and agrees with the appearances, must be accepted. This attitude also explains the Epi-

---

223 Woltjer (1877) 123, quoting Epic. *Hdt.* 74 with scholion: "Non perspicuum est quam formam terrae tribuerit poeta, quam Stoici sphaeram esse contendebant. Epicurus de mundorum forma dixit: Ἔτι δὲ τοὺς κόσμους οὔτε ἐξ ἀνάγκης δεῖ νομίζειν ἕνα σχηματισμὸν ἔχοντας, ἀλλὰ (καὶ διαφόρους αὐτοὺς ἐν τῇ ι β Περὶ φύσεώς φησιν) οὓς μὲν γὰρ σφαιροειδεῖς, καὶ ᾠοειδεῖς ἄλλους, καὶ ἀλλοιοσχήμονας ἑτέρους· οὐ μέντοι πᾶν σχῆμα ἔχειν. Ergo *potest* terra sphaericam habere formam ...". For Epicurus' views concerning the shape of the sun see Aëtius II 22.4.

cureans' indifference towards the astronomical proofs of the earth's sphericity: although they might accept some of the observations underlying these proofs, the proofs themselves depend on conceptual models, which vainly pretend to capture the essence of the phenomena they are meant to explain, when other theories agree with the appearances just as well.[224] Furley's attempt to defend the Epicureans is misguided on two accounts: in the first place he assumes wrongly, as most scholars before him, that the Epicureans were committed to the earth's flatness, and in the second place he misinterprets Epicurus' opposition to mathematical astronomy and its partisans as directed against specific theories, rather than its method and assumptions. In reality, as Wasserstein (1978) recognized,[225] Epicurus viewed the cosmos as a series of unconnected phenomena, each to be explained separately in agreement with the appearances here with us. To force them all into a preconceived system, whether one based on a spherical or on a flat earth, is to fall back into myth.

---

224 Concerning the Epicureans' insistence on 'agreement with the phenomena' see Chapter Two, esp. p. 19 f., pp. 32 ff., and p. 56 f. above.
225 Wasserstein (1978) 490–494; see pp. 32 ff. above.

CHAPTER 5

General Conclusions

Having reached the end of this exploration of Epicurean meteorology, I will now conclude by summarizing the main findings.

In Chapter Two I have examined Epicurus' method of multiple explanations from several different perspectives. First I dealt with the view that Epicurus held all alternative explanations to be true at the same time. This view seems to follow from Epicurus' claim that non-contestation establishes truth and his use of non-contestation to support each one of his alternative explanations. However, although the inference seems to be sound, and Epicurus may have actually asserted the truth of alternative explanations, in fact this claim holds good only insofar as the phenomenon under consideration is viewed as an instance of a general type, and only insofar as the explanations are safe from being falsified by closer observation, and even then Epicurus does not seem to set much value on this *truth*, for, when it comes to proving the fundamental physical theories, only *singular truths* qualify. Then I argued that Diogenes of Oenoanda's claim that some explanations are more plausible than others is a departure from Epicurus and Lucretius for whom all alternative explanations are equally true, and I suggested that Diogenes may have used this licence in order to be able to embrace the findings of contemporary astronomy without actually rejecting Epicurus' multiple explanations. In this connection I also examined Bailey's claim that in astronomy Lucretius usually presents the views of the mathematical astronomers first, 'as though he really preferred these.' Although Bailey's observation seems correct, his interpretation is not. Instead, by presenting these explanations first Lucretius simply acknowledges the predominant position these explanations had among his contemporaries, a predominance which he sets out to undermine by showing that other explanations are equally possible. Next, we found that the individual explanations given in Epicurus' *Letter to Pythocles* and the meteorological and astronomical portions of Lucretius' DRN to all likelihood derive from a doxographical work resembling Aëtius' *Placita*, but, in contrast to Aëtius, combining the reported *doxai* with explanatory analogies. Finally, I compared Epicurus' use of multiple explanations with Aristotle's and Theophrastus'. It turns out that their occasional use of multiple explanations, and their epistemological justifications for this use differ from Epicurus' and Lucretius' method in several important respects, but may have been a source of inspiration. In this context I also pointed out that the pervading use of multiple explanations in the *Syriac meteorology*, a meteorological treatise preserved in Syriac and Arabic and commonly ascribed to Theophrastus, while closely resembling Epicurus' method, differs greatly from the occasional use of multiple explanations in the uncontested works of Theophrastus. For this reason the possibility that the Syriac meteorology is Epicurean rather than Theophrastean should be taken into consideration.

In Chapter Three I compared Epicurus' *Letter to Pythocles* and Lucretius' *DRN* VI with a number of meteorological works and passages as to the range and subdivision of subjects included, and as to the order in which these subjects are dealt with. This led me to conclude, among other things, that the range of subjects covered by the *Syriac meteorology* may well be complete (except for the omission of the rainbow), and that the meteorological accounts of Lucretius, Epicurus and the *Syriac meteorology* belong to a tradition, inaugurated by Aristotle, which connected earthquakes with atmospherical phenomena. Comparison of *DRN* VI with a number of meteorological and paradoxographical works showed that Lucretius' treatment of exceptional local phenomena is unique, and may well be his own invention, rather than something derived from Epicurus. Examining the often observed correspondence in the order of subjects between the *Syriac meteorology*, book III of Aëtius' *Placita*, book VI of Lucretius' *De rerum natura*, and the second part of Epicurus' *Letter to Pythocles*, I found that the order of all four works can be reduced to an original order, from which each of the four texts deviates in its own manner. This reconstructed 'original order' led me in turn to investigate the relations these four texts bear to each other and to this 'original', as far as the order of subjects is concerned. In this connection I also re-examined the *Syriac meteorology*, looking for clues that might link it more closely to either Theophrastus and the Peripatos in general, or to Epicurus. While the general character of the work is much closer to Epicurus than to Theophrastus' undisputed works, certain details preclude an exclusively Epicurean origin. Interestingly, these un-Epicurean details are concentrated in just a few passages, some of which are of a markedly different character from the rest of the treatise. This seems to suggest that the treatise has a mixed origin, being mostly Epicurean but with an admixture of Peripatetic views.

In Chapter Four I examined the claim, made by many modern scholars, that Epicurus and Lucretius believed the earth to be flat. After a thorough examination of every possibly relevant passage in their works I arrived at the following conclusions. Although the Epicureans assumed a parallel downward motion (with the exception of Lucretius' cosmogony, which may be a *corpus alienum*), they never concluded from this that the earth must be flat. Moreover, the Epicureans' failure to engage with the available astronomical evidence for the earth's sphericity does not mean they were ignorant of the underlying observations, but could equally well be construed as a rejection of their evidential status, which in the end always depends on certain assumptions. This is in line with the Epicureans' general distrust of astronomical models, as repeatedly expressed in Epicurus' *Letter to Pythocles* and the fragments of book XI of his *On nature*.

Throughout the main chapters of this book, several themes recur. One such theme, considered in Chapters Two and Three, is the authorship and identity of the *Syriac meteorology*. Another theme, also dealt with in Chapters Two and Three, is Lucretius' treatment of exceptional local phenomena. A third theme, common to Chapters Two and Four, concerns the Epicureans' attitude to mathematical astronomy: their mistrust of arbitrary assumptions and physical and conceptual models as a means to arrive at the exclusive truth of hypotheses, and their persistent habit of considering every phenomenon in isolation, leading to the acceptance of every possible theory, instead of combining their accounts into unified theories so as to weed out impossible individual explanations. A fourth theme, broached in Chapters Two, Three and Four, regards Lucretius' relation to Epicurus: is the *De rerum natura*, as far as its philosophical content is concerned, based solely on Epicurus' own writings, or did Lucretius incorporate later intellectual developments as well? In the course of this book three passages present themselves that have been, or could be, viewed as departures or developments from Epicurus. The first is Lucretius' rejection of centripetal downward motion in *DRN* I 1052–1093, discussed on p. 181 ff. above. Although its arguments and conclusion agree well with orthodox Epicureanism, it is commonly believed that Epicurus himself did not engage with Stoicism. Attempts to explain the passage as an attack on other philosophers, whom Epicurus could have criticized himself, fail to convince. So, unless we challenge the dogma that Epicurus ignored the Stoics, the passage must be considered post-Epicurean. The second passage is Lucretius' account of a number of exceptional local phenomena in *DRN* VI 608 ff. Although the account is not incompatible with orthodox Epicurean doctrine, it is unlikely that Lucretius should have derived it from Epicurus. The third and final passage is Lucretius' cosmogonical account in *DRN* V 449–508. Here at last we have a passage that is definitely *in*compatible with Epicurean orthodoxy. However, before we start accusing Lucretius of heterodoxy, it must be noted that the passage is also incompatible with passages of Lucretius himself: the cosmogony appears to be a foreign body which Lucretius inadvertently incorporated in his otherwise orthodox account of Epicurean physics.

A very interesting aspect of Epicurean meteorology, touched upon in all three chapters, is its ambivalent relation to Peripatetic astronomy and meteorology. On the one hand, it is clear that the Peripatos has exerted an immense positive influence on Epicurus' treatment of these subjects. Firstly, Epicurean meteorology derives most of its explanations from doxography, a genre rooted in Peripatetic philosophy. Secondly, the organization of its subject matter closely matches Aëtius' *Placita* and probably derives from a Peripatetic work (note e.g. the inclusion of earthquakes with atmospherical phenomena, and the *prēstēr* being appended to thunder, lightning and thunderbolts). Thirdly, the Epicurean use of multiple explanations has clear antecedents in the works of Aristotle and Theophrastus, from which Epicurus may well have drawn his inspiration. The Peripatetic influence on Epicurus' method of multiple explanations might be even greater, if the Syriac meteorology, which also makes extensive use of multiple explanations, is indeed a genuine work of Theophrastus', which in my view is still open to debate. On the other hand, Epicurean meteorology is strongly opposed to Aristotle's astronomical and meteorological views. In the first place, Epicurus' combination of atmospherical and astronomical phenomena under the single heading of τὰ μετέωρα and his subjection of both types of phenomena to a single explanatory method implies a rejection of Aristotle's sharp division of the sublunary and supralunary realms. In the second place, Epicurus' method of multiple explanations is opposed not just to the use of single explanations, but to the very idea of a unified explanatory principle, such as we find in the works of most of the natural philosophers, including Aristotle. In this light we must also view Epicurus' rejection of mathematical astronomy, to which Aristotle too was committed. It is ironic that the weapons for Epicurus' attack on Peripatetic astronomy and meteorology (viz. multiple explanations and doxography) should have been provided by the very same school.

APPENDIX 1

Multiple Explanations in Epicurus' *Letter to Pythocles*

Ch.	Subject	Number of alternative explanations	Possibility	Grammatical conjunctions	Inexhaustivity of the list
1	Introduction	–			
2	Method	–			
3	Definition of 'cosmos'	–			
3a	motion of its boundary	2	ἐνδέχεται	ἢ Α ἢ Β	–
3b	shape of its boundary	2+	ἐνδέχεται	ἢ Α ἢ Β	ἢ οἵαν δήποτε …
4	Number and origin of cosmoi	1			
5	Formation of the heavenly bodies	1			
6	Size of the heavenly bodies	1			
7	Risings and settings	2	δύνασθαι	καὶ Α ⟨καὶ⟩ Β	–
8	Motions of the heavenly bodies	3	οὐκ ἀδύνατον	Α ἢ Β, εἶτα Γ	–
9	Turnings of the sun and moon	4(+)	ἐνδέχεται	Α, ὁμοίως δὲ καὶ Β ἢ καὶ Γ ἢ καὶ Δ	πάντα τὰ τοιαῦτα καὶ τὰ τούτοις συγγενῆ …
10	Phases of the moon	3+	δύναιντ' ἂν	καὶ Α καὶ Β, ἔτι τε καὶ Γ	καὶ κατὰ πάντας τρόπους …
11	Light of the moon	2	ἐνδέχεται	ἐνδέχεται ⟨μὲν⟩ Α, ἐνδέχεται δὲ Β	–
12	The face in the moon	2+	δύναται	καὶ Α καὶ Β	καὶ ὅσοι ποτ' ἂν τρόποι …
13	Eclipses of sun and moon	2	δύναται	καὶ Α καὶ Β	–
14	The regularity of the periods	1			
15	Length of nights and days	2	–	καὶ Α καὶ Β	–
16	Weather signs	2	δύνανται	καὶ Α καὶ Β	–

270   APPENDIX 1

(cont.)

Ch.	Subject	Number of alternative explanations	Possibility	Grammatical conjunctions	Inexhaustivity of the list
17A	Clouds	3+	δύναται	καὶ A καὶ B καὶ Γ	καὶ κατ' ἄλλους δὲ τρόπους πλείους ... οὐκ ἀδυνατοῦσι ...
17B	Rain	3	δύναται	ᾗ μὲν A, ᾗ δὲ B, ἔτι τε Γ	–
18	Thunder	5	ἐνδέχεται	καὶ A καὶ B καὶ Γ καὶ Δ καὶ E	–
19	Lightning	8+	–	καὶ A καὶ B καὶ Γ καὶ Δ ἢ E ἢ Z καὶ H καὶ Θ	καὶ κατ' ἄλλους δὲ πλείους τρόπους ...
20	Why lightning precedes thunder	2	–	καὶ A καὶ B	–
21	Thunderbolts	2+	ἐνδέχεται	καὶ A καὶ B	καὶ κατ' ἄλλους δὲ τρόπους πλείονας ἐνδέχεται ...
22	Whirlwinds	2	ἐνδέχεται	καὶ A καὶ B	–
23	Earthquakes	2+	ἐνδέχεται	καὶ A ⟨καὶ⟩ B	καὶ κατ' ἄλλους δὲ πλείους τρόπους ...
24	Subterranean winds	3	–	A καὶ B, τὸ δὲ λοιπὸν Γ	–
25-a	Hail	2	–	καὶ A καὶ B	–
25-b	Round shape of hailstones	2	οὐκ ἀδυνάτως ἔχει	A καὶ B	–
26	Snow	3+	ἐνδέχεται	καὶ A καὶ B καὶ Γ	καὶ κατ' ἄλλους δὲ τρόπους ἐνδέχεται ...
27A	Dew	2	–	καὶ A καὶ B	–
27B	Hoar-frost	1 (?)			
28	Ice	2	–	καὶ A καὶ B	–
29-a	The rainbow	2	–	A ἢ B	–
29-b	Round shape of the rainbow	2	–	A ἢ B	–
30-a	The halo around the moon	3	–	[καὶ] A ἢ B ἢ καὶ Γ	–
30-b	Circumstances leading to a halo	2	–	ἤτοι A ἢ B	–
31	Comets	3	–	ἤτοι A ἢ B ἢ Γ	–

Ch.	Subject	Number of alternative explanations	Possibility	Grammatical conjunctions	Inexhaustivity of the list
32	Revolution of the stars	3+	–	οὐ μόνον A, ἀλλὰ καὶ B, ἢ καὶ Γ	καὶ κατ' ἄλλους δὲ πλείονας τρόπους τοῦτο δυνατὸν ...
33	Planets	2	ἐνδέχεται	ἐνδέχεται μὲν καὶ A, ἐνδέχεται δὲ καὶ B	–
34	Lagging behind of certain stars	3	–	καὶ A καὶ B καὶ Γ	–
35	Shooting stars	3+	δύνανται	καὶ A καὶ B καὶ Γ	καὶ ἄλλοι δὲ τρόποι ... εἰσιν.
36	Weather signs from animals	1			
37	Conclusion				

APPENDIX 2

Multiple Explanations in Lucretius' *DRN* V and VI

Multiple explanations in DRN V *509–770 (on astronomy)*

Lines		Subject	Number of explanations
509–533	1.	Motions of the stars	5
534–563	2.	Immobility of the earth	1
564–591	3.	Size of the sun, moon and stars	1
592–613	4.	Source of the sun's light and heat	3
614–649	5.	Turnings of the sun, moon and planets	2
650–655	6.	Causes of nightfall	2
656–679	7.	Causes of dawn	2
680–704	8.	Varying lengths of day and night	3
705–750	9.	Phases of the moon	4
751–761	10.	Solar eclipses	3
762–770	11.	Lunar eclipses	3

Multiple explanations in DRN VI *(on meteorology)*

Lines	Subject	Number of explanations
96–159	Thunder	9
160–218	Lightning	4
219–422	Thunderbolts	5
423–450	Whirlwinds (*prēstēres*)	2
451–494	Clouds	5
495–523	Rain	3
524–526	Rainbow	-
527–534	Snow, wind, hail, hoar-frost, ice	-
535–607	Earthquakes	4
608–638	Constant size of the sea	5
639–702	Etna	1 (+ 1 subsidiary expl.)
712–737	The Nile flood	4
738–839	Poisonous exhalations	2

Lines	Subject	Number of explanations
840–847	Temperature in wells	1
848–878	Spring of Hammon	1 (+ 1 subsidiary expl.)
879–905	Spring which kindles tow	1
906–1089	Magnets	1 (+ 2 subsidiary expl.)
1090–1286	Diseases	2

APPENDIX 3

General Structure of the *Syriac Meteorology*

Loc.		Subject	Number of explanations
1		Thunder	
	1.2–23	Causes of thunder	7
	1.24–38	How clouds can produce noise	1
2		Lightning	4
3		Thunder without lightning	3
4		Lightning without thunder	2
5		Why lightning precedes thunder	2
6		Thunderbolts	
	6.2–9	Their nature	1
	6.10–16	Their fineness and penetration	1
	6.16–21	Their causes	2
	6.21–28	Necessary conditions	2
	6.28–36	Their escape from the cloud	2
	6.36–41	Reasons for their downward motion	2
	6.41–67	Why clouds burst at the bottom	1 (+ 1 subsidiary expl.)
	6.67–74	Why they are more frequent in spring	1
	6.74–85	Why more frequent in high places	2
	6.85–91	Their effects	1
7		Clouds	
	7.2–5	Causes of clouds	2
	7.5–9	Causes of air condensation	2
	7.9–27	Reasons for the clouds floating on air	3
	7.27–29	Causes of clouds turning into water	2
8		Rain	
	8.2	Causes of heavy rain	1
	8.3–4	Causes of continuous rain	1
9		Snow	
	9.2–8	Causes of snow	2
	9.8–11	Reasons for the whiteness of snow	1
10		Hail	
	10.2–3	Causes of hail	1
	10.3–6	Causes of the hailstone being round	3

Loc.	Subject	Number of explanations
11	Dew	1
12	Hoar-frost	
12.2	*Causes of hoar-frost*	*1*
12.2–6	*Reasons for the whiteness of hoar-frost*	*1*
13	Winds	
13.2–3	*Their nature*	*1*
13.3–6	*Their origin (from above and below)*	*3*
13.7–18	*Wind from below*	*1*
13.18–21	*Wind from above*	*2*
13.21	*Wind moving sideways*	*1*
13.22	*Causes of strong winds*	*1*
13.23	*Causes of continuous winds*	*1*
13.24–27	*Causes of hot and cold winds*	*2*
13.27–32	*Winds arising from high and low places*	*1*
13.33–42	*The wind called 'WRS (Euros?)*	*2*
13:43–54	*The prēstēr*	
(43–45)	Its nature	1
(45–47)	Its causes	2
(47–54)	Its effects on ships	2
14	Halo	
14.2–13	*Account of the halo*	*1*
14.14–29	*Thunderbolts not the work of God*	-
15	Earthquakes	
15.2–16	*Causes of earthquakes*	*4*
15.16–21	*Influence of wind*	*1*
15.22–25	*Why some places don't have earthquakes*	*3*
15.26–35	*Types of earthquakes*	*3*

Bibliography

Abel, K. (1974), 'Zone', *Paulys Realencyclopädie der classischen Altertumswissenschaft*, Supplementband XIV, columns 989–1188.
Alfieri, V.E. (1953), *Atomos idea: l'origine del concetto dell' atomo nel pensiero greco*, Florence.
Algra, K.A. (1988), 'The Early Stoics on the Immobility and Coherence of the Cosmos', *Phronesis* 33, 155–180.
——— (1993), "Posidonius' Conception of the Extra-cosmic Void: The Evidence and the Arguments." *Mnemosyne* 46.4, 473–505.
——— (1995), *Concepts of Space in Greek Thought*, (*Philosophia Antiqua* vol. 65), Leiden.
——— (ed.) (1999), *The Cambridge History of Hellenistic Philosophy*, Cambridge.
——— (2001), *Epicurus en de zon, Wiskunde en fysica bij een Hellenistisch filosoof*, Amsterdam.
——— (2002), 'Zeno of Citium and Stoic Cosmology: some notes and two case studies', in Th. Scaltsas & A.S. Mason (eds.) *The Philosophy of Zeno*, Larnaca, 157–183.
——— (2003), 'Zeno of Citium and Stoic Cosmology: some notes and two case studies', *Elenchos* 24, 9–32.
Algra, K.A., M.H. Koenen & P.H. Schrijvers (eds.) (1997), *Lucretius and his intellectual background*, Amsterdam-Oxford-New York.
Allen, J. (2001), *Inference from Signs. Ancient Debates about the Nature of Evidence*, Oxford.
Arnim, J. von (ed.) (1903–1924), *Stoicorum Veterum Fragmenta*, Bd. I–III, Leipzig 1903–1905; Bd. IV with indices by M. Adler, Leipzig.
Arrighetti, G. (1973), *Epicuro, Opere*, Torino 1960, 1973².
——— (1975), 'L'Opera "Sulla natura" e le lettere di Epicuro a Erodoto e Pitocle', *Cronache Ercolanesi* 5, 39–52.
Asmis, E. (1984), *Epicurus' Scientific Method*, Ithaca / London.
——— (1999), 'Epicurean epistemology', in Algra (1999), 260–294.
Bailey, C. (1926), *Epicurus, The Extant Remains*, Oxford.
——— (1928), *The Greek Atomists and Epicurus*, Oxford 1928, repr. New York 1964.
——— (1947), *Titi Lucreti Cari De Rerum Natura*, ed. with prolegomena, critical apparatus, translation and commentary, 3 vols., Oxford.
Bakker, F.A. (2010), *Three Studies in Epicurean Cosmology*, Quaestiones Infinitae 64, Diss. Utrecht University, Zutphen, freely accessible via Utrecht University Repository http://dspace.library.uu.nl/handle/1874/188108.
——— (2013), 'Aëtius, Achilles, Epicurus and Lucretius on the Phases and Eclipses of the Moon', *Mnemosyne* 66.4–5, 682–707.

Baltussen, H. (1998), 'The Purpose of Theophrastus' *de Sensibus* Reconsidered', *Apeiron* 31 (2), 167–199.

Barnes, J. (1977), review of Reale (1974) in *Classical Review* 27, 40–43.

———— (1989), 'The size of the sun in antiquity', *Acta Classica Univ. Scient. Debrecen.* XXV, 29–41.

Bénatouïl, T. (2003), 'La méthode épicurienne des explications multiples', in T. Bénatouïl, V. Laurand & A. Macé (ed.), *Etudes épicuriennes, Cahiers philosophiques de Strasbourg* 15 (2003), 15–47.

Bergsträsser, G. (1918), *Neue meteorologische Fragmente des Theophrast, arabisch und deutsch*, Heidelberg.

Bollack, J. & A. Laks (1978), *Epicure à Pythoclès: Sur la cosmologie et les phénomènes météorologiques*, Cahiers de Philologie de Lille, vol. 3, Villeneuve d'Ascq.

Bowen, A.C. & R.B. Todd (2004), *Cleomedes' Lectures on Astronomy. A Translation of The Heavens with an Introduction and Commentary*, Berkeley / Los Angeles / London.

Brennan, T. (2000), Review of: K.A. Algra, J. Barnes et al. (edd.), *The Cambridge History of Hellenistic Philosophy* (Cambridge 1999), in BMCR 2000.09.11, http://bmcr.brynmawr.edu/2000/2000-09-11.html.

Brieger, A. (1884), *Urbewegung der Atome und die Weltentstehung bei Leucipp und Demokrit*, Halle.

Brown, P.M. (1984), *Lucretius, De Rerum Natura I*, edited with introduction, commentary & vocabulary, Bristol.

Brown, R.D. (1982), 'Lucretius and Callimachus' in *Illinois Classical Studies* 7 (1982) 77–97, repr. in M.R. Gale (ed.), *Oxford Readings in Classical Studies: Lucretius*, Oxford 2007, 328–350.

Burnet, J. (1892), *Early Greek Philosophy*, London.

Costa, C.D.N. (1984), *Lucretius, De Rerum Natura V*, edited with introduction and commentary, Oxford.

Capelle, W. (1912a), 'Μετέωρος—μετεωρολογία', *Philologus* 71, 414–448.

———— (1912b), 'Das Proömium der Meteorologie', *Hermes* 47, 514–535.

———— (1913), 'Zur Geschichte der meteorologischen Litteratur', *Hermes* 48, 321–358.

———— (1935), 'Meteorologie', *Paulys Realencyclopädie der classischen Altertumswissenschaft*, Supplementband VI, colums 315–358.

Chalmers, A. (1997), 'Did Democritus ascribe weight to atoms?', *Australasian Journal of Philosophy* 75.3, 279–287.

———— (2009), *The Scientist's Atom and the Philosopher's Stone*, Dordrecht.

Cherniss, H. (1957), 'Plutarch: Concerning the Face Which Appears in the Orb of the Moon.', in H. Cherniss & W.C. Helmbold, *Plutarch: Moralia, Volume XII*, with an English translation, Cambridge (Mass.) / London, 1–223.

———— (1976), 'Plutarch: On Stoic Self-Contradictions', in H. Cherniss, *Plutarch: Mora-*

lia, Volume XIII, Part 2, with an English translation, Cambridge (Mass.) / London, 367–603.
Codoñer Merino, C. (1979), *L. Annaei Senecae Naturales Quaestiones*, Madrid.
Conroy, D.P. (1976), *Epicurean Cosmology and Hellenistic Astronomical Arguments*, unpublished PhD thesis, Princeton University.
Corcoran, T.H. (1971/2), *Seneca: Naturales Quaestiones*, 2 vols., Cambridge (Mass.) / London.
Cornford, F.M. (1937), *Plato's cosmology: the Timaeus of Plato*, London.
Coutant, V. (1971), *Theophrastus. De igne. A post-Aristotelian view of the nature of fire. Edited with introduction, translation and commentary*, Assen (Netherlands).
D'Ancona, C. (2015), 'Aristotle and Aristotelianism' in Kate Fleet, Gudrun Krämer, Denis Matringe, John Nawas, Everett Rowson (eds.), *Encyclopaedia of Islam*, THREE (2007–) (Brill Online 2015), accessed 30 October 2015 via Radboud University http://referenceworks.brillonline.com.ru.idm.oclc.org/entries/encyclopaedia-of-islam-3/aristotle-and-aristotelianism-COM_0170.
Daiber, H. (1992), 'The *Meteorology* of Theophrastus in Syriac and Arabic Translation', in Fortenbaugh & Gutas (1992), 166–293.
De Lacy, P. and E. (1978), *Philodemus, On methods of inference* (ed. 2), Naples.
De Sanctis, D. (2012), 'Utile al singolo, utile a molti: il proemio dell'*Epistola a Pitocle*', *Cronache Ercolanesi* 42, 95–109.
Dicks, D.R. (1970), *Early Greek Astronomy to Aristotle*, London.
Diels, H.A. (1879), *Doxographi Graeci*, Berlin 1879, 1965^4.
Diels, H.A. & W. Kranz (eds.) (1952), *Die Fragmente der Vorsokratiker*, 3 vols. Berlin.
Dover, K.J. (1968), *Aristophanes: Clouds*, Oxford.
Dreyer, J.L.E. (1906), *History of the Planetary Systems from Thales to Kepler*, Cambridge 1906 (republished as *A History of Astronomy from Thales to Kepler*, New York 1953, and, with the original title, New York 2007).
Drossaart Lulofs, H.J. (1955), 'The Syriac translation of Theophrastus' Meteorology' in *Autour d'Aristote. Recueil d'études de philosophie ancienne et médiévale offert à Monseigneur A. Mansion*, (Bibliothèque philosophique de Louvain 16), Louvain 1955, 433–449.
Dyroff, A. (1899), *Demokritstudien*, Munich.
Edelstein, L. & I.G. Kidd (1972), *Posidonius, I. The Fragments*, Cambridge 1972, 1989^2.
Eichholz, D.E. (1965), *Theophrastus, De lapidibus, edited with introduction, translation and commentary*, Oxford.
Ernout, A. (1920), *Lucrèce, De la nature. Texte établie et traduit*, Paris.
Ernout, A. & L. Robin (1925–1928), *Lucrèce. De rerum natura. Commentaire exégétique et critique*, Paris 1925–1928, 1962^2.
Evans, J. (1998), *The History and Practice of Ancient Astronomy*, New York.
Fortenbaugh, W.W. & D. Gutas (eds.) (1992), *Theophrastus: His Psychological, Doxo-*

graphical and Scientific Writings, Rutgers University Studies in Classical Humanities, vol. V, New Brunswick (N.J).

Fortenbaugh, W.W., P.M. Huby, R.W. Sharples & D. Gutas (eds.) (1992), *Theophrastus of Eresus: Sources for his life, writings, thought and influence*, Pt. I (*Philosophia antiqua* vol. 54, 1), Leiden (= FHS&G).

Fowler, D.P. (2002), *Lucretius on Atomic Motion. A commentary on* De Rerum Natura *book two, lines 1–332*, Oxford.

Fowler, R.L. (2000), 'P. OXY. 4458: POSEIDONIOS', *Zeitschrift für Papyrologie und Epigraphik* 132, 133–142.

Furley, D.J. (1955), 'Aristotle: On the Cosmos', in E.S. Foster & D.J. Furley, *Aristotle: On Sophistical Refutations, On coming-to-be and Passing-away, On the Cosmos*, Cambridge (Mass.) / London 1955, 331–409.

——— (1966), 'Lucretius and the Stoics', in *Bulletin of the Institute of Classical Studies*, 13, 13–33; repr. in Furley (1989b) 183–205.

——— (1971), 'Knowledge of Atoms and Void in Epicureanism', in J.P. Anton & G.L. Kustas (eds.), *Essays in Ancient Greek Philosophy*, Albany 1971, 607–619, repr. in Furley (1989b) 161–171.

——— (1976), 'Aristotle and the Atomists on Motion in a Void', in P.K. Machamer & R.J. Turnbull (eds.), *Motion and Time, Space and Matter*, Columbus (Ohio) 1976, 83–100, repr. in Furley (1989b) 77–90.

——— (1978), 'Lucretius the Epicurean: on the history of man', in *Lucrèce*, Entretien sur l'Antiquité Classique 24, Geneva 1978, 1–37, repr. in Furley (1989b) 206–222.

——— (1981), 'The Greek Theory of the Infinite Universe', in *Journal of the History of Ideas* 42, 571–585, repr. in Furley (1989b) 1–13.

——— (1983), 'Weight and Motion in Democritus' Theory', in *Oxford Studies in Ancient Philosophy* 1, 193–209, repr. in Furley (1989b) 91–102.

——— (1986), 'The Cosmological Crisis in Classical Antiquity', in *Proceedings of the Boston Area Colloquium in Ancient Philosophy* 2, 1–19, repr. in Furley (1989b) 223–235.

——— (1989a), 'The Dynamics of the Earth: Anaximander, Plato, and the Centrifocal Theory', in Furley (1989b), 14–26.

——— (1989b), *Cosmic Problems*, Cambridge 1989.

——— (1996), 'The Earth in Epicurean and Contemporary Astronomy' in Gabriele Giannantoni and Marcello Gigante (eds.), *Epicureismo greco e romano: Atti del congresso internazionale, Napoli, 19–26 maggio 1993*, Naples 1996, vol. 1, 119–125.

——— (1999), 'Cosmology', in Algra (1999), 412–451.

Gale, M.R. (2009), *Lucretius, De Rerum Natura V*, edited with translation and commentary, Oxford.

Gambetti, S. (2015), 'Anonymous, On the Nile (647)' in Ian Worthington (ed.), *Brill's New Jacoby* (Brill Online 2015), accessed 30 October 2015 via Radboud University http://referenceworks.brillonline.com.ru.idm.oclc.org/entries/brill-s-new-jacoby/anonymous-on-the-nile-647-a647.

Garani, M. (2007), *Empedocles Redivivus: Poetry and Analogy in Lucretius*, Oxford.

Giussani, C. (1896–1898), *T. Lucreti Cari De rerum natura libri sex. Revisione del testo, commento e studi introduttivi*, 4 vols., Torino.

Goethe, A. (1987), *M. Tullii Ciceronis De natura deorum libri tres*, Leipzig.

Gottschalk, H.B. (1965), Review of: Wagner & Steinmetz (1964), in *Gnomon* 37, 758–762.

——— (1998), 'Theophrastus and the Peripatos', in J.M. van Ophuijsen & M. van Raalte (eds.), *Theophrastus: Reappraising the Sources*, Rutgers University Studies in Classical Humanities, vol. VIII, New Brunswick (New Jersey) 1998, 281–298.

Gross, N. (1989), *Senecas naturales quaestiones. Komposition, Naturphilosophische Aussagen und ihre Quellen*, Palingenesia 27, Stuttgart.

Guthrie, W.K.C. (1965) *A history of Greek Philosophy*, vol. 2: *The Presocratic Tradition from Parmenides to Democritus*, Cambridge.

Hahm, D.E. (1977), *The Origins of Stoic Cosmology*, Columbus.

Hankinson, R.J. (1999a), *Cause and Explanation in Ancient Greek Thought*, Oxford 1999.

——— (1999b), 'Explanation and causation', in Algra (1999), 479–512.

Heath, sir Th. (1913), *Aristarchus of Samos, the ancient Copernicus. A history of Greek astronomy together with Aristarchus' treatise on the size and distance of the sun and moon, a new greek text with translation and notes*, Oxford 1913, repr. New York 1981.

——— (1932), *Greek Astronomy*, London 1932, repr. New York 1991.

Hine, H.M. (1981), *An Edition with Commentary of Seneca, Natural Questions, Book Two*, New York.

——— (2002), 'Seismology and Vulcanology in Antiquity?', in C.J. Tuplin & T.E. Rihl (eds.), *Science and Mathematics in Ancient Greek Culture*, Oxford 2002.

Hübner, W. (2002), 'Der *descensus* als ordnendes Prinzip in der 'Naturalis historia' des Plinius', in C. Meier, *Die Enzyklopädie im Wandel vom Hochmittelalter bis zur frühen Neuzeit*, Munich 2002, 25–41.

Hultsch, F.O. (1897), 'Bion (11)', *Paulys Realencyclopädie der classischen Altertumswissenschaft*, Band III, colums 485–487.

Jacob, Ch. (1983), 'De l'art de compiler à la fabrication du merveilleux. Sur la paradoxographie grecque', *Lalies* 2 (Actes de Sessions de Linguistique et de Littérature: Thessolonique, 24 Août–6 Septembre 1980), Paris 1983, 121–140.

Jakobi, R. & W. Luppe (2000), 'P. Oxy. 4458 col. 1: Aristoteles redivivus', *Zeitschrift für Papyrologie und Epigraphik* 131, 15–18.

Jones, H.L. (1917–1932), *The Geography of Strabo*, with an English translation, 8 vols., Cambridge (Mass.) / London.

Jürss, F. (1994), 'Wissenschaft und Erklärungspluralismus im Epikureismus', *Philologus* 138.2, 235–251.

Kahn, C.H. (1960), *Anaximander and the Origins of Greek Cosmology*, New York.

Kany-Turpin, J. (1997), 'Cosmos ouvert et épidémies mortelles dans le *De rerum natura*', in Algra, Koenen & Schrijvers (1997), 179–185.

Kaufmann, G. (1894), 'Antipodes', *Paulys Realencyclopädie der classischen Altertumswissenschaft*, Band I, colums 2531–2533.

Kechagia, E. (2010), 'Rethinking a Professional Rivalry: Early Epicureans against the Stoa', *The Classical Quarterly*, 60.1, 132–155.

Kenney, E.J. (1971), *Lucretius, De Rerum Natura book III*, ed. with introduction and commentary, Cambridge.

Kidd, I.G. (1992), 'Theophrastus' *Meteorology*, Aristotle and Posidonius', in Fortenbaugh & Gutas (1992), 294–306.

Kirk, G.S. & J.E. Raven (1957), *The Presocratic Philosophers*, Cambridge 1957, 2nd edition revised by M. Schofield, 1983.

Konstan, D. (1972) 'Epicurus on "Up" and "Down" (Letter to Herodotus § 60)', *Phronesis* 17, 269–278.

——— (2014), 'Epicurus' in E.N. Zalta (ed.), *The Stanford Encyclopedia of Philosophy* (Summer 2014 Edition), accessed 30 October 2015 http://plato.stanford.edu/archives/sum2014/entries/epicurus/.

Kroll, W. (1930), *Die Kosmologie des Plinius*, Breslau.

Lachenaud, G. (1993), *Plutarque: Oeuvres morales t. XII2: Opinions des Philosophes*, Paris.

Lambinus, D. (1563), *Titi Lucretii Cari De rerum natura libri sex*, (text with commentary), Paris.

Lasserre, F. (1966), *Die Fragmente des Eudoxos von Knidos*, Berlin.

Lee, H.D.P. (1952), *Aristotle: Meteorologica*, with an English translation, Cambridge (Mass.) / London.

Leonard, W.E. (1921), *On the Nature of Things by Lucretius*, a metrical translation, London.

Leven, K.-H. (ed.) (2005), *Antike Medizin. Ein Lexikon*, Munich.

Lewis, C.T. & C. Short (1879), *A Latin dictionary*, Oxford.

Liepmann, H.C. (1885), *Mechanik der Leucipp-Democritschen Atome*, Berlin.

Long, A.A. & D.N. Sedley (1987), *The Hellenistic Philosophers*, 2 vols., Cambridge.

Louis, P. (1982), *Aristote, Météorologiques. Texte établi et traduit*, 2 vols., Paris.

Lück, W. (1932), *Die Quellenfrage im 5 und 6 Buch des Lukrez*, Ohlau in Schles.

Mansfeld, J. (1986), 'Diogenes Laertius on Stoic Philosophy', in G. Giannantoni (ed.), *Diogene Laerzio, Storico del Pensiero Antico*, Elenchos 7, 297–382.

——— (1989), 'Chrysippus and the *Placita*', *Phronesis* 34, 311–342, repr. in Mansfeld & Runia (2009b) 125–160.

——— (1990), 'Doxography and Dialectic. The Sitz im Leben of the Placita', in Haase, W. & H. Temporini, *ANWR* III 36.3, Berlin / New York 1990, 3056–3229.

——— (1991), 'Two attributions', *The Classical Quarterly*, 41.2 (1991) 541–544.

——— (1992a), 'A Theophrastean excursus on God and nature and its aftermath in Hellenistic thought', *Phronesis* 37, 314–335.

——— (1992b), 'Physikai doxai and Problemata physika from Aristotle to Aëtius (and Beyond)', in Fortenbaugh & Gutas (1992), 63–111, repr. in Mansfeld & Runia (2009b) 33–98.

——— (1992c), 'ΠΕΡΙ ΚΟΣΜΟΥ: A Note on the History of a Title', *Vigiliae Christianae* 46.4 (Dec., 1992), 391–411.

——— (1994), 'Epicurus Peripateticus', in A. Alberti (ed.), *Studi di filosofia antica. Realtà e ragione*, Florence 1994, 29–47, repr. in Mansfeld & Runia (2009b) 237–254.

——— (2005), 'From Milky Way to Halo, Aristotle's *Meteorologica*, Aëtius, and Passages in Seneca and the *Scholia* on Aratus', in A. Brancacci (ed.), *Philosophy and Doxography in the Imperial Age*, Firenze 2005, 23–58.

——— (2013), 'Doxography of Ancient Philosophy' in E.N. Zalta (ed.), *The Stanford Encyclopedia of Philosophy* (Winter 2013 Edition), accessed 30 October 2015 http://plato.stanford.edu/archives/win2013/entries/doxography-ancient/.

Mansfeld, J. & D.T. Runia (1997), *Aëtiana, the Method and Intellectual Context of a Doxographer, vol. 1: The Sources* (Philosophia Antiqua vol. 73), Leiden 1997.

——— (2009a), *Aëtiana, the Method and Intellectual Context of a Doxographer, vol. 2 (in two parts): The Compendium* (Philosophia Antiqua vol. 114), Leiden 2009.

——— (2009b), *Aëtiana, the Method and Intellectual Context of a Doxographer, vol. 3: Studies in the doxographical Traditions of Greek Philosophy* (Philosophia Antiqua vol. 118), Leiden 2009 [my references to the articles in this collection use the original page numbers, which are included in the text of the present edition].

Masi, F.G. (2015), 'The Method of Multiple Explanations. Epicurus and the Notion of Causal Possibility', *Aitia II, Avec ou sans Aristote*, Louvain, Louvain-La-Neuve (2015), 37–63.

Mejer, J. (1978), *Diogenes Laertius and his Hellenistic Background*, Hermes Einzelschr. 40, Wiesbaden.

——— (1992), 'Diogenes Laertius and the transmission of Greek Philosophy,' in W. Haase (ed.), *ANRW II* 36.5, Berlin / New York 1992, 3556–602.

Merrill, W.A. (1907), *T. Lucreti Cari De Rerum Natura libri sex*, New York / Cincinnati / Chicago.

Milton, J.R. (2002), 'The Limitations of Ancient Atomism', in C.J. Tuplin & T.E. Rihl (eds.), *Science and Mathematics in Ancient Greek Culture*, Oxford 2002.

Montarese, F. (2012), *Lucretius and his sources: a study of Lucretius, De rerum natura I 635–920*. Sozomena 12, Berlin / Boston.

Montserrat, J.M. & L. Navarro (1991), 'The Water Cycle in Lucretius', *Centaurus* 34, 289–308.

Munro, H.A.J. (1864), *T. Lucreti Cari De rerum natura libri sex*, with notes and a translation, Cambridge 1864, 1893[4].

Naas, V. (2002), *Le projet encyclopédique de Pline l'Ancien*, Rome 2002.

O'Brien, D. (1981), *Theories of Weight in the Ancient World, Vol. 1 'Democritus: Weight and Size'*, Paris.
O'Keefe, T.S. (2000), 'Epicurus', *The Internet Encyclopedia of Philosophy*, accessed 10 November 2015 http://www.iep.utm.edu/epicur/.
Oxford Latin Dictionary, ed. P.G.W. Glare, Oxford 1968–1982.
Pajón Leyra, I. (2011), *Entre ciencia y maravilla. El género literario de la paradoxografía griega*. Monografías de Filología Griega 21, Zaragoza.
Parroni, P. (2002), *Seneca. Ricerche sulla natura*, Milan.
Partsch, J. (1909), 'Des Aristoteles Buch "Über das Steigen des Nil"', *Abhandlungen der Königlichen Sächsischen Gesellschaft der Wissenschaften, philologisch-historische Klasse* 27, 551–600.
Pease, A.S. (1955/8), *M. Tulli Ciceronis de natura deorum*, 2 vols., Cambridge (Mass.).
Podolak, P. (2010), 'Questioni pitoclee', *Würzburger Jahrbücher für die Altertumswissenschaft* 34, 39–80.
Raalte, M. van (2003), 'God and the Nature of the World: the 'Theological Excursus' in Theophrastus' Meteorology', *Mnemosyne* 56.3, 306–342.
Reale, G. (1974), *Aristotele: Trattato sul cosmo per Alessandro*, Naples.
Reale, G. & A.P. Bos (1995), *Il trattato sul cosmo per Alessandro attribuito ad Aristotele*, Milan.
Regenbogen, O. (1940), 'Theophrastos', *Paulys Realencyclopädie der classischen Altertumswissenschaft*, Supplementband VII (1940), colums 1354–1562.
Reitzenstein, E. (1924), *Theophrast bei Epikur und Lucrez* (Orient und Antike 2), Heidelberg.
Rist, J.M. (1972), *Epicurus. An Introduction*, Cambridge.
Roos, A.G. & G. Wirth (eds.) (1967/8), *Flavii Arriani quae exstant omnia*, 2 vols., Leipzig.
Rösler, W. (1973), 'Lukrez und die Vorsokratiker: doxographische Probleme im I. Buch von "De Rerum Natura"', *Hermes* 101 (1973), 48–66, repr. in C.J. Classen (ed.), *Probleme der Lukrezforschung* (Olms Studien, Bd. 18), Hildesheim / Zürich / New York (1986), 57–73.
Rouse, W.H.D. & M.F. Smith (1982), *Lucretius: On the Nature of Things*, with an English translation, Cambridge (Mass.) / London 1982, 1992.
Runia, D.T. (1989), 'Xenophanes on the Moon: a Doxographicum in Aëtius', *Phronesis* 34, 245–269, repr. in Mansfeld & Runia (2009b) 99–124.
―――― (1992), 'Xenophanes or Theophrastus? An Aëtian *Doxographicum* on the Sun', in Fortenbaugh & Gutas (1992), 112–140.
―――― (1997a), 'Lucretius and Doxography' in Algra, Koenen & Schrijvers (1997), 93–103, repr. in Mansfeld & Runia (2009b) 255ff.
―――― (1997b), Art. 'Doxographie', in *Der neue Pauly*, Bd. 3 (1997), columns 803–806.
Sambursky, S. (1959), *Physics of the Stoics*, London.

Schenkeveld, D.M. (1991), 'Language and style of the Aristotelian *De mundo* in relation to the question of its inauthenticity', *Elenchos* 12, 221–255.

Schepens, G. & K. Delcroix (1996), 'Ancient Paradoxography: Origin, Evolution, Production and Reception', in Pecere, O. & A. Stramaglia (eds.), *La letteratura di consumo nel mondo greco-latino*, Cassino 1996, 373–460.

Schmidt, J. (1990), *Lukrez, der Kepos und die Stoiker: Untersuchungen zur Schule Epikurs und zu den Quellen von "De Rerum Natura"*, Frankfurt am Main [revised version of Schmidt's 1975 dissertation *Lukrez und die Stoiker: Quellenuntersuchungen zu De rerum natura*].

Schrijvers, P.H. (1978), 'Le regard sur l'invisible. Etude sur l'emploi de l'analogie dans l'oeuvre de Lucèce', in O. Gigon (1978), *Lucrèce. Huit exposés suivis de discussions. Entretiens de la Fondation Hardt* XXIV, Genève 1978, 77–114, repr. in Schrijvers (1999) 183–213.

——— (1999), *Lucrèce et les sciences de la vie*, Leiden.

Sedley, D.N. (1976), 'Epicurus and the mathematicians of Cyzicus', *Cronache Ercolanesi* 6, 23–54.

——— (1982), 'On Signs', in J. Barnes et al. (eds.), *Science and Speculation: Studies in Hellenistic Theory and Practice*, Cambridge / Paris 1982, 239–272.

——— (1998a), *Lucretius and the transformation of Greek wisdom*, Cambridge 1998.

——— (1998b), 'Epicureanism' in E. Craig (ed.), *Routledge Encyclopedia of Philosophy*, London 1998, 2005, pp. 340–350.

——— (2013), 'Lucretius' in E.N. Zalta (ed.), *The Stanford Encyclopedia of Philosophy* (Fall 2013 Edition), accessed 30 October 2015 http://plato.stanford.edu/archives/fall2013/entries/lucretius/.

Sharples, R.W. (1985), 'Theophrastus on the Heavens', in J. Wiesner (ed.), *Aristoteles Werk und Wirkung*, Bd. 1 Aristoteles und seine Schule, Berlin 1985, 577–593.

——— (1998) (with contributions by D. Gutas), *Theophrastus of Eresus: Sources for his life, writings, thought and influence, Commentary, vol. 3.1, Sources on Physics (texts 137–223)* (*Philosophia Antiqua* vol. 79), Leiden.

Smith, M.F. (1993), *Diogenes of Oinoanda: The Epicurean Inscription*, Naples.

——— (2003), *Supplement to Diogenes of Oinoanda: The Epicurean Inscription*, Naples.

Sorabji, R. (1988), *Matter, Space and Motion: Theories in Antiquity and Their Sequel*, Ithaca (NY).

Spoerri, W. (1959), *Späthellenistische Berichte über Welt, Kultur und Götter: Untersuchungen zu Diodor von Sizilien*, Basel.

Steinmetz, P. (1964), *Die Physik des Theophrastos van Eresos*, Palingenesia 1, Bad Homburg.

Striker, G. (1974), 'Κριτήριον τῆς ἀληθείας', in *Nachrichten der Akademie der Wissenschaften zu Göttingen*, I. Philologisch-Historische Klasse 2, 48–110.

——— (1996), 'Κριτήριον τῆς ἀληθείας', English translation of the former, in G. Striker, *Essays on Hellenistic epistemology and ethics*, Cambridge 1996, 22–76.

Strohm, H. (1937), 'Zur Meteorologie des Theophrast', *Philologus* 92, 249–268 & 403–428.

Taub, L.C. (2003), *Ancient Meteorology*, London / New York.

——— (2009) 'Cosmology and meteorology', in J. Warren (ed.), *The Cambridge Companion to Epicureanism*, Cambridge 2009, 105–124.

Thesaurus Linguae Latinae, Munich 1894-present.

Thomson, J.O. (1948), *History of Ancient Geography*, Cambridge 1948, 1965³.

Todd, R.B. (1990), *Cleomedis Caelestia* (Μετέωρα), Leipzig.

Usener, H. (1858), 'Analecta Theophrastea' (1858), in H. Usener, *Kleine Schriften, erster Band: Arbeiten zur griechischen Philosophie und Rhetorik—Grammatische und textkritische Beiträge*, Berlin 1912, Osnabrück 1965², 50–87.

——— (1887), *Epicurea*, Leipzig 1887, repr. Rome 1963.

Vallance, J. (1988), 'Theophrastus and the Study of the Intractable: Scientific Method in *De lapidibus* and *De igne*', in W.W. Fortenbaugh & R.W. Sharples (eds.), *Theophrastean Studies on Natural Science, Physics and Metaphysics, Ethics, Religion and Rhetoric*, New Brunswick / Oxford 1988, 25–40.

Vanotti, G. (2007), *Aristotele, Racconti Meravigliosi*, Milano.

Verde, F. (2013), 'Cause epicuree', *Antiquorum Philosophia* 7, 127–142.

Vlastos, G. (1975), *Plato's Universe*, Oxford 1975.

Wagner, E. & P. Steinmetz (1964), *Der syrische Auszug der Meteorologie des Theophrast* (Akademie der Wissenschaften und der Literatur. Abhandlungen der geistes- und sozialwissenschaftlichen Klasse, Wiesbaden, Jahrgang 1964, nr. 1).

Warren, J. (2007), 'Lucretius and Greek Philosophy' in Stuart Gillespie & Philip Hardie (eds.), *The Cambridge Companion to Lucretius*, Cambridge 2007, 19–32.

Wasserstein, A. (1978), 'Epicurean Science', *Hermes* 106, 484–494.

Wehrli, F. (ed.) (1969), *Straton von Lampsakos*, Die Schule des Aristoteles: H.5, Basle / Stutgart 1969².

Wenskus, O. & L. Daston (2000), 'Paradoxographoi', in *Der neue Pauly*, Bd. 9 (2000), columns 309–314.

Wilson, M. (2013), *Structure and Method in Aristotle's* Meteorologica. *A More Disorderly Nature*, Cambridge.

Wöhrle, G. (1985), *Theophrasts Methode in seinen botanischen Schriften*, Amsterdam.

Wolff, M. (1989), 'Hipparchus and the Stoic Theory of Motion', in J. Barnes & M. Mignucci (eds.), *Matter and Metaphysics*, Naples 1989, 471–545.

Woltjer, J. (1877), *Lucretii philosophia cum fontibus comparata*, Groningen.

Zeller, E. (1879), *Die Philosophie der Grieche*, Leipzig 1879³.

Ziegler, K. (1949), 'Paradoxographoi', *Paulys Realencyclopädie der classischen Altertumswissenschaft*, Band 18, columns 1137–1166.

Index Locorum

Achilles
Isagoge
4	169n34, 194n88
9	189n77, 193n87

Aelian
Natura animalium
V 8	120
XII 2	121n120

Aëtius
Placita (ed. Diels *DG*; book II: M&R 2)
I 3.18	217
I 4	224–228, 235
I 12.4	193n88
I 12.5	215n146
I 12.6	217
I 12.7	208n133
I 18.4	208n133
II	61–62, 160
II 8.1	166n22, 167n24, 222n165, 248
II 8.2	166n22, 249n206
II 9.4	196n104
II 13.15	59, 227n172
II 17a.1	194n92
II 20.14	227n172
II 22.4	59
II 23.5	53n136
II 23.7	194n92
II 24.9	52, 233n180
II 29.7	50–51, 53n138
III	61–62, 78, 84–87, 108, 127–129, 137–138, 140, 142–143, 155–160
III 0	85n31
III 1	130
III 2	90n50, 107, 130
III 3	132–133
III 3.13	133n141
III 4.1	69n185
III 4.4	69n185
III 4.5	144
III 5	86, 134–136
III 6	86, 134–136
III 7	134–136
III 8.2	85n32
III 9–14	103, 131
III 10.1	165n12, 169n34
III 10.2	165n14
III 10.3	165n17
III 10.4	167n25
III 10.5	167n25
III 12.1	249n207
III 12.2	249n208
III 15	104, 136
III 15.7	166n19
III 15.8	165n17
III 15.11	144
III 16	136–137
III 17	136–137
III 18	86, 134, 135–136
IV 1	86–87, 116–117, 119, 126, 136–137
IV 1.4	69n185
IV 1.7	168n30
V 30.6	252n213

Alexander
In Aristotelis Meteorologica (Hayduck)
p. 103.30–31	254
p. 179.1–5	80n14

Anaxagoras
fr.A1 D-K	167n24, 248
fr.A12 D-K	218n152
fr.A42 D-K	167n24, 218n152+154+157
fr.A67 D-K	222n165, 248
fr.A71 D-K	218n152
fr.A87 D-K	167n24
fr.B15 D-K	218n157

Anaximander
fr.A10 D-K	165n14
fr.A11 D-K	165n14–15
fr.A26 D-K	165n15
fr.B5 D-K	165n14

INDEX LOCORUM   287

Anaximenes
    fr.A6 D-K   165n17
    fr.A7 D-K   165n17
    fr.A17 D-K   69n185
    fr.A20 D-K   165n17

Antigonus
Historiarum mirabilium collectio
    12   117, 120
    121–123   117
    122   117
    123   117, 119n115
    129–165   114n106, 117, 122
    129.2   117, 123
    132   83n26
    144   117, 123
    148   110n91, 117, 123
    152a–b   117, 120
    166–167   117, 119

Apollonius
Historiae Mirabiles
    9   120
    23   123

Apuleius
De deo Socratis 1.14–30   54
De mundo 17.17 ff.   121n119

Archytas
    fr.A24 D-K   183n66

Aristophanes
Nubes
    369–371   118
    375–394   118
    395–407   118, 131n135
    403–407   62
    1278–1295   118

Aristotle
Analytica posteriora
    I 8, 75b21–36   115n108
    I 13, 78b4–11   53n138
    I 31, 87b39–88a1   51n129
    II 2, 90a15–18   51n129
De anima
    I 2, 405a19 ff.   123n126
    II 4, 415b28–416a9   196n100

De caelo
    I 2, 269a18–19   195n98
    I 3, 270b6–25   57n149
    I 8, 277a28–29   149n178
    I 8, 277b5–9   149n178
    I 9, 278b14–16   57n149
    I 9, 279a7–17   200n115
    I 9, 279a12–17   196n103
    II 1, 284a12–14   57n149
    II 1, 284b3–5   57n149
    II 2, 285b22–27   166n20
    II 3, 286a10–13   57n149
    II 6, 288a17–22   149n178
    II 7, 289a11–35   202n122
    II 11, 291b18–21   53n138
    II 12, 292b32–293a1   53n138
    II 13–14   82n21, 104
    II 13, 293b25–30   166n20
    II 13, 294a28–32   165n12
    II 13, 294b13–30   165n17, 167n24–
                             25, 177n58, 218,
                             235
    II 13, 295a13–14   218n154
    II 13, 295b10–16   165n15
    II 14, 296b28–297a2   196n101
    II 14, 297a8–298b20   168n31
    II 14, 297a12–b18   178
    II 14, 297a12–19   218n155
    II 14, 297b17–21   179
    II 14, 297b24–298a10   169–175, 237n188
    II 14, 297b24–31   51n129, 174, 243
    II 14, 297b31–298a10   174
    III 2, 300b8   217
    IV   195n98
    IV 1, 308a14–24   178, 195n99
    IV 3–5, 310a16–313a13   149n177
    IV 3, 310a30–35   149n176
De divinatione per somnum
    1, 462b28–29   52n131
De generatione animalium
    I 8, 326a8   217
    II 6, 742b17   217
De generatione et corruptione
    I 10, 336a32–b24   53n136
    I 10, 337a8   53n136
Metaphysica
    VI 2, 1027a20–26   115
    VIII 4, 1044b9–15   51n129
    XII 5, 1071a16   53n136

288   INDEX LOCORUM

Metaphysica (cont.)
  XII 6, 1071b32   217
  XII 8, 1073b17 ff.   53n136
  XII 8, 1074a38–b14   57n149
Meteorologica
  I 1, 338b1–3   79
  I 2, 339a19–21   79–80
  I 3, 339b7–9   82n21
  I 3, 339b16–340a18   202n122
  I 3, 340b35–36   168n31
  I 3, 341a12–31   65–66
  I 4–8   99
  I 4, 341b2–3   94n62
  I 4, 341b6–13   149n176
  I 4, 341b34–35   94n62
  I 4, 341b36–342a13   65
  I 4, 342a13–27   148n174, 198n109
  I 6–7   90n50
  I 6, 343b5   94n62
  I 7, 344a5–b4   65–66
  I 8, 345b1–9   243n195
  I 9, 346b22–347a8   82n19
  I 9, 346b24   82n21
  I 9, 346b30–31   69
  I 13   134n151
  I 13, 349b3–8   82n19
  I 13, 349b8   115
  I 13, 350b36–351a18   114, 115, 116, 122
  I 13, 351a14–16   116, 123
  II 2, 354b28–34   82n19
  II 2, 354b34–355a32   196n100
  II 2, 355b20–32   116
  II 3, 356b22–357a2   82n19
  II 3, 359a18–b22   114, 115, 116, 122
  II 4–6   134n151
  II 5, 362a33–b33   82n21
  II 5, 362b16–18   251n211
  II 7, 365a14–15   81, 105
  II 7, 365a20–37   177n58
  II 7, 365b1–6   65
  II 8, 365b29–366a5   39n106
  II 8, 366a3–5   81
  II 8, 367a1–11   82n20, 116
  II 8, 367b20–22   51n129
  II 9–III 1   135n151
  II 9, 369a20–30   148n174, 198n109
  III 1, 371a8–18   120n, 132n141
  III 2–6   86n34
  III 6, 378a13 ff.   81n17, 123
  IV   80
Physica
  II 2, 193b22 ff.   44n119
  III 4, 203b20–22   182n65
  III 4, 203b25–30   22n52
  IV 1, 208b9–22   149n176–177
  IV 1, 210a3–6   149n176
  IV 5, 212b14–17   196n103
  IV 8–9, 214b12–217b28 196n103
  IV 9, 217b11–12   214n143
  VIII 1, 252a34   217

[Aristotle]
De mirabilibus auscultationibus
  34–40   117, 119
  53–57   117, 122
  102   117, 120
De mundo
  2, 391b9–392a32   84n27
  2, 392a32–b5   83n25, 84n27
  2, 392b5–13   84n27
  3, 392b14–394a6   84n27
  4   82–84, 100–103
  4, 394a7–8   83
  4, 394b7–395a10   134n151
  4, 394b13   134n151
  4, 395a11–24   135n151
  4, 395a21–24   132n141
  4, 395a32–b3   86n34
  4, 395b8–9   90n50
  4, 395b17–18   83n27
  4, 395b19–23   117
  4, 395b26–30   117, 120
  4, 395b30   117
  4, 396a17   83n27
  4, 396a27–32   84n27
Liber de inundacione Nili
  2–4   115n110
Problemata
  X 66, 898b   251n212

Arius Didymus (ed. Diels *DG*)
  fr.14a   87
  fr.23   189n77, 192–193,
                           197, 199, 203
  fr.31   178n61
  fr.32   52n133
  fr.33   194n93

INDEX LOCORUM

Augustine
Enarratio in Psalmos
10, 3 — 54

Cicero
Academica
II 82 — 176n53
De divinatione
II 103 — 181n63
De fato
22 — 215n146
46 — 176n52
De finibus
I 18–20 — 176n52–53, 215n147

De natura deorum
I 24 — 221–223
I 30–39 — 57n149
I 52 — 32n78, 152n189
II 13–15 — 124–125
II 25.7–26.1 — 112n94
II 41 — 194n93
II 115 — 193n87
II 116 — 178
II 117 — 194n88
Lucullus 123.7 — 177n59, 195n99

Cleomedes (ed. Todd)
I 1 — 104
I 1, 39–149 — 185
I 1, 91–92 — 193n87
I 1, 112–122 — 184n69, 185
I 1, 158–192 — 178n61
I 1, 258–261 — 194n90
I 3, 76–4, 17 — 53n135
I 5–8 — 104
I 5, 11–13 — 176
I 5, 30–44 — 175, 241
I 5, 30–37 — 176, 241
I 5, 37–44 — 253n214
I 5, 44–49 — 175, 253
I 5, 47–48 — 250
I 5, 49–54 — 175
I 5, 54–56 — 175
I 5, 114–125 — 175
I 7, 7–48 — 169n38
I 7, 49–110 — 169n35, 238n190
I 7, 71–78 — 254–255
I 8, 79–82 — 194

II 1–3 — 44n117
II 1, 1–413 — 176n53 + 55
II 1, 2–5 — 176
II 1, 136–139 — 220n161
II 1, 426–466 — 52n133–134, 173n46, 176, 241
II 2, 19–30 — 243n196
II 3, 68 ff. — 44n117
II 4, 1 — 54
II 4, 21–32 — 53n138
II 6, 60–108 — 243n196

Democritus
fr.A39 D-K — 218n152
fr.A61a D-K — 212n138, 219n158
fr.A61b D-K — 212n140, 219n158
fr.A85 D-K — 218n152
fr.A87 D-K — 218n152
fr.A90 D-K — 218n152
fr.A94 D-K — 167n25
fr.A96 D-K — 249n208
fr.A97 D-K — 65
fr.A98 D-K — 63–65, 146n170
fr.A99 D-K — 69n185

Diogenes of Apollonia
fr.A11 D-K — 166n22, 222n165, 248
fr.A16c D-K — 167n24

Diogenes Laërtius
II 8 — 167n24
II 9.1–3 — 248
IV 58.4–6 — 167n28
V 44 — 91n54, 145n164
VII 123 — 119n115
VII 132 — 44n119
VII 138 — 77n5
VII 144 — 53n136, 77n5
VII 145 — 53n138, 169n34, 194n92
VII 151–154 — 90–91, 107, 140n155
VII 152 — 77n5
VII 154.4–6 — 133n141
VII 155.9 — 53n136
VIII 25 — 166n18
VIII 48 — 166n18, 169n34
VIII 54–56 — 166n21

Diogenes Laërtius (cont.)
   IX 21 — 166n18
   IX 30 — 167n25
   IX 33.6–8 — 249n207
   IX 47 — 76
   X — 2, 92
   X 4 — 63n167
   X 28 — 109n87
   X 31–34 — 13n33
   X 34 — 14n35–37, 15n40

Diogenes of Oenoanda (ed. Smith)
   fr.13 I 11–13 — 240, 256
   fr.13 III 2–13 — 37–38, 41–42, 242
   fr.14 — 42n114
   fr.66 — 52, 233, 242
   fr.67 — 185n73
   fr.98.8–11 — 42n114

Empedocles
   fr.A58 D-K — 166n22, 249n206
   fr.B35 D-K — 166n22

Epicurus
Epicurea (ed. Usener)
   266 —
   276 — 21n52
   280 — 211–213, 228, 234
   281 — 215n146
   293 — 176n52, 215, 234
   297 — 109n86
   346 — 181n63
   351 — 176n54
   376 — 146n170
      — 14n35

Epistula ad Herodotum
   38–44 —
   38 — 15n39, 17
   40 — 14n37, 15n39
   41–42 — 17
   43 — 256
   45 — 209n135
   46–50 — 127, 256
   51 — 18n46
   55–56 — 14
   60 — 18
   61 — 149n179, 164, 256
   67 — 149n179, 214n144
    — 206n130

68 — 127
73 (scholion) — 256
74 (scholion) — 262n223
76–77 — 32n78, 152n189
78–80 — 10
80 — 14n38, 35, 37, 57n150

83 — 127
Epistula ad Menoeceum
134 — 152
Epistula ad Pythoclem
1 [84] — 1, 92, 95
2 [85–88] — 10
2 [86] — 19n49, 20, 24, 26, 28, 33n84, 58n152

2 [87] — 19n49, 25, 35, 37, 57n150
2 [88] — 15n41, 19n48, 25
3 [88] — 15n41, 19n48
4 [89–90] — 224, 235
5 [90–91] — 11n22, 226, 235
6 [91] — 10n14, 35n90, 236, 257
7 [92] — 10n12, 15n41, 19, 25n61, 33n80, 42, 55, 241, 258
8 [92–93] — 55
8 [93] — 201n118
9 [93] — 19n48–49, 45, 47n123, 55, 56, 59n156, 66n174, 247n202
10 [94] — 35, 37, 37n99, 55, 57n150, 58n152
11 [94–95] — 11n19, 33n81, 55
11 [95] — 19n48, 36
12 [95–96] — 55
12 [95] — 19n49
13 [96–97] — 12, 55, 243, 258
14 [97] — 32n78, 55
15 [98] — 36, 55, 57n150
16–36 [98–116] — 100–103
16 [98] — 19n48
17–35 [99–115] — 127–130, 137–140, 143–144
17–30 [99–111] — 139–140, 154, 156–159
17 [99–100] — 151n185

INDEX LOCORUM   291

18 [100]	10n13, 33n83, 35n, 58n152	*Hipp. Aër.* 8.6	69n184, 150n183
19 [101–102]	33n83, 35n93, 63n166	**[Galen] (ed. Kühn)**	
21 [103–104]	33n83	*Definitiones medicae*	
23 [105–106]	105, 107, 134–135	19.391	113n96
24 [106]	134–135		
29 [109–110]	102, 134	**Geminus**	
30 [110–111]	102, 134	*Isagoge*	
31 [111]	55, 94n62, 107, 130	VI 7–8	173n46
		VI 9	168n33
32 [112]	19n49, 49–50, 55		
33 [112–113]	55	**Herodotus**	
33 [113]	32n78, 33n85, 56–57, 58n152	*Historiae*	
		II 19–27	119n114
34 [114]	53n, 55, 57n150	II 22	251n212
35 [114–115]	55, 94n62, 107, 130	II 31.1	222
36 [115–116]	32n78	III 104	166n23
Περὶ φύσεως (ed. Arr.)		IV 31.2	222
XI	95, 144, 163, 180, 236, 240, 260, 266	IV 42	166n23
		Hipparchus	
		In Arati et Eudoxi Phaenomena	
XI frs.22 & 23	235n182	I 3, 6–7	172n44
XI fr.38	47n123		
XI fr.42	104, 235n182, 256	**Hippocrates**	
XII	95, 144, 156, 243	*De aëre*	
XIII	95, 144, 156	8.7	69n185
XIV & XV	60n161		
Ratae sententiae		**Lactantius**	
1	32n78	*Divinae institutiones*	
24	14n37	III 24	167n27
Eudoxus (ed. Lasserre)		**Leucippus**	
fr.75a+b	171n42	fr.A1 D-K	167n25, 218n152 + 154 + 157, 249n207
fr.124.84–85	245n201		
fr.288	168n30		
		fr.A26 D-K	167n25
Galen (ed. Kühn)		fr.A27 D-K	222n165, 249n207
Nat. Fac.			
I 14 = 2.45.4–52.2	109n86		
Hipp. Acut.		**Lucretius**	
15.429–430	112n96, 113n97	*De rerum natura*	
Hipp. Nat. Hom.		I 73	204n126
15.117	113n98	I 170	229n176
Hipp. Epid.		I 231	201
17a.1–2	113n96	I 235	198
17a.10	113n98, 119–120n115	I 334–345	17n45
17a.12–13	113n96–97	I 358–369	213

De rerum natura (cont.)

I 402–409	127	V 204–205	221–223, 257
I 437–439	207n131	V 422–431	21n52
I 514	229n176	V 449–508	223–235, 240, 256, 266
I 565–576	214n143	V 509–770	10, 21, 42–58, 60, 96, 97, 109, 146, 272
I 635–920	60		
I 665–674	214n142		
I 782–797	214n142	V 509–533	44, 46
I 951–1051	181–186	V 525	201
I 984–997	208–209	V 526–533	21, 28, 43n116, 47, 210
I 1045	198		
I 1051	199	V 534–563	10, 44, 46, 97, 104, 164, 196, 231, 233, 235, 256
I 1052–1113	164, 181–210		
I 1052–1093	233–234, 266		
I 1058–1064	194n91	V 534	233
I 1062–1063	230n178	V 564–591	10, 44, 46, 236–239, 257–258
I 1068–1075	190		
I 1089–1091	200	V 592–613	44–45, 46
I 1094–1101	190	V 614–649	45, 46, 239
I 1102–1113	203–205	V 621–636	239, 256
I 1114–1117	127	V 621–622	40–41, 58
II 62–250	184, 210–220, 234	V 650–704	189, 258
II 62–111	209n135	V 650–679	10n12, 19n47, 42, 241–242
II 83–85	214		
II 100–108	214n143	V 650–655	33n80, 45, 46, 233
II 184–250	256		
II 184–215	184, 208, 211, 228, 234	V 656–679	33n80, 45, 46, 52
		V 680–704	45, 46, 53
II 184–186	188n74	V 682–695	42
II 190	214	V 691–693	45n121, 247n203
II 203–215	148, 198	V 694–695	46–47
II 205	214	V 705–750	33n83, 43, 46, 53–55
II 216–250	214–215, 234, 256		
II 235–237	207n131	V 705–714	33n81–82, 43
II 1048–1089	256	V 713–714	43, 47, 53n139, 240, 256
II 1066	230, 233		
II 1090–1104	32n78, 152n189	V 727–730	43, 47, 53–54, 58
II 1133–1135	205	V 751–770	43, 48
III 9–12	3n5, 144n161	V 751–761	46, 48–52
III 346	229n176	V 762–770	33n83, 46, 48–52
III 370–373	41, 188n75	V 762–767	33n82
III 453	246	V 762–764	242–245, 258–259
IV 54–216	18n46	V 771–1457	109
IV 404–413	220, 257	V 1183–1240	32n78
IV 436–437	246	V 1183–1193	125, 152n189
IV 515	246	V 1189–1193	96
V 55–56	3n5	V 1204–1210	152n189
V 156–234	32n78	VI 48–90	125
V 203	111n92	VI 50–79	32n78

VI 59–61	110n88	VI 848–878	32, 111, 116, 122–123
VI 96–607	153–154, 156		
VI 96–159	10n13, 12n28, 33n83, 63n166, 143	VI 850	110, 111
		VI 879–905	110, 111, 116, 118, 122–123
VI 121–131	61	VI 890–894	116, 118, 122–123
VI 160–218	12n29, 33n83	VI 906–1089	111, 116, 123, 129, 136
VI 219–422	33n83, 141		
VI 379–422	32, 131–132, 151	VI 910	110, 111
VI 423–450	132–133	VI 956	112n95
VI 463–464	151n185	VI 1056	110, 111
VI 470–475	111	VI 1090–1286	111, 112–113, 116, 124–125, 129, 136
VI 483–494	112n95		
VI 503–505	111	VI 1093–1096	124n128
VI 510–512	151n185	VI 1098–1102	112
VI 524–526	102, 139	VI 1103–1118	112–113
VI 527–534	97, 127, 134, 139	VI 1106–1109	245–253
VI 535–607	12n30, 97	VI 1110–1113	251–253
VI 605–607	204n128	VI 1115–1117	111, 124
VI 608–1286	109, 116–117, 136, 266	VI 1119–1132	113
		VI 1133–1137	113
VI 608–638	110–111, 116, 118, 136	VI 1138–1286	113
		VI 1237	40n108
VI 608–609	98		
VI 639–702	12n26, 111, 116, 119, 134, 136	**Manilius**	
		Astronomica	
		I 215–235	169, 173–174
VI 639	111	I 167	229n175
VI 703–711	22–23, 31, 111	IV 758–759	251n212
VI 703–704	24		
VI 712–737	111, 116, 119, 136	**Paradoxographus Florentinus**	
VI 712–713	30n73, 110, 111		
VI 720	246	11	110n91, 123
VI 722	251n212	19	123
VI 738–839	111, 116, 119–121, 136	**Paradoxographus Vaticanus**	
		13	120
VI 738	119	36	121n119
VI 740–746	124n128		
VI 746–748	111, 116, 120	**Parmenides**	
VI 749–755	111, 116, 120	fr.A1 D-K	166n18
VI 753–754	32	fr.A44 D-K	166n18–19
VI 756–759	111, 116, 121		
VI 762–766	32	**Philodemus**	
VI 762	120n115	*De signis*	
VI 769–780	113n99	47.3–8 (De Lacy)	254–255, 259
VI 781–817	113n99		
VI 834	246	**Philostratus**	
VI 840–905	129, 136	*Vita Apollonii*	
VI 840–847	10n16, 111–112, 116, 121–122	II 10	120

Plato
Phaedo
97d	162, 167
108e–109a	162, 167
110b	162

Timaeus
40c–d	47n123, 52n131
62c–63e	195n98
62c–63a	178
63a	195n99
63b1–3	202n121

Pliny
Naturalis Historia
I	88
II	79, 87
II i–xxxvii 1–101	88
II iv 11.6	229n175
II viii 51	243n196
II xvii 81	53n135
II xxii–cxiii 89–248	87–89, 100–103
II xxviii 98	102
II xxxi 99	102
II xxxviii 102	88n44
II l 133–134	133n141+145
II lx 150–151	102
II lxii 153	88n45
II lxiii 154	88n45
II lxv 164.1–5	169–175
II lxvi 166.9–11	117, 118
II lxviii 172	222
II lxxi–lxxvii 177b–187a	169–175
II lxxv 183	255
II lxxvii 186–187a	167n29, 168n33
II lxxx 189.2–3	251n212, 252n213
II xcv–xcviii 206b–211	115, 120
II xcv 207.1–7	123
II xcv 207.9–208.10	117, 120
II xcv 208.4–5	117, 121n119
II xcv 208.9–10	120
II cvi 224b–234	115, 117, 121, 122
II cvi 227.4–5	117, 123
II cvi 228.1–6	117, 123
II cvi 228.1–3	110n91
II cvi 228.6–10	117, 122
II cvi 233.1–2	112n94, 117, 121
II cviii–cx 235–238	115
II cx 236–238	117, 119
IV xvi 104	168n33
V xix 81	121n120
X xiv 30	120n117
XXXI xviii 21	120n116, 228n173

Plutarch
Adversus Colotem
3, 1108e–f	63n166

De audiendis poëtis
11, 31d	201n119

De communibus notitiis
46, 1084e	201n120

De facie in orbe lunae
7, 924a4–6	194n91
7, 924b5–6	207n132
11, 926b4–7	207n132
13, 927c	194n88
25, 940c	194n92

De Stoicorum repugnantiis
41, 1053a	201n120
42, 1053e	194n88
44, 1054b–1055a	200n114
44, 1054b	175n51
44, 1054c	199–200
44, 1055a	193n87

Pomponius Mela
De situ orbis
II 37	123
II 43	110n91

Ptolemy
Almagest
I 3, 11–12	53n134
I 4, 14–16	169–175
II 6	167n29

Geographia
I 4, 2	174n50

Tetrabiblos
II 2, 56	251n212, 252n213

Seneca
Naturales quaestiones
	35, 76, 89–90, 100–103
I 2–13	86n34
II 1.1–2	89n48
II 1.3	104–105
II 1.5	90n51, 104
II 10.4	119
II 13.1–14.1	194n88, 198

II 24.1	194n88, 198	**Simplicius**	
II 26.4–6	119	*In Aristotelis De caelo* (ed. Heiberg)	
II 30.1	119	267.30–34	208n133, 212
II 42	132	269.4–6	208n133, 211n137
II 58.2	194n88		
III	87, 115	284.28–285.2	183n67, 185
III 4–8	117	495.22–23	245n201
III 14	165n12	569.5–9	212, 219n158
III 20	117, 122	712.27–29	212n140, 219n158
III 21	117, 120		
III 25–26	117, 122	*In Aristotelis Physica* (ed. Diels)	
III 26.1	30n72	291.21–292.21	44n119
IVa	87, 90, 117, 119	467.26–35	183n66
IVa 2.18.1–2	251n212		
IVa 26–27	117, 121–122	**Stobaeus**	
IVb	90	*Eclogae physicae*	
V 13.1–3	133n145, 133n141	19.4	192n83
V 14.4	117, 119	21.5	178n61
VI 4.1	119	25.5	194n93
VI 13.1	69, 146n169	29.2.4–7	133n141
VI 13.3–4	112n94, 117, 121–122	39	86–87, 128, 136, 137, 138
VI 20	144, 146n170	*Stoicorum veterum fragmenta* (ed. von Arnim)	
VI 20.1	63–64	I 99	192n83
VI 20.2	65n170	I 100	194n88
VI 20.5–7	95	I 119	51n129, 52n131
VI 20.5	59, 64	I 120	51n129, 194n93
VI 20.7	12, 38	I 276	169n34
VI 21.1	39	I 501	194n92
VI 27–28	117, 124n128, 124	I 504	194n92–93
VI 28	120, 124	I 528	124
VII 4	90n49	I 535	201n119
VII 21.1	90n49	I 542	53n136
VII 22.1.1–3	90n49, 99	II 421	194n92
VII 23.1	194n88	II 434	194n88
		II 527	178n61
Sextus Empiricus		II 535a	183n67, 185
Adversus mathematicos		II 539	175n51
VII 203–211	13n33	II 549	193n87
VII 211–216	15–17	II 550	193n87
VII 211	14n35–36	II 551	200n113
VII 212	15n43	II 554	193n87
VII 213–214	15n41, 16	II 555b	169n34, 174n88
VII 214	15	II 557	178n61
VII 215	15n44	II 571	193n88
VII 216	14n36	II 572	194n92
X 221–222	207n131	II 579	201n120
		II 593	194n92
		II 648	169n34

Stoicorum veterum fragmenta (ed. von Arnim) (*cont.*)

II 650	52n131, 53n136+138, 169n34, 194n92	6.36–41	148
		6.41–48	148–149, 152
		7	149–152
		7.2–9	150
		7.27–29	150
II 651	53n136	13	132–136, 139, 141, 149, 152
II 658	194n92		
II 663	194n92	13.33–54	131
II 676	51n129	13.43–54	132–133, 138–139, 146–147, 150
II 677	194n92		
II 678	51n129	14a = 14.1–13	102, 131, 134–136, 138–139
II 683	52n133		
II 690	194n92	14b = 14.14–29	131–132, 139, 142, 151–153
II 703	133n141		
II 806d	201n120	14b.14–17	152
III 642	119n115	14b.25–26	131
		14b.25–29	152
		15	105, 107, 135

Strabo

I 1.12	169n37		
I 1.20.18–27	169, 171, 174–175, 260	**Thales**	
		fr.A14 D-K	165n12
I 1.20	44n119	fr.A15 D-K	165n12
I 2.24–28	252n213	fr.A22 D-K	123n126
I 4.4	168n32		
II 5.2	44n119	**Theon of Smyrna**	
II 5.8	168n32–33	*Expositio*	
II 5.10.3–5	169n36	III 2–3	154–158
II 5.38–42	173n46		
V 4.5	120n115	**Theophrastus**	
XII 8.17	119n115	*De causis plantarum*	
XIII 4.14	120n115, 121n119	I 17.5	67n178
XIV 1.11	120n115	*De igne*	
XIV 1.44	119–120n115	1.4–11	67–68
XV 1.24	251n212	1.8–9	133, 146–147
XVI 1.27	121n120	52–56	68n183
XVI 2.13	123	56	196n100
		De lapidibus	
Strato (ed. Wehrli)		3, 1–3	68
fr.50	208n133, 211n137	*De sensibus*	
fr.51	208n133	61	217
fr.52	208n133, 212–213	*De ventis*	
fr.54	208n133	5.1–5	69n184, 150n183
fr.55	208n133	53	133n142, 147
		Historia plantarum	
Syriac meteorology (ed. Daiber)		III 1.4–5	68n183
1	142–143	III 2.2	68n183
1.18–20	62	*Fragments* (ed. FHS&G)	
6	131, 141	159	145n165
6.36–67	148, 152	166	81n16

186A–194	145n166	IX 2	54
195	69–70, 146n169	IX 3.1–3	53n135
211A	69n184, 150n183		
211B	69, 150–151, 152	**Xenophanes**	
211C	69n184, 150n183	fr.A32 D-K	52n134
		fr.A33 D-K	52n134
Vitruvius		fr.A38 D-K	52n134
De architectura		fr.A40 D-K	52n134
VI 1.1	253n214–215	fr.A41a D-K	233n180
IX 1.3	53n136	fr.A46 D-K	69n185

General Index

Aëtius' *Placita*
  On Epicurus' multiple explanations 59
  Pseudo-Plutarch and Stobaeus 84–87
Aëtius' *Placita* III (+ IV 1)
  General account and scope 84–87
  Milky Way, comets and shooting stars 99, 107–108, 130, 155
  Rainbows, haloes, rods and mock suns 86, 134–137
  The Earth as a whole 85, 103–104, 131, 136–137, 155
Analogy 34–37, 61–62, 63, 66–70, 72–73, 122, 142, 146, 237, 264
Aristotle's *Meteorology* I–III
  General account and scope 79–82
  Milky Way, comets and shooting stars 81, 99, 107–108, 130, 137, 146–147, 158
  Multiple explanations in 65–67
  Rainbow, haloes, rods and mock suns 99
[Aristotle's] *De mundo*
  Authorship 82, 83n26
  Comets and shooting stars 84n27, 107
  General account and scope of ch.4 82–84
Aristotle's (?) *Liber de Nilo* 115–116, 119n114
Attestation 14–15, 17, 30
  See also non-attestation

Bion of Abdera 167

Callimachus 114, 122, 126
Clinamen: *see* swerve
Contestation 15–16, 18, 32
  See also non-contestation
Cosmic whirl 166n22, 210, 216–219, 226, 227, 234–235
Cosmogony 223–235, 256–257, 265, 266
Cosmology
  Centrifocal 166, 168, 177–181, 186, 192–193, 196–197, 202–203, 204n127, 205–214, 216–219, 223–224, 228–235, 239–240, 256–257, 266

  Parallel-linear 164, 177–179, 203, 204n127, 208, 210–211, 214–216, 218–220, 223–224, 228, 230, 232–235, 240, 256–257, 262, 265

Democritus
  Acknowledged by Lucretius 40–41, 45, 58
  Distinguishing astronomy and meteorology 76
  Influence on Epicurus 2, 63, 73
  Multiple explanations 5, 8, 63–65, 73, 146
Diseases 109, 111–113, 116, 124–125, 245, 251
Doxography
  Definition of 5n7, 8
  Epicurus' and Lucretius' dependence on 58–62
  Multiple explanations and 37, 58–62, 64, 71–72, 74–75, 126, 155, 157, 160, 264, 267
  Peripatetic origins of 60, 75, 155–156, 160, 267
  See also Aëtius' *Placita*; Theophrastus' Φυσικαὶ δόξαι

Earth
  Flat 6, 34, 162–168, 170–173, 175–179, 181, 220, 222–223, 235–239, 241, 244–245, 247, 249–250, 253, 255–263, 265
  Problems pertaining to the whole 82, 84–88, 90, 95, 97, 103–104, 131, 136–137, 155
  Size of the earth 82n21, 89, 104, 176
  Spherical 44, 162–179, 181, 222–223, 233, 237–238, 241, 243–244, 249n206, 253, 255, 257–263, 265
  Supported underneath 44, 165–167, 179, 235, 256
  See also inclination
Earthquakes
  Causing plagues 124

GENERAL INDEX

Explanation   38–40, 42n114, 59, 63–65, 70, 73, 81, 95, 121, 135, 146
Inclusion in meteorology   81, 97–98, 104–105, 107–108, 135–136, 158–159, 265, 267
Type-differentiation   12
Epicurean epistemology
   Applied to astronomy and meteorology   19–20
   Applied to fundamental physical theories   17–18
   Principles of   4, 13–15
   Sextus Empiricus on   15–17
Epicurean physics
   Principles of   3–4, 17–18
Epicurus, *Letter to Pythocles*
   Authenticity of   1
   Classification of comets and shooting stars   94–95, 99, 107, 108, 130, 139, 158
   Distinguishing astronomy and meteorology   94
   General account and scope of   92–95
   Rainbows, haloes, rods and mock suns   94, 102, 108, 134, 140
   Summary nature of   1, 2, 35, 62, 95, 142, 144
Eratosthenes   169–170, 238, 255
Etna
   See volcanoes

Heraclitus   52, 60
Hydrological cycle   82

Inclination
   Of the Cosmos   166n22, 246–250, 253n215
   Of the Earth   86, 103, 249–250

Lucretius' *De rerum natura*
   Book VI, general account and scope of   95–98
   Classification of comets and shooting stars   97, 99
   Division of atmospherical and terrestrial phenomena   97–98
   Privileging astronomers' views   42–58

Rainbows, haloes, rods and mock suns   102, 108, 134, 139
Separating astronomy and meteorology   96–97
(Un)orthodoxy of   3, 191–192, 235, 257, 266

Magnets   96, 106, 109–112, 116, 123, 126, 129, 136
Mathematical astronomy
   Deification of heavenly bodies   57
   Diogenes of Oenoanda's attitude towards   5, 37, 41–42, 74, 242, 264
   Endorsed by Plato, Aristotle and Stoics   51–58
   Epicurean attitude towards   5, 6, 33, 42–58, 74–75, 77, 260, 263, 264, 266, 267
   Physical models in   47, 266
   Physics and   44–45, 77n5
Meteorology, ancient
   Definition and scope   76–78
   Milky Way, comets an shooting stars   81, 90, 92, 94–95, 99, 107–108, 130, 137, 146–147, 155, 158
   Rainbows, haloes, rods and mock suns   86, 87, 94, 99, 102, 107–108, 134–140
   Range and order of subjects   100–103
Metrodorus of Chios   52
Mirabilia
   Explanation of   29–31, 114–115, 117, 119–123, 125–127, 158
   Listed in meteorological works   114–118, 122, 125–127, 157, 158
   Lucretius' treatment of   6, 29–31, 78, 109–113, 126–127, 157–158, 265
   See also paradoxography
Multiple explanations
   Degrees of probability   5, 13, 37–42, 74, 242, 264
   (In)exhaustivity   11, 25, 56n143, 269–271
   Multiple vs. single explanations   6, 8, 10–11, 20–21, 26, 27, 34, 36, 38, 56–58, 65, 69–70, 72, 146, 242, 267

Mutual exclusivity or inclusivity   11–12, 13, 22, 247
Number of explanations   10–11, 30n74, 46, 55–56, 64, 66–67, 70, 72, 126, 269–275
Subsidiarity of explanations   12, 30n74, 272–274
Truth value   12–13, 21–31
Type-differentiation   12, 65, 73

Nile, summer flooding of   22, 29–31, 64n169, 86–87, 90, 96, 102, 103, 109–111, 113, 115–117, 119, 121–123, 125–126, 128–129, 136–138, 157, 159, 167–168, 272
Non-attestation   14–15, 17, 30
   See also attestation
Non-contestation   13–20, 23–31, 33–36, 74, 264
   See also contestation

Obliquity of the zodiac   45, 247–248

Paradoxography   6, 78, 113–126, 157–159, 265
   See also mirabilia
Philodemus   2–3, 254–255, 259, 261
Pliny's *Natural History* 11
   Comets and shooting stars   87, 99, 107
   Earth as a whole   87, 103–104
   General account and scope   87–89
   On *mirabilia*   114, 115, 119–123
   Rainbows, haloes, rods and mock suns   87, 102
Posidonius   53n138, 77, 79, 169, 170
Prester (whirlwind)   67–68, 132–133, 138–140, 146–147, 159, 267
Principle of plenitude   21–24, 28–31, 74, 210
Pythagoras & Pythagoreans   165n12, 166, 170, 183
Pytheas of Marseille   168, 170, 173, 175, 261

Risings and settings   8, 10, 11, 19, 42, 55, 58, 93, 176, 178, 180, 220n, 241–242, 257–259, 269

Seneca's *Natural Questions*
   Comets and shooting stars   90, 99, 107
   General account and scope   89–90
   On Epicurus' multiple explanations   38–39, 59, 63–64
Size of the sun   10, 11, 34, 36, 43–44, 46, 93, 236, 239, 244–245, 257–259, 269, 272
Stoic Meteorology in D.L. 7.151–154
   Comets and shooting stars   107–108
   General account and scope   90–91
   Rainbows, haloes, rods and mock suns   108
Swerve (clinamen)   3, 176, 214–215, 220
Syriac meteorology
   Authorship   5, 6, 8–9, 70–73, 74–75, 78, 108
   Character   71
   Completeness   106–108
   General account and scope   5, 70–72, 91–92
   Multiple explanations in   5, 8–9, 70–75
   Rainbows, haloes, rods and mock suns   102, 107–108, 134–139
   Theological excursus   71n194, 128–129, 131–132, 138, 139, 151–153
   Use of analogies   35, 62

Terrestrial phenomena   11, 92, 95, 97–98, 106, 109–113, 115, 117–118, 123, 125, 136–138, 144, 157
   See also mirabilia
Terrestrial waters   81–82, 86–87, 90, 114–115, 122–123, 157
   See also mirabilia
Theophrastus
   Doxography   60, 68–69, 74, 75, 155, 160, 267
   Multiple explanations   67–70
   Φυσικαὶ δόξαι   60n158, 154n191, 155–156, 160
   See also Syriac meteorology

Void
   Existence of   15–17, 196, 200, 202

Infinity of   3, 21, 31, 74, 181–185, 191, 193, 196, 199, 200, 205–210, 234
Lightness due to admixture of   213
Volcanoes   12, 82, 84, 116, 119
Vortex: *see* cosmic whirl

Wasserstein, Abraham   8n1, 32–34, 244, 263
Whirlwinds: *see* prester

Zones   82n21, 103, 104, 221–223